Superoxide Ion: Chemistry and Biological Implications

Volume II

Oxygen Radicals in Biology

Author

Igor B. Afanas'ev, D.Sc.
Chief
Laboratory of Chemical Kinetics
All-Union Vitamin Research Institute
Moscow, Soviet Union

CRC Press
Taylor & Francis Group
Boca Raton London New York

CRC Press is an imprint of the
Taylor & Francis Group, an **informa** business

First published 1991 by CRC Press
Taylor & Francis Group
6000 Broken Sound Parkway NW, Suite 300
Boca Raton, FL 33487-2742

Reissued 2018 by CRC Press

A Library of Congress record exists under LC control number: 88022277

ISBN 13: 978-1-138-50706-7 (hbk)
ISBN 13: 978-0-203-71016-6 (ebk)

Visit the Taylor & Francis Web site at http://www.taylorandfrancis.com and the CRC Press Web site at http://www.crcpress.com

PREFACE TO VOLUME II

This book is a continuation of the first volume of *Superoxide Ion: Chemistry and Biological Implications,* and it is concerned with the problems of production and reactions of the superoxide ion and other oxygen radicals in biological processes. By contrast with chemical experiments, it is virtually impossible to produce the superoxide ion alone in biological systems; moreover, since the *in vitro* studies are always carried out to model real *in vivo* processes, the attempt to study the role of the superoxide ion in some biological systems without other oxygen species is not only unrealistic but is actually not very fruitful.

Therefore, in this book we consider all the processes in which the participation of oxygen radicals, i.e., superoxide ion, perhydroxyl radical, hydroxyl radicals, and active iron-oxygen intermediates formed in the Fenton reaction during the oxygenation of iron ions ("crypto-hydroxyl" radicals, perferryl and ferryl ions), is suggested. There is now an essential change in understanding of the role of oxygen radicals in biological processes: if earlier they were considered only as some damaging species, at present, their important regulatory role is becoming evident. Both damaging and regulatory functions of oxygen radicals will be discussed here. Since superoxide ion is a precursor of all other oxygen species, their reactions should be considered as a consequence of the formation of superoxide ion.

This book consists of five chapters. The questions of classification and reactivity of oxygen radicals and other oxygen metabolites are briefly discussed in Chapter 1. The three subsequent chapters are concerned with the modes of production of oxygen radicals in biological systems. In Chapter 5, the interaction of oxygen radicals with naturally and biologically active compounds, cellular components and cells, and the reactions of oxygen radicals in biological systems are discussed. There is a big difference in conducting oxygen radical studies in chemistry and in biology. As we have seen (Volume I), chemists actually use only a few methods for superoxide production (radiolysis, photolysis, electrochemical reduction of dioxygen, and alkali metal superoxides) while in biological systems, superoxide ion is formed as a result of numerous redox processes in which natural substrates, enzymes, xenobiotics, cellular components, and whole cells take part. Therefore, much attention is given to studying the mechanisms of superoxide production in cells and cellular components.

The difference in the experimental techniques used in biological and chemical investigations should also be stressed. Again, in chemical studies, workers do not apply many specific methods for detection and identification of oxygen radicals: they are ESR and optical spectroscopy, pulse radiolysis and flash photolysis, electrochemical technique, and the usual chemical methods based on the identification of reaction products. All these methods are also applied in biological systems for the identification of oxygen radicals and the investigation of their reactions. However, in addition to these methods, numerous biochemical, physiological, immunological, and pathological experiments are carried out in which important information concerning the role of oxygen radicals in physiological and pathological processes is obtained.

In this book an attempt was made to describe if not all, at least most biological processes in which oxygen radicals are produced. Due to the great number of such works, it is impossible (and even pointless, because it can be correctly done only by scientists working in specific experimental fields) to consider the experimental details of these works. However, I tried to compile a special index of experimental techniques applied in cited works which I hope will be useful to readers. Furthermore, since there is great growth in the oxygen radical studies in biology, about 400 papers were published during the compilation of Volume II. Therefore, I included short discussions on these works in Additions to all chapters.

As the reader can see, Volume II does not contain all data concerning oxygen radicals in biology. Thus many questions such as the role of oxygen radicals in pathological processes which we wanted to discuss in this Volume (see Preface to Volume I) are not considered.

The reason is obvious: too many works in this field, especially concerned with pathobiological aspects of this problem, appear in the current literature. I hope that these questions, including the studies of oxygen radicals in organs and a whole organism, the role of oxygen radicals in free radical pathologies (ischemia, cancer, inflammation, environmental-associated diseases, etc.), and the role of antiradical, antioxidative, and chelating drugs in the treatment of these diseases, will be discussed in subsequent work.

Professor I. B. Afanas'ev

THE AUTHOR

Igor B. Afanas'ev, D.Sc., Professor, is Chief of the Laboratory of Chemical Kinetics, All-Union Vitamin Research Institute, Moscow.

Dr. Afanas'ev graduated in 1958 from Moscow Chemico-Technological Institute, received his Candidate Sc. and D.Sc. degrees (Chemistry) in 1963 and 1972, respectively. He became a Professor of Chemical Kinetics and Catalysis in 1982.

Dr. Afanas'ev has presented lectures and reports at many national and international conferences on chemistry, biology, and medicine of free radicals, including "Oxygen Radicals in Chemistry and Biology", Munich, 1983; "Superoxide and Superoxide Dismutase in Chemistry, Biology, and Medicine", Rome, 1985; and International Conference on Medical, Biochemical, and Chemical Aspects of Free Radicals, Kyoto, 1988; etc.

Dr. Afanas'ev is a Chairman of the U.S.S.R. annual seminar on oxygen radicals in chemistry and biology. He has published more than 130 research papers, among them about 30 papers in English. His current research interests include studying the reactions of oxygen radicals in chemical and biological systems and developing new antioxidative drugs for combating free radical pathologies.

TABLE OF CONTENTS

Chapter 1

WHO IS THE VILLAIN?

(Classification and Reactivities of Oxygen Species in Biological Systems)

There is now abundant literature concerning the role of reactive oxygen species in biological processes. However, authors do not always clearly determine even hypothetical structures of these compounds. In part, it is due to the confusing terminology when such terms as "free oxygen radicals", "oxygen-centered free radicals", "reactive oxygen species" (ROS), "active oxygen", "oxygen metabolites", and "partly reduced oxygen" are applied indiscriminately. We have already discussed the structures of various oxygen radicals in connection with the mechanism of the superoxide-driven Fenton reaction (Volume I, Chapter 9) and will again consider the mechanisms of their formation in subsequent chapters. However, it seems necessary to discuss the questions of classification and some disputable questions of the structures of oxygen radicals at the beginning of this volume.

It is certain that all reduced oxygen species, excluding H_2O, are reactive intermediates which may mediate important physiological processes and simultaneously may induce the cellular damage leading to various pathologies. Hence, the most general terms for them are "reactive oxygen species" or "oxygen metabolites", which include free radicals (superoxide ion $O_2^{\bar{\cdot}}$, perhydroxyl $HO_2\cdot$, and hydroxyl $HO\cdot$), hydrogen peroxide, singlet oxygen, and oxygenated transition metal complexes (crypto-hydroxyl radicals, ferryl FeO^{2+}, and perferryl FeO_2^{2+} ions). Thus oxygen radicals, which are the principal subject of this book, are only a part of a whole body of ROS. However, there are important reasons to consider biological processes mediated by oxygen radicals and oxygenated transition metal complexes separately from those promoted by other ROS.

As is known, free radicals are species possessing an unpaired electron. The occurrence of an unpaired electron, as a rule, but not always, sharply increases the reactivity of a compound especially in hydrogen atom abstraction reactions, in the addition to double bonds of organic compounds, and in other reactions. The enhanced reactivity of free radicals is a basis of free radical hypothesis in biology, explaining the development of many pathological processes by the destructive action of free radicals on cellular components (mitochondria, microsomes, nuclei, lipid membranes, etc.), and on the cells themselves. Thus, in accord with this hypothesis, free radicals are our major villains.

However, here some difficulties begin in both the classification and the understanding of the role of oxygen radicals, other free radicals, and other oxygen species in biological processes. Oxygen radicals are not the only free radicals forming in cells. For example, it is well known that such a destructive process as lipid peroxidation is mediated by lipid alkyl and peroxy radicals. Superoxide ion, hydroxyl radicals, and oxygenated transition metal complexes are important initiators of lipid peroxidation (see Chapter 5). However, their role is insignificant in the subsequent stages (propagation and termination) of this process. Within the limits of free radical hypothesis, all free radicals should be considered. This point of view is shown in the term "oxygen-centered" free radicals, which includes both oxygen and peroxy free radicals. Nonetheless, the distinctions in the modes of formation and chemical properties of peroxy and oxygen radicals which make their roles different in physiological and pathological processes justify separate consideration of both types of free radicals.

Singlet oxygen and hydrogen peroxide are frequently united with oxygen radicals in studying the effects of ROS on cells and cellular components. There have been many discussions about the possibility of singlet oxygen forming in biological systems. After the original enthusiasm in 1970 to 1980, no further light was shed on its role in biological processes until about 1985. At present, the formation of singlet oxygen in cells cannot be completely excluded. However, the main question remains to be decided: is the formation of this highly reactive species of real significance in pathobiological processes? The difficulties are evident when we compare this question with the study of the role of other highly reactive oxygen species, hydroxyl radicals: although there is much evidence of their production being stimulated by xenobiotics (Chapter 3), during lipid peroxidation (Chapter 5), or during oxidative burst in phagocytes (Chapter 4), in many cases doubts remain concerning their real significance in damaging processes (see below).

Another nonradical oxygen metabolite, hydrogen peroxide, is undoubtedly an effective damaging agent, the effects of which are not always easily separated from those of oxygen radicals because hydrogen peroxide is formed during the dismutation of superoxide ion. There is an important theoretical question: is hydrogen peroxide able to exhibit destructive effects, without being converted into hydroxyl (or "crypto-hydroxyl") radicals? It is true that hydrogen peroxide is capable of reacting with organic compounds without apparent decomposition into hydroxyl radicals, for example, in epoxydation of olefins. It is also supposed that the catalysis of hydrogen peroxide decomposition by peroxidases and especially catalase proceeds by a nonradical pathway. However, the dissociation of radical and nonradical pathways of hydrogen peroxide transformation in biological systems remains a very difficult problem. But it is certainly wrong to assume (as is sometimes maintained in the literature) that the hydrogen peroxide-dependent damage is always a free radical process, if there is no evidence of oxygen radical formation (by ESR* spectroscopy, the spin-trapping technique, or with the aid of oxygen radical scavengers).

Now we can return to the title of this chapter, "Who is the Villain?", and try to estimate the activity of oxygen radicals in biological systems. However, it should be stressed that in contrast to other oxygen radicals, the role of the superoxide ion cannot be confined only to destructive processes. Surprisingly, this question has been studied insufficiently, although the role of the superoxide ion in physiological processes is probably far more important than it seems now. For example, the superoxide ion is apparently able to mediate one-electron transfer processes in mitochondria; especially, its actual role in so-called cyanide-insensitive respiration in plant mitochondria is of great interest. Furthermore, the superoxide ion may activate or be a substrate of certain enzymes. Low, nontoxic concentrations of superoxide ions produced by cells may also mediate the transfer of activation signals. Other potential physiological functions of the superoxide ion are considered through the book.

There is no doubt that, as follows from its chemical properties (Volume I), the superoxide ion is not a very reactive damaging agent; however, its role as a precursor of other more active oxygen radicals is certainly important. It is also very important that the superoxide ion not only initiates the formation of hydrogen peroxide and active oxygen species originating from it, but, due to its relatively long lifetime, it can transport a free valence (an unpaired electron) far from the formation site and produce there hydroxyl radicals by a site-specific mechanism. Current studies show that there is another mode of indirect damaging activity of the superoxide ion, independent of the transition metal-catalyzed production of active oxygen species. Since the superoxide ion may activate enzymes or be a messenger of signals between cells, both an increase and a decrease in its concentration will disturb its normal functions and stimulate the disruption of extra- and intracellular processes.

In contrast to the superoxide ion, hydroxyl radicals are extremely reactive free radicals which practically unselectively react with any native compound. Therefore, hydroxyl radicals are frequently considered as main candidates for the role of a "major villain" in destructive biological processes. Indeed, the damage by hydroxyl radicals has been shown for lipid membranes, proteins, enzymes, DNA, carbohydrates, etc. (Chapter 5). However, it is their high activity that casts doubts on the real importance of hydroxyl radicals in free radical damage because free hydroxyl radicals can react practically only with neighboring molecules. As we will see, in some cases it is possible to dissociate the formation of hydroxyl radicals

*Appendix 2 at the end of this volume contains a list of abbreviations used.

and the damage to lipid membranes or enzymes which therefore must be mediated by other active oxygen species. Therefore, even if the hydroxyl radical formation in biological systems is proven (for example, by the spin-trapping technique), it is not sufficient to ascertain their participation in the damaging process. Because of this, other data and, in the first place, the use of hydroxyl radical scavengers (mannitol, ethanol, DMSO, formate, dimethyl-thiourea, etc.) are needed to show the participation of free hydroxyl radicals in some destructive processes.

There are three types of oxygen radical-mediated mechanisms of cellular damage which are realized with active oxygen species other than the superoxide ion or free hydroxyl radicals, namely, the site-specific mechanism of hydroxyl radical formation, the damage mediated by oxygenated iron complexes, and the damage mediated by crypto-hydroxyl radicals. We will not consider here in detail the reactions responsible for the formation of these oxygen species because they were discussed in Volume I (Chapter 9) and will be further considered in subsequent chapters. The subject of this discussion is the features and distinctions of these active oxygen species.

The formation of such oxygenated iron complexes as perferryl and ferryl ions was proposed in lipid peroxidation even before the detection of superoxide ions in biological systems (Chapter 5). These iron-oxygen complexes may formally be considered as iron-substituted peroxy (Fe–OO·) and hydroxy (Fe–O·) free radicals. It is difficult to estimate the reactivities of these ions, although *a priori* ferryl ion must be a much more active species.

The concept of "crypto-hydroxyl" radicals has been the subject of many discussions. It is explained for two principal reasons: first, at the beginning, there was no constructive proposal concerning the structure of crypto-hydroxyl radicals; it was simply suggested that, similar to free hydroxyl radicals, they are formed in the Fenton reaction, but have a different reactivity. Second, it was concluded that all the data indicating the formation in the Fenton reaction of oxygen radicals other than free hydroxyl radicals may be explained on the basis of the site-specific mechanism of the formation of these radicals (Volume I, Chapter 9).

As is known, the site-specific mechanism of hydroxyl radical formation suggests that the decomposition of hydrogen peroxide into hydroxyl radicals (the Fenton reaction) proceeds at the biological polymeric molecules-targets (proteins, carbohydrates, DNA, etc.) which adsorbed the catalysts of this reaction, transition metal ions (iron or copper). Because of this, classical hydroxyl radical scavengers (mannitol, ethanol, formate, etc.) are not able completely or partially to protect biological molecules against oxygen radical damage, or the reactivities of these scavengers in the reaction with hydroxyl radicals become different from those in solution. Since the last phenomenon is a major reason for the introduction of the crypto-hydroxyl radical concept, one may suggest that this becomes obsolete.

The site-specific mechanism of hydroxyl radical production in biological systems was confirmed experimentally. However, this fact obviously cannot be sufficient to deny the other mechanism which is mediated by crypto-hydroxyl radicals. Again, we have two reasons to admit the formation of these oxygen species. First, the Fenton reaction is an inner-sphere electron transfer process, and therefore hydrogen peroxide forms some complexes with ferrous ion, before decomposing into hydroxyl radicals; for example,

$$H_2O_2 + Fe^{2+} \rightleftarrows Fe(OOH)^+ + H^+ \tag{1}$$

$$Fe(OOH)^+ \rightarrow FeO^+ + HO\cdot \tag{2}$$

Hence, we must admit that the Fenton reaction does produce intermediates other than the hydroxyl radical which may be a reactive oxygen species. Second, it was shown that oxygen species with reactivity other than that of hydroxyl radicals are formed in the Fenton reaction in the absence of polymeric compounds; these data obviously cannot be explained with the aid of a site-specific mechanism. Taking all these facts into account, it seems that crypto-hydroxyl radicals really mediate some biological free radical processes. It is of interest that a crypto-hydroxyl radical may be identical to a ferryl ion, as the latter can be formed during the decomposition of a peroxy precursor (instead of Reaction 2):

$$Fe(OOH)^{+} \rightarrow FeO^{2+} + HO^{-} \quad (3)$$

In conclusion, one may assume that there is no ''main villain'', because all oxygen radicals, despite their different reactivities in various free radical processes, may contribute to oxidative damaging processes in cells. It is a consequence of very complicated relationships of oxygen radicals in cells in which not only the chemical reactivity of a free radical, but also its lifetime, the place of formation, the ability to traverse through lipid membranes and the cytosol, and the capacity to participate in some indirect activation and deactivation processes are of great importance.

Chapter 2

NONENZYMATIC AND ENZYMATIC PRODUCTION OF OXYGEN RADICALS IN BIOLOGICAL SYSTEMS

I. INTRODUCTION

Production of oxygen radicals in biological systems may be tentatively divided into four categories: (1) nonenzymatic production during the oxidation of natural substrates; (2) enzymatic production catalyzed by reductases, dehydrogenases, and other enzymes; (3) the drug and xenobiotic stimulation of superoxide formation; and (4) the production of oxygen radicals by whole cells. The last two categories will be considered in subsequent chapters, while the first ones are discussed in this chapter.

All oxygen radicals whose physical and chemical properties were considered in Volume I, namely, superoxide ion, hydroxyl radical, and perhydroxyl radical, are formed in biological systems. However, in addition to these free radicals with definite structures, there are enigmatic reactive species named "crypto-hydroxyl" radicals. These species were also discussed in Volume I in connection with a mechanism of the Fenton reaction. Crypto-hydroxyl radicals play a very important role in biological processes and also will be considered below.

As we will see, in most cases the superoxide ion is a main precursor of all oxygen radicals. It can be formed in enzymatic processes and during the oxidation of various natural reductants. Other oxygen radicals are always formed nonenzymatically:

$$O_2^{\bar{\cdot}} + H^+ \rightleftarrows HOO\cdot \text{ (perhydroxyl radical)} \tag{1}$$

$$O_2^{\bar{\cdot}} + Fe^{3+} \rightarrow O_2 + Fe^{2+} \tag{2}$$

$$Fe^{2+} + H_2O_2 \rightarrow Fe^{3+} + HO\cdot + HO^- \text{ (hydroxyl radical)} \tag{3}$$

$$Fe^{2+} + H_2O_2 \rightarrow \text{"crypto-hydroxyl" radical} \tag{4}$$

Thus, the superoxide ion is a precursor of the perhydroxyl radical formed in Reaction 1 and hydroxyl and crypto-hydroxyl radicals formed in Reactions 3 and 4. In these cases genuine catalysts of the hydrogen peroxide decomposition are Fe^{2+} ions (the Fenton Reaction 3), which are formed by Reaction 2. In addition to the superoxide-driven Fenton reaction (Equations 2 and 3), there may be other pathways to hydroxyl and crypto-hydroxyl radicals. For example, ferric ions can be reduced by ascorbate or some other natural substrates:

$$AH^- + Fe^{3+} \rightarrow AH\cdot + Fe^{2+} \tag{5}$$

However, it is usually believed that Reaction 2 is the most important pathway to active oxygen radicals.

II. NONENZYMATIC PRODUCTION OF THE SUPEROXIDE ION

In Volume I (Chapter 3) we discussed the conditions which favor the one-electron reduction of dioxygen by organic and inorganic electron donors, including hydroquinones, pyridinium compounds, iron and copper ions and complexes, etc. Similar processes probably proceed in cells which contain many oxidizable organic and inorganic compounds. The ability of ascorbate, α-tocopheryl anion, dihydroflavins, and metalloporphyrins to reduce dioxygen to a superoxide ion was already discussed in Volume I.

It seems that thiols may be an important source of the superoxide ion in biological systems. In 1974, Misra[1] had shown that many thiols, including dithiothreitol, β-mercaptoethanol, ethyl mercaptan, and glutathione induced the SOD-inhibitable reduction of nitroblue tetrazolinium (NBT) under aerobic conditions. On these grounds, the formation of $O_2^{\cdot-}$ during the autoxidation of thiols was proposed. In subsequent works the nonenzymatic and enzymatic oxidation of cysteine and glutathione has been studied. Saez et al.[2] obtained the DMPO spin adducts of cysteinyl and hydroxyl radicals in autoxidation of cysteine in the presence of DMPO. Suppression of the DMPO-OH formation by catalase and the transformation of it into the DMPO-C_2H_4OH spin adduct in the presence of ethanol confirms the production of genuine hydroxyl radicals. The DMPO-OOH spin adduct was not found, possibly due to a low rate of the reaction of $O_2^{\cdot-}$ with DMPO. However, the fact that SOD stimulated the formation of both DMPO-cysteine and DMPO-OH spin adducts indicated the formation of the superoxide ion in the oxidation of cysteine.

It has been shown[3,4] that the oxidation of thiols (including cell membrane thiol groups) is catalyzed by transition metal ions.

$$RSH + M^{n+} \rightarrow RS\cdot + M^{(n-1)+} + H^+ \tag{6}$$

$$RS\cdot + RSH \rightarrow RSSR^{\cdot-} + H^+ \tag{7}$$

Thiyl radicals are apparently incapable of transferring an electron to dioxygen (Reaction 8) and are added to O_2 (Reaction 9).

$$RS\cdot + O_2 \not\rightarrow RS^+ + O_2^{\cdot-} \tag{8}$$

$$RS\cdot + O_2 \longrightarrow RSOO\cdot \tag{9}$$

(However, in the case of cysteine and glutathione, RSOO· radicals are instable and are converted to RSO·.)[5] The superoxide ion is apparently formed via Reaction 10 or as a result of the dioxygen reduction by metal ions:

$$RSSR^{\cdot-} + O_2 \rightarrow RSSR + O_2^{\cdot-} \tag{10}$$

$$M^{(n-1)+} + O_2 \rightarrow M^{n+} + O_2^{\cdot-} \tag{11}$$

In addition to nonenzymatic oxidation, thiols are oxidized to free radicals by peroxidases. Thus, cysteine was oxidized to a free radical (identified as a DMPO spin adduct) by horseradish peroxidase (HRP) under aerobic and anaerobic conditions (however, in the last case only in the presence of hydrogen peroxide).[6] Excluding the enzymatic production of RS·, the oxidation process includes the same stages as a nonenzymatic one. Under the same conditions, L-cysteine sulfinic acid, a metabolite of the oxidative degradation of cysteine, also produced a free thiyl radical, but only via nonenzymatic oxidation.

Similar to the cysteine oxidation, the DMPO-glutathione spin adduct was obtained in the oxidation of glutathione catalyzed by HRP,[7] lactoperoxidase (in the presence of phenols),[8] and prostaglandin H synthase or ram seminal vesicle microsomes.[9] One may expect that all these enzymes must catalyze the superoxide production by Reactions 10 and 11. Indeed, the HRP-catalyzed oxidation of glutathione was accompanied by O_2 uptake; moreover, GSSG was formed over twice the initial hydrogen peroxide concentration, which indicates the participation of dioxygen in the one-electron oxidation of GSH. In contrast to these enzymes, glutathione peroxidase oxidizes glutathione via a two-electron pathway without the production of thiyl radicals. It is of interest that at the initial stage of glutathione oxidation by HRP, glutathione peroxidase stimulates the formation of thiyl radicals. The mechanism of stimulation is uncertain, although it was proposed that it may be caused by the oxidation of the seleno prosthetic group.[9]

The NADH free radical (NAD·) reduces dioxygen with a diffusion-controlled rate constant ($2 \cdot 10^9 \ M^{-1} \ s^{-1}$).[10] Therefore, one may expect that NADH autoxidation will produce a superoxide ion in biological systems (that also follows from comparison of the one-electron reduction potentials of NADH and O_2). However, it seems that this reaction proceeds with a reasonable rate only in the presence of catalysts. For example, phenazine methosulfate (PMS) catalyzed the SOD-inhibitable reduction of NBT by NADH that was interpreted as evidence of the formation of a superoxide ion during the NADH oxidation.[11] However, SOD did not affect the reduction of cytochrome c in the same system. On these grounds, it was proposed that superoxide ion was formed during the reoxidation of the NADH-reduced PMS. Picker and Fridovich[12] have further studied the mechanism of the PMS action on the superoxide production by NADH. They concluded that $O_2^{\overline{\cdot}}$ is formed only in the presence of NBT as a result of the reduction of dioxygen by NBTH· radicals. However, this conclusion contradicts the results obtained by Bielski et al.[13] who found that this reaction has a very small rate.

Superoxide production by NADH oxidation is accelerated in the presence of catalase[14] which apparently catalyzes the oxidation of NADH to a free radical:

$$\text{NADH} \xrightarrow[\text{H}_2\text{O}_2]{\text{Catalase}} \text{NAD}\cdot + \text{H}_2\text{O} \qquad (12)$$

$$\text{NAD}\cdot + \text{O}_2 \longrightarrow \text{NAD}^+ + \text{O}_2^{\overline{\cdot}} \qquad (13)$$

The reaction requires Mn^{2+} or phenols, whose role in superoxide production is unclear. Formation of the superoxide ion also increased under illumination with visible light in the presence of tricine[15] or with ultraviolet light[16] and in the presence of lactate dehydrogenase.[17] In the latter case, it was proposed that lactate dehydrogenase (LDH) catalyzed the oxidation by forming an active complex with NADH.

$$\text{LDH} + \text{NADH} \rightleftarrows \text{LDH·NADH} \qquad (14)$$

$$\text{LDH·NADH} + \text{O}_2^{\overline{\cdot}} + \text{H}^+ \rightarrow \text{LDH·NAD}\cdot + \text{H}_2\text{O}_2 \qquad (15)$$

$$\text{LDH·NAD}\cdot + \text{O}_2 \rightarrow \text{LDH·NAD}^+ + \text{O}_2^{\overline{\cdot}} \qquad (16)$$

It was shown in Volume 1 (Chapter 10) that iron(II) porphyrins are not good catalysts of one-electron reduction of dioxygen to superoxide ion, since their dioxygen adducts do not dissociate to form $O_2^{\overline{\cdot}}$ and iron(III) porphyrins. An exclusion is iron(II) porphyrin with a thiolate ligand which becomes capable of producing a superoxide ion as a result of the dissociation of its six-coordinate dioxygen complex.[18] However, in contrast to porphyrins, hemoglobins are oxidized more easily to form superoxide ion. This process may be of importance as the methemoglobin formed cannot be oxygenated and is therefore physiologically inactive. The same is true for the autoxidation of myoglobin.

Misra and Fridovich[19] have shown that superoxide ion is formed during the autoxidation of shark hemoglobin (which is oxidized more rapidly than human hemoglobin) under physiological conditions. In the case of mammalian hemoglobin, the superoxide ion was formed only at a high concentration of potassium phosphate (1 M).[20] Superoxide production was measured by the SOD-inhibitable oxidation of epinephrine and the SOD-inhibitable reduction of cytochrome c. Superoxide ion was also produced during the autoxidation of isolated α- and β-chains of human hemoglobin,[21] in the autoxidation of oxyhemoglobin,[22] and in the autoxidation of soluble trypsin-cleaved microsomal ferrocytochrome b_5.[23]

There are various pathways of the stimulation of superoxide production during the autoxidation of hemoglobin. The irradiation of human oxyhemoglobin with white light of low intensity induced the photodissociation of oxyhemoglobin to form methemoglobin and the superoxide ion (determined via the SOD-inhibitable cytochrome c reduction).[24]

$$HbO_2 \xrightarrow{h\nu} metHb + O_2^{\cdot -} \quad (17)$$

The α-chain of human oxyhemoglobin is about three to four times more effective as a superoxide generator than the β-chain.[25] It was found[26,27] that the decomposition of oxyhemoglobin to metHb and $O_2^{\cdot -}$ was accelerated by phenylhydrazine (a hemolytic agent). The superoxide ion was identified on the basis of its ESR spectrum using the rapid freezing method. However, in this system $O_2^{\cdot -}$ was formed as a result of one-electron reduction of dioxygen by phenyldiazine (an intermediate of the oxidation of phenylhydrazine) and not due to the release from oxyhemoglobin.[28]

Autoxidation of hemoglobin and myoglobin is accelerated by anions; this reaction may proceed by a reductive displacement mechanism:

$$HbO_2 + L^- \rightarrow MetHb^+L^- + O_2^{\cdot -} \quad (18)$$

However, a more elaborate scheme was recently proposed for a complete explanation of experimental data.[29] The accelerating effect of electron donors on hemoglobin oxidation and superoxide production was also shown. It has been proposed[30] that δ-aminolevulinic acid (a heme precursor) in the enolic form may transfer an electron to oxyhemoglobin, inducing the production of metHb, H_2O_2, and $O_2^{\cdot -}$. Similarly, glutathione accelerated the autoxidation of human hemoglobin, but in this case it was shown that glutathione is unable to reduce oxyhemoglobin and the superoxide ion is formed in the direct reaction with glutathione.[31]

Iron hemoproteins (hemoglobin and myoglobin) are not only natural iron-containing complexes which are capable of producing the superoxide ion. It was found that ferredoxin (a non-heme iron protein) and putidamonooxin (an oligomeric conjugated iron-sulfur protein and the dioxygen-activating component of 4-methoxybenzoate monooxygenase) also generate the superoxide ion. Thus, the superoxide ion was detected (on the basis of its ESR spectrum and the SOD-inhibitable oxidation of epinephrine) in the autoxidation of bacterial ferredoxin by dithionite[32,33] and the oxidation of bacterial and spinach ferredoxin by ferredoxin-NADH reductase.[34] Superoxide production increased with rising pH. It was concluded that the superoxide ion was formed as a result of a direct electron transfer from ferredoxin and not from flavoprotein to dioxygen.[35] Ferredoxin also increased the oxygen uptake by isolated chloroplasts.[36] SOD inhibited its effect, which indicated the formation of superoxide ion. On the basis of kinetic analysis, it was proposed[37] that the oxidation of reduced ferredoxin by dioxygen proceeds via a two-step mechanism with the formation of the reduced ferredoxin-O_2 complex as an intermediate:

$$Fd(II) + O_2 \rightleftarrows Fe(II)O_2 \quad (19)$$

$$Fe(II)O_2 \rightarrow Fd(III) + O_2^{\cdot -} \quad (20)$$

Putidamonooxin stimulated the formation of superoxide ion which was detected by the reduction of modified cytochrome c and the SOD-inhibitable formation of lactoperoxidase Compound III.[38] However, a part of the total electron flux channeled by pytidamonooxin to dioxygen (which corresponds to one-electron reduction of dioxygen) was only about 6%.

Recently, it was shown[39,40] that glycated human albumin and glycated polylysine reduced NBT and cytochrome c. The reaction was partially inhibited by SOD. In addition, glycated polylysine in the presence of lactodehydrogenase stimulated the SOD-inhibitable oxidation of NADH. It was proposed that superoxide ion may be formed as a result of one-electron oxidation of the ketoamine form of glycated proteins.

III. ENZYMATIC PRODUCTION OF THE SUPEROXIDE ION

A. XANTHINE OXIDASE

In 1968 McCord and Fridovich[41] showed that xanthine oxidase catalyzed the reduction of cytochrome c under aerobic conditions. The reaction was inhibited by bovine erythrocyte carbonic anhydrase, which contained superoxide dismutase as an admixture. These findings may be considered the first reliable evidence of the generation of a free superoxide ion in biological systems, although some facts indicating the possibility of its formation by this enzyme were obtained much earlier.[42] Now, xanthine oxidase is undoubtedly the most popular superoxide generator used in numerous *in vitro* experiments.

From 1968 to 1970, decisive evidence of superoxide production by xanthine oxidase was obtained. Formation of superoxide ion during the reoxidation of reduced xanthine oxidase by dioxygen and the steady-state oxidation of xanthine by xanthine oxidase has been shown by the ESR method using ordinary and ^{17}O-enriched dioxygen.[43,44] Arneson[45] applied SOD-inhibitable chemiluminescence for superoxide detection in the xanthine-xanthine oxidase and the acetaldehyde-xanthine oxidase systems. Later on, the spin-trapping technique (using DMPO as a spin trap)[46,47] and SOD-inhibitable lucigenin-dependent chemiluminescence[48] were applied for detecting $O_2^{\cdot-}$ in the xanthine-xanthine oxidase system.

Enhancement of chemiluminescence stimulated by xanthine oxidase with rising pH led to a proposal that under these conditions, a new intermediate, $H_2O_4^-$, is formed. However, this phenomenon was later rightly explained by the accumulation of the superoxide ion due to its relative stability at elevated pH values.[49]

The mechanism of the production of $O_2^{\cdot-}$ and H_2O_2 by xanthine oxidase was developed by Olson et al.[50] This mechanism is based on a proposal that the one-electron transfer equilibriums between redox centers of xanthine oxidase (one molybdenum, one FAD, and two Fe-S centers) are rapid and are governed by their reduction potentials. This conclusion was recently confirmed by the pulse radiolysis study of xanthine oxidase.[51] During the oxidation of reduced xanthine oxidase, six electrons may be transferred to dioxygen. It was found that reoxidation exhibits a biphasic character, with the rate of the first fast stage being approximately ten times greater than that of the slow second stage. A simplified scheme of reoxidation is presented below.[52]

$$\mathrm{XO(6)} \xrightarrow{O_2} \begin{matrix}\mathrm{XO(4)}\\+\\H_2O_2\end{matrix} \xrightarrow{O_2} \begin{matrix}\mathrm{XO(2)}\\+\\H_2O_2\end{matrix} \xrightarrow{O_2} \begin{matrix}\mathrm{XO(1)}\\+\\O_2^{\cdot-}\end{matrix} \xrightarrow{O_2} \begin{matrix}\mathrm{XO(0)}\\+\\O_2^{\cdot-}\end{matrix}$$

(Here, XO(n) is reduced xanthine oxidase containing n electrons). In accord with this scheme, two $O_2^{\cdot-}$ are produced for each enzyme molecule reoxidized.[53] The individual rate constants are estimated to be 35 s^{-1}, 33 s^{-1}, 20 s^{-1}, and 0.9 s^{-1}, respectively.

It is believed that in all steps except the last, electrons are transferred to dioxygen from fully reduced flavin which is rapidly regenerated during the first two steps as a consequence of intramolecular electron transfer from the other redox centers of xanthine oxidase: iron-sulfur centers and molybdenum. However, when XO(2) was formed, fully reduced flavin cannot be regenerated after the transfer of the first electron to dioxygen, because the last electron remains in the iron-sulfur centers due to their high affinity for electrons.[50] Because of this, superoxide ion is released in the two last steps of reoxidation. This mechanism seems to contradict the previous conclusion[54] that $O_2^{\overline{\cdot}}$ is formed in the initial steps of the oxidation of xanthine oxidase and that H_2O_2 is produced only by dismutation. This apparent discrepancy, however, is explained by different levels of the reduction of xanthine oxidase at different xanthine and dioxygen concentrations. Thus, it was found that when the concentration of dioxygen is greater than that of xanthine, only XO(2) and XO(0) states exist, and oxidation occurs largely by a one-electron transfer. As a result, under such conditions the superoxide ion is a principal product of the oxidation of xanthine oxidase.[52]

The above mechanism of production of the superoxide ion by xanthine oxidase was confirmed in experiments with lumazine.[55] It was proposed that the substitution of xanthine by a low-turnover substrate, lumazine, must increase the univalent reduction of dioxygen due to a decrease in the steady-state reduction level of xanthine oxidase. Indeed, it was found that univalent flux increased from 25% in the case of xanthine to 75% in the case of lumazine (at 5 μM substrate, pH 8.1).

B. PEROXIDASES, CATALASE, PGH SYNTHASE, LIPOXYGENASE, AND OTHER ENZYMES

It has been shown that the superoxide ion is formed during the oxidation of various substrates catalyzed by peroxidases, catalase, PGH synthase, and lipoxygenase. Thus, the oxidation of dihydroxyfumarate by horseradish peroxidase (HRP) stimulated the formation of superoxide ion, identified on the basis of its ESR spectrum.[32] However, in this case the possibility of nonenzymatic production of superoxide ion cannot be excluded, since the superoxide ion may be formed as a result of the reduction of dioxygen by the dihydroxyfumarate (DHF) radical cation.[56]

$$\mathrm{DHF} + \mathrm{H_2O_2} \xrightarrow{\mathrm{HRP}} \mathrm{DHF}^{\dot{+}} + \mathrm{H_2O} \qquad (21)$$

$$\mathrm{DHF}^{\dot{+}} + \mathrm{O_2} \longrightarrow \text{dioxosuccinate} + \mathrm{O_2^{\overline{\cdot}}} \qquad (22)$$

It has also been shown[57] that HRP oxyperoxidase (Compound III) decomposed to produce superoxide ion, which was detected on the basis of the SOD-inhibitable oxidation of epinephrine.

Medeiros et al.[58] reported the inhibitory effect of SOD on the chemiluminescence produced during the oxidation of reduced glutathione by HRP. On these grounds, it was proposed that the superoxide ion formed in this system takes part in the production of photoemissive species.

However, the most important substrates which are oxidized by peroxidases to form the superoxide ion are probably NADH and NADPH. It was proposed[59,60] that the superoxide production by HRP in the presence of NADH is due to the reduction of dioxygen by NAD·, which possibly arises in the NADH oxidation by oxygenated intermediates of peroxidase (Compounds I and II).

$$\mathrm{HRP} + \mathrm{H_2O_2} \rightarrow \text{Compound I} \qquad (23)$$

$$\text{Compound I} + \mathrm{NADH} \rightarrow \text{Compound II} + \mathrm{NAD\cdot} \qquad (24)$$

$$\text{Compound II} + \mathrm{NADH} \rightarrow \mathrm{HRP} + \mathrm{NAD\cdot} \qquad (25)$$

$$\mathrm{NAD\cdot} + \mathrm{O_2} \rightarrow \mathrm{NAD^+} + \mathrm{O_2^{\overline{\cdot}}} \qquad (26)$$

The superoxide ion is also formed in the oxidation of NAD dimers by HRP, apparently by the same mechanism.[61]

It has been recently found[62] that PGH synthase and soybean lipoxygenase catalyze superoxide production during the oxidation of arachidonic and linoleic acids, respectively, in the presence of NADH or NADPH. It is believed that superoxide production is a result of enzyme hydroperoxidase activity and that $O_2^{\overline{\cdot}}$ is formed by the same mechanism as in the case of HRP. As was already pointed out, the superoxide ion is formed by the one-electron oxidation of NADH catalyzed by catalase.[14]

Rotilio et al.[63] concluded that the ability of pig kidney diamine oxidase to initiate the sulfite oxidation and the aerobic reduction of cytochrome c points to the formation of the superoxide ion in this system. This conclusion is possibly supported by the inhibitory effect of copper chelators.[64] Sulfite oxidation was also initiated by ferredoxin-NADP oxidase.[65] It is believed that in this case both flavoprotein and iron-sulfur protein are responsible for superoxide production.

2-Nitropropane dioxygenase from *Hausenula mrakii* catalyzes denitrification of 2-nitropropane by dioxygen:

$$2\ CH_3CH(NO_2)CH_3 + O_2 \rightarrow 2\ CH_3COCH_3 + 2\ HNO_2 \qquad (27)$$

It was found[66] that SOD significantly inhibited nitrite production. The reaction was also inhibited by cytochrome c, epinephrine, NADH, and tyron. SOD also inhibited the concomitant reduction of cytochrome c and the oxidation of epinephrine. It was supposed that Reaction 27 was mediated by the superoxide ion which was formed at the iron or flavin moiety of the enzyme.

SOD inhibited the oxygenase reaction catalyzed by ribulose-1,5-diphosphate carboxylase isolated from spinach leaves that indicates the superoxide production in this enzymatic process.[67] Fontecave et al.[68] have recently found that flavin reductase from an enzyme system from *Escherichia coli* catalyzed the superoxide production. The superoxide ion was identified as the DMPO-OOH spin adduct and was detected via the cytochrome c reduction. It was proposed that the superoxide ion generated by flavin reductase inactivates ribonucleotide reductase.

It is of interest that superoxide ion may be formed in the reactions catalyzed by SOD in the presence of strong electron acceptors. Thus, in the presence of tetranitromethane, SOD catalyzed the oxidation of hydrogen peroxide by dioxygen, evidently due to a high rate of Reaction 30.[69]

$$Cu^{2+}\text{–}SOD + HO_2^- \rightleftarrows Cu^{+}\text{–}SOD + O_2^{\overline{\cdot}} + H^+ \qquad (28)$$

$$Cu^{+}\text{–}SOD + O_2 \rightarrow Cu^{2+}\text{–}SOD + O_2^{\overline{\cdot}} \qquad (29)$$

$$O_2^{\overline{\cdot}} + C(NO_2)_4 \rightarrow O_2 + \cdot C(NO_2)_3 + NO_2^- \qquad (30)$$

C. PRODUCTION OF THE SUPEROXIDE ION BY MITOCHONDRIA

Formation of the superoxide ion in mitochondria was first shown by Loschen et al.,[70] who found that washed mitochondrial membranes from beef heart mitochondria oxidized epinephrine in the presence of antimycin A. SOD inhibited epinephrine oxidation. It was proposed that the site of superoxide production must be placed between rotenone- and antimycin-sensitive sites. Later on, superoxide production by mitochondrial membranes was shown for mitochondria isolated from beef and rat heart,[71-75] rat brain,[76] normal and neoplasmic tissues,[77] and potato tubers.[78] In whole mitochondria, superoxide production is difficult to register due to the presence of matrix SOD. Nonetheless, superoxide ion was detected in intact rat heart mitochondria in the presence of ATP on the basis of the SOD-inhibitable epinephrine oxidation.[72] Similarly, superoxide ion was detected in isolated rat heart mitochondria with the use of tyron (4,5-dihydroxybenzene-1,3-disulfonic acid) as a spin trap[79] and in rat skeletal muscle mitochondria on the basis of epinephrine oxidation.[80] It was found[79] that (3-chlorophenyl)hydrazone of carbonyl cyanide (an uncoupler of oxidative phosphorylation) increased the rate of superoxide production.

Submitochondrial particles from beef and rat heart mitochondria produced superoxide ion in the presence of antimycin A at a rate of about 5.3 nmol/min/mg protein (superoxide production was measured as the SOD-inhibitable reduction rate of cytochrome c).[71] In mitochondria from whole testis, seminiferous tubules, and Leydig cells, the rate of superoxide production in the presence of various respiratory inhibitors was substantially smaller and was equal to 0.27 to 1.67 nmol/min/mg protein.[81]

Superoxide formation apparently increases in tumor mitochondria.[82] Indeed, it was recently shown[83] that the rate of superoxide production by submitochondrial particles from ascites hepatomas was 1.5 to 2 and 10 times higher than the rates of superoxide production by submitochondrial particles from bovine heart and rat liver, respectively. This fact was explained by a decrease in the SOD activity in malignant tissues.

A very important question is the identification of the site of superoxide production in mitochondrial membrane. Recently, Boveris and Cadenas[84] discussed proposals made in previous works and concluded that there are two main sources of mitochondrial superoxide ion: the $O_2^{\cdot-}$ generator acting in Complex III between the rotenone- and antimycin-sensitive sites and NADH dehydrogenase. The existence of two different sites of superoxide production follows from the different effects of respiratory chain inhibitors. It was found that $O_2^{\cdot-}$ production by NADH dehydrogenase was stimulated by rotenone and was not affected by cyanide,[73,75,78] while $O_2^{\cdot-}$ production by the ubiquinone-cytochrome b site is inhibited by rotenone and cyanide.[75,78] In contrast to rotenone, the effect of cyanide is hard to explain with the aid of the classic scheme of electron transport, but it can be explained by the Mitchell proton-motive hypothesis.[78] Turrens and Boveris[75] concluded that NADH dehydrogenase is a quantitatively less active $O_2^{\cdot-}$ generator than the ubiquinone-cytochrome b site, since the fully reduced NADH dehydrogenase produced 0.90 ± 0.07 nmol $O_2^{\cdot-}$ per minute per milligram protein, while the ubiquinone-cytochrome b site generated 1.85 ± 0.20 nmol $O_2^{\cdot-}$ per minute per milligram protein at pH 7.4. In contrast to this, Takeshige and Minakami[73] did not find the superoxide production by the ubiquinone-cytochrome b site. Recently, Nohl[85] concluded that mitochondria contain one more superoxide generator, the exogenous NADH oxidoreductase, which is not involved in energy-linked respiration.

The presence of antimycin A, a specific inhibitor of mitochondrial electron transfer between b and c_1 cytochromes, together with succinate, is an obligatory condition of superoxide production by mitochondria. However, other specific inhibitors such as 2-*n*-nonyl-4-hydroxyquinoline *N*-oxide and funiculosin may also be used for the stimulation of superoxide production.[86]

Three compounds, namely, ubisemiquinone, ubiquinol, and cytochrome b, are considered potential generators of the superoxide ion in Complex III. One-electron oxidation of ubisemiquinone by dioxygen seems to be a very important route to the superoxide ion.[74,75,87,88] This conclusion is supported by the existence of a linear relationship between the ubiquinone and hydrogen peroxide content in mitochondria and by production of the superoxide ion by isolated Complex I or III.[84] Recently,[89] ubisemiquinone was accepted as a single generator of the superoxide ion in Complex III due to the inability of ubiquinol and reduced cytochrome b to catalyze the H_2O_2 production in intact rat heart mitochondria in the presence of antimycin A and myxothiazol (an inhibitor of the reduction of the Rieske iron-sulfur center). Within the limits of the Mitchell hypothesis, it was assumed that $O_2^{\cdot-}$ production is a consequence of autoxidation of the unstable ubisemiquinone in the so-called center *o* of Q cycle under the condition that the cytochrome b reoxidation in center *i* is inhibited. However, Nohl and Jordan[90,91] found that dioxygen does not affect the ESR spectrum of ubisemiquinone formed in rat heart mitochondria in the presence of myxothiazol or antimycin A. In the authors' opinion, this fact indicates an importance of cytochrome b as a superoxide generator.

It is obvious that, in the first place, the efficiency of superoxide production by ubisemiquinone, cytochrome b, or some other potential $O_2^{\overline{\cdot}}$ generators is determined by their redox potentials. Boveris and Cadenas[84] concluded that all reduced carriers are expected to be oxidized by dioxygen since its reduction potential $E_o(H_2O/O_2)$ is equal to 0.82 V. However, this conclusion bears no relation to the production of superoxide ion by one-electron oxidation of respiratory carriers as the one-electron reduction potential of dioxygen $E_o(O_2^{\overline{\cdot}}/O_2) = -0.16$ V (Volume I, Chapter 1). Nohl and Jordan[91] believed that low-potential cytochrome b_{566} is the only component in the antimycin-rotenone section of mitochondrial respiratory chain having a reduction potential (-0.050 V) close to the one-electron reduction potential of dioxygen, and therefore cytochrome b is the most probable mitochondrial superoxide generator.

However, a direct measurement of the equilibrium for Reaction 31 in aprotic media[92] showed that $K_{31} = 2.85$ for ubiquinone Q_9:

$$O_2^{\overline{\cdot}} + Q_9 \rightleftarrows O_2 + Q_9^{\overline{\cdot}} \tag{31}$$

Therefore, the one-electron reduction potentials of ubiquinones and dioxygen in aprotic media are practically equal, and ubisemiquinone must be a thermodynamically even more effective superoxide generator than cytochrome b. Indeed, in contrast to data obtained by Nohl and Jordan,[91] ubisemiquinone was readily oxidized by dioxygen in DMF.[92] Nonetheless, there is one more fact which must be taken into account when ubisemiquinone and cytochrome b are compared as superoxide generators. It is well known that the ubiquinone pool in mitochondria is much greater than the content of other electron carriers. Therefore, it may be expected that an excess of ubiquinone will shift the equilibrium for Reaction 31 to the right and suppress the efficiency of superoxide production by ubiquinone.

In contrast to the above proposals, Forman and Kennedy[76,93] concluded that primary mitochondrial dehydrogenases rather than electron carriers are real generators of the superoxide ion. (Thermodynamically, primary dehydrogenases having more negative redox potentials must certainly be more favorable superoxide producers.) These authors showed that superoxide production depends on the nature of the dehydrogenase. Thus, it was found that superoxide ion was formed in rat liver mitochondria with succinate dehydrogenase and dihydroorotate dehydrogenase and in rat brain mitochondria with dihydroorotate dehydrogenase, but it was not formed in rat brain mitochondria with succinate dehydrogenase. Furthermore, at least in rat liver mitochondria, ubiquinone and cytochrome b were not superoxide generators, since blocking the ubiquinone reduction by thenoyltrifluoroacetone and removing cytochrome b from mitochondrial particles with Triton® left the rate of superoxide production unchanged.

Henry and Vignais[94] have shown that plasma membrane vesicles isolated from *Paracocous denitrificans* cells (a free-living bacterium possessing an aerobic respiratory chain resembling the mitochondrial chain) oxidized epinephrine. The reaction was inhibited by SOD (60%). It was proposed that the site of superoxide production is the low-potential cytochrome b.

D. PRODUCTION OF THE SUPEROXIDE ION BY MICROSOMES

It has been shown that rat liver microsomes stimulated the NADPH-dependent SOD-inhibitable oxidation of epinephrine and the reduction of NBT[95-97] as well as the formation of lactoperoxidase Compound III.[98] These findings were interpreted in favor of formation of the superoxide ion by microsomes. There are two potential generators of superoxide ion in microsomes: NADPH-cytochrome P-450 reductase and cytochrome P-450. Superoxide production catalyzed by NADPH-cytochrome P-450 reductase has been shown on the basis of the SOD-inhibitable oxidation of epinephrine to adrenochrome[95,99,100] and neotetrazolium to formazon;[101] later on, superoxide ion was detected in this system with the aid of spin-trapping agents DMPO and PBN.[102,103] It was found that both prosthetic groups of the enzyme, FAD and FMN, appear to be equally effective in superoxide production.[103]

The use of NBT reduction as an assay of $O_2^{\cdot-}$ production by the reductase turns out to be insufficient despite the inhibition of this reaction by SOD, because superoxide production may be in this case a consequence of one-electron reduction of dioxygen by the NBT free radical.[104]

$$NBT \xrightarrow{\text{reductase}} NBT^{\cdot} \tag{32}$$

$$NBT^{\cdot} + O_2 \longrightarrow NBT_{ox} + O_2^{\cdot-} \tag{33}$$

Production of the superoxide ion during autoxidation of cytochrome P-450 apparently was first shown by Sligar et al.[105] and Sligar and Debrunner.[106] These authors found that low-level chemiluminescence from the autoxidation of substrate-bound cytochrome $P\text{-}450_{cam}$ (from the monooxygenase system of *Pseudomonas putida*) was inhibited by SOD. Bartoli et al.[107] tried to estimate a ratio of superoxide production by NADPH-cytochrome P-450 reductase and cytochrome P-450 in rat liver microsomes. They compared superoxide production in microsomes from 1-day-old rats (in which NADPH-cytochrome P-450 reductase is mostly responsible for formation of the superoxide ion) and from adult rats and concluded that in adult rats about 50% $O_2^{\cdot-}$ is formed by the reductase.

However, subsequent results do not agree with this conclusion. It was found[108,109] that cytochrome P-450 greatly increased production of the superoxide ion and hydrogen peroxide (measured by succinylated cytochrome c reduction and the formation of Compound III of lactoperoxidase) in a reconstituted monooxygenase system (NADPH-cytochrome P-450 reductase + the substrate of *O*-dealkylation, 7-ethoxycoumarin). It was proposed that superoxide ion was formed during the dissociation of the oxycomplex of cytochrome P-450. Donating the second electron to oxycytochrome P-450 after incorporation of cytochrome b_5 into phospholipid vesicles containing NADPH-cytochrome P-450 reductase and cytochrome P-450 decreased superoxide production.[110] It was concluded[109] that all H_2O_2 was formed by dismutation of superoxide ion, since the $O_2^{\cdot-}/H_2O_2$ ratio was equal to 2. Stoichiometry of the $O_2^{\cdot-}$, H_2O_2, and H_2O formation in rabbit liver microsomes was later determined by Zhukov and Archakov.[111]

A predominant role of cytochrome P-450 in microsomal superoxide production was supported by the results obtained by Morehouse et al.[112] These authors showed that superoxide production is only a small fraction of the total NADPH-dependent reduction capacity of NADPH-cytochrome P-450 reductase. Thus, it was found that contrary to the acetylated cytochrome c reduction by xanthine oxidase, which was completely inhibited by SOD, the reduction of acetylated cytochrome c by NADPH-cytochrome P-450 reductase was inhibited by SOD only to a small degree. Close results were obtained for the detergent-solubilized NADPH-cytochrome P-450 reductase.[113]

In comparison with the purified reductase, microsomes produced a greater amount of superoxide ion (about 1.5%),[112] which indicates a more important role of cytochrome P-450 in superoxide production. Recently, superoxide production by rat liver microsomes was studied with the aid of a highly specific test for detection of $O_2^{\cdot-}$: the lucigenin-dependent chemiluminescence.[48]

E. PRODUCTION OF THE SUPEROXIDE ION BY CELL NUCLEI

Bartoli et al.[114] found that nuclei isolated from Ehrlich-Lettre ascites tumor cells oxidized epinephrine in the presence of NADPH. As the epinephrine oxidation was inhibited by SOD, these authors concluded that tumor cell nuclei are able to produce superoxide ion. This process is supposedly catalyzed by nuclear NADPH-cytochrome c (P-450) reductase. Later on, it was shown[115] that the superoxide ion (identified by its DMPO-spin adduct) is generated by hamster hepatic nuclei (supposedly by flavoprotein FAD-monooxygenase). Peskin et al.[116,117] have shown that nuclear membranes from hepatoma 22a ascites cells oxidized epinephrine; the reaction was inhibited by SOD. It was proposed that superoxide production was catalyzed by an enzyme different from NADPH-cytochrome c reductase as this enzyme was able to utilize both NADH and NADPH and was sensitive to cyanide.

Using the SOD-inhibitable reduction of succinoylated cytochrome c as an assay for detecting the superoxide ion, Yusa et al.[118] studied superoxide production by porcine lung and liver nuclei. Superoxide production was NADH- and NADPH-dependent in the case of liver nuclei and, in general, NADPH-dependent in the case of lung nuclei. Hyperoxia enhanced the generation of superoxide ion in nuclei: NADH-dependent superoxide production increased from 0 to 2.21 nmol/min/mg protein for lung nuclei and from 0.16 to 1.34 nmol/min/mg protein for liver nuclei, and NADPH-dependent superoxide production increased from 0.20 to 1.20 nmol/min/mg protein for liver nuclei and remained constant (0.45 nmol/min/mg protein) for lung nuclei when dioxygen concentration increased from 0 to 100%. Generation of superoxide ion by rat liver nuclei was also determined on the basis of SOD- and catalase-inhibitable nickling in nuclear DNA in the presence of NADPH.[119]

IV. PRODUCTION OF HYDROXYL, CRYPTO-HYDROXYL, AND SITE-SPECIFIC RADICALS

While there are a number of enzymatic and nonenzymatic pathways to the superoxide ion in biological systems, it seems there is only one route to hydroxyl radicals: the Fenton or Fenton-like reactions. These reactions (the reduction of hydrogen peroxide by transition metal cations) were discussed in detail in Volume I (Chapter 9).

$$H_2O_2 + M^{n+} \rightarrow HO\cdot + HO^- + M^{(n+1)+} \qquad (34)$$

Also considered there were the reductants of transition metal ions which are able to recycle $M^{(n+1)+}$ to M^{n+}; among them, the superoxide ion is apparently the most important species initiating the superoxide-driven Fenton reaction:

$$O_2^{\overline{\cdot}} + F^{3+} \rightarrow O_2 + Fe^{2+} \qquad (2)$$

$$Fe^{2+} + H_2O_2 \rightarrow Fe^{3+} + HO\cdot + HO^- \qquad (3)$$

In many cases, the Fenton and Fenton-like reactions generate active oxygen radicals similar, but not identical, to free hydroxyl radicals. These radicals can be "crypto-hydroxyl" radicals which probably are oxy or peroxy iron complexes. Other species which are different from free hydroxyl radicals are site-specific hydroxyl radicals. It is supposed that in the case when iron ion binds to protein or another compound, the production of hydroxyl radicals occurs only close to the site of the iron ion fixation that changes the apparent reactivity of the hydroxyl radicals formed.

If the Fenton and Fenton-like reactions are really the most important pathways to hydroxyl radicals or hydroxyl-like species in biological systems, then the HO· production must chiefly depend on the presence of transition metal ions, since hydrogen peroxide is always present in all cellular components. Therefore, the search for the "physiological" metal ion (iron ion) catalyst of the Fenton reaction becomes one of the key questions in studying hydroxyl radical production. However, at the beginning, it is important to consider the possible pathways of the reduction of ferric ions and complexes in biological systems.

A. ENZYMATIC AND NONENZYMATIC REDUCTION OF FERRIC IONS AND COMPLEXES

It was already pointed out that the superoxide ion is probably the most important reductant of ferric ions. Therefore, one may propose that enzymes catalyzing the one-electron reduction of dioxygen must also stimulate the HO· production when iron ions or complexes are available. Indeed, as early as 1970, Beachamp and Fridovich[120] proposed that hydroxyl radicals are responsible for the production of ethylene from methional catalyzed by xanthine oxidase. After this work, the ability of xanthine oxidase to produce hydroxyl and "crypto-hydroxyl" radicals in the presence of adventitious or specially added iron ions and complexes has been shown by many authors. Thus, Fong et al.[121] determined HO· production using the oxidation of reduced cytochrome c. Richmond et al.[122] and Motohashi and Mori[123] applied hydroxylation of salicylate to the detection of hydroxyl radicals.

The presence of iron ions or complexes is an obligatory condition for the production of active oxygen species by xanthine oxidase. It was concluded[124] that the most effective catalyst of the production of free hydroxyl radicals is Fe(EDTA) (see also Volume I, Chapter 9). Without EDTA, other than free hydroxyls, active oxygen radicals are largely formed. They can be "crypto-hydroxyl" radicals, i.e., oxy or hydroperoxy iron complexes. However, as was mentioned, there is another explanation of different reactivity of these species, namely, the site-specific formation of hydroxyl radicals. Both concepts are frequently used as interchangeable ones, but it may not be true. Thus, Winterbourn and Sutton[124] believe that in the absence of EDTA xanthine oxidase generates oxygen species (which can be named "crypto-hydroxyl" radicals) having a structure different from hydroxyl radicals, as a result of the interaction of superoxide ion with iron ions bound to the enzyme. However, Aruoma et al.[125] concluded that these species can be the same hydroxyl radicals, but due to chelating iron ions by a substrate (deoxyribose), their formation becomes "site-specific", which changes, respectively, the apparent reactivity of hydroxyl radicals.

Another enzyme which is able to catalyze the HO· formation is ferredoxin-ferredoxin:NADP$^+$ oxidoreductase. It was found that under aerobic conditions, this enzyme stimulates the formation of DMPO-OH spin adduct.[126] As in the case of xanthine oxidase, the presence of chelators (DTPA) was an obligatory condition.

Microsomal NADPH-cytochrome P-450 reductase and rat liver microsomes generated hydroxyl radicals which were detected as DMPO-OH spin adducts[99,127,128] or as active species which induced the formation of ethylene from methional or 2-keto-4-thiomethylbutyric acid, the peroxidation of lysosomal membranes,[135] the formation of methane from DMSO,[129] and the generation of low-level chemiluminescence.[130] All these processes proceeded only in the presence of iron ions or complexes. Ingelman-Sundberg and Ekstrom[131] and Ingelman-Sundberg and Hagbjork[132] have shown that cytochrome P-450 catalyzed hydroxylation of aniline and destruction of deoxyribose. Since these processes were inhibited by SOD, catalase, and hydroxyl radical scavengers, it was concluded that cytochrome P-450 is able to produce hydroxyl radicals.

The HRP-H_2O_2-NADPH system is apparently able to produce hydroxyl radicals in the presence of Fe(III)EDTA, since this system stimulated the formation of MDA-like products during the deoxyribose destruction.[133] It is of interest that HO· production was accelerated by porphyrins (uroporphyrin I, hematoporphyrin, and hematoporphyrin derivative); in this case deoxyribose destruction also occurred under anaerobic conditions. Therefore, it was assumed that radical anions of porphyrins may replace superoxide ion in the reduction of Fe(III)EDTA:

$$HRP + H_2O_2 \rightarrow \text{Compound I} \tag{35}$$

$$\text{Compound I} + NADPH \rightarrow \text{Compound II} + NADP\cdot \tag{36}$$

$$NADP\cdot + O_2 \rightarrow NADP^+ + O_2^{\overline{\cdot}} \tag{37}$$

$$NADP\cdot + \text{porphyrin} \rightarrow NADP^+ + \text{porphyrin}^{\overline{\cdot}} \tag{38}$$

$$\text{porphyrin}^{\overline{\cdot}} + Fe(III)EDTA \rightarrow \text{porphyrin} + Fe(II)EDTA \tag{39}$$

In addition to the superoxide ion, ferric ions and complexes may be reduced by other native reductants such as ascorbate, thiols, NADPH, and NADH. Ascorbate-dependent production of hydroxyl and crypto-hydroxyl radicals is supposed to be an initiation step of nonenzymatic lipid peroxidation, and will be considered in subsequent chapters. The accelerating effect of thiols (*N*-acetylhomocysteinylglycine, glutathione, or dithiothreitol) on the production of hydroxyl radicals in the H_2O_2 decomposition catalyzed by Fe(III)EDTA and Cu^{2+} has also been shown.[134] Hydroxyl radicals were detected by way of the mannitol-inhibited hydroxylation of salicylic acid and oxidation of tryptamine. It was suggested that in the case of the Cu^{2+}-catalyzed decomposition of hydrogen peroxide the formation of site-specific hydroxyl radicals is possible. This conclusion was based on a decrease in the inhibitory effect of mannitol in this system.

Besides iron and copper ions, the catalysts of the decomposition of hydrogen peroxide may be cobalt ions. Thus, it has been shown[136] that hydrogen peroxide decomposition in the presence of Co^{2+} ions induced the degradation of deoxyribose by hydroxyl radicals (in the presence of EDTA) and by "site-specific" hydroxyl radicals (in the absence of EDTA).

B. PHYSIOLOGICAL METAL ION CATALYSTS OF THE FENTON REACTION

The exclusively important role of iron as a catalyst of the Fenton reaction stimulated study of various iron proteins as potential promoters of hydroxyl radical production in biological systems. Sadrzadeh et al.[137] proposed that hemoglobin may react as a biological Fenton catalyst stimulating HO· production in the presence of superoxide ion or hydrogen peroxide. The same suggestion was made by Benatti et al.,[138] who found that under aerobic conditions, ascorbate reacted with methemoglobin to produce hydroxyl radicals. However, Gutteridge[139] questioned the possibility of HO· escaping from the iron center of hemoglobin through the protein core into solution. He proposed that hydrogen peroxide may degrade hemoglobin and release iron ions, which are the real catalysts of the Fenton reaction. This conclusion was confirmed in subsequent works[140-142] for both hemoglobin and myoglobin. However, oxyhemoglobin, methemoglobin, and metmyoglobin are possibly able to produce active oxygen species (crypto-hydroxyl radicals) in reaction with low concentrations of hydrogen peroxide.[141,143]

More important physiological catalysts of hydroxyl radical production are probably the iron-storage and iron-transporting proteins. It was proposed that ferritin,[144,145] hemosiderin (an iron-storage protein formed by lysosomal modification of ferritin),[145,146] lactoferrin,[147,148] and transferrin[148] are effective *in vivo* catalysts of the Fenton reaction. Thus, hydroxyl radicals were detected by hydroxylation of salicylate into dihydroxybenzoates, promoting the deoxyribose degradation, and the formation of DMPO-OH spin adducts in the reactions of ferritin and hemosiderin with the superoxide ion generated by the hypoxanthine-xanthine oxidase system or ascorbate.[145,146] All these processes were inhibited by hydroxyl radical scavengers (mannitol, formate, and Tris®) and an iron chelator, desferrioxamine. The catalytic effect of iron-storage proteins is apparently bound with the release of iron after reducing ferric ions into ferrous ions by reductants (superoxide ion, ascorbate, etc.). For example, Biemond et al.[144] have shown that activated polymorphonuclear leukocytes mobilized iron from horse spleen ferritin. The iron release was inhibited by SOD, which indicated the participation of superoxide ion in this process.

However, the ability of transferrin and lactoferrin to catalyze the HO· production was recently questioned.[148,149] It was found[149] that neither protein affected the ESR spectra of the DMPO-OH spin adduct formed in the xanthine-xanthine oxidase reaction. In contrast to this, two other iron-binding proteins, uteroferrin and bovine spleen purple acid phosphatase, were able to increase the hydroxyl radical production in this system. It was proposed that the difference in the catalytic activity of iron-binding proteins is determined by their ability to be reduced by the superoxide ion. As the reduction potential of transferrin is equal to (−0.28) to (−0.40) V, $O_2^{\overline{\cdot}}$ is unable to reduce Fe^{3+}-transferrin to Fe^{2+}-transferrin. The same is probably true for lactoferrin. In the authors' opinion, the production of hydroxyl radicals by uteroferrin and bovine spleen purple acid phosphatase is due to more positive values of their reduction potentials. It is also believed that previous results indicating the HO· production in the presence of transferrin and lactoferrin are explained by the use of DTPA for removing iron traces, since this chelator cannot do it quantitatively.[149]

In general, similar results were obtained by Aruoma and Halliwell.[148] In this work, the ability of lactoferrin and transferrin to stimulate HO· production was measured by their effects on deoxyribose degradation. It was found that iron-loaded lactoferrin and transferrin did not affect deoxyribose degradation by the hypoxanthine-xanthine oxidase system or by the ascorbate-hydrogen peroxide system at pH 7.4, while apo-lactoferrin and apo-transferrin even manifest inhibitory effects. It was concluded that since both proteins are not even approaching full iron loading *in vivo*, they must be *in vivo* antioxidants at physiological pH. Previous contradicting results are believed to be explained by using inadequate methods of iron loading or the application of chelating agents. At the same time, both lactoferrin and transferrin were able to stimulate the HO· production at acidic pH values (4 to 5) (depending on their degree of iron loading) that may be of importance in ischemic tissues where microenvironments of sufficiently low pH may exist.

V. ADDITIONS

Tajima and Shikama[150] recently determined the overall stoichiometry of superoxide formation during the oxidation of oxymyoglobin. There are recent studies of oxygen radical production by xanthine oxidase. Storch and Ferber[151] have studied the production of the superoxide ion by xanthine oxidase and microsomal NADPH oxidase, using the detergent-amplified lucigenin-dependent chemiluminescence. In contrast to cytochrome c reduction, the lucigenin-dependent chemiluminescence was completely inhibited by SOD. Radi et al.[152] concluded that both the superoxide ion and hydroxyl radicals are able to induce the luminol-dependent chemiluminescence in the reactions catalyzed by xanthine oxidase. Direct production of hydroxyl radicals by xanthine oxidase has been shown with the aid of the spin-trapping technique.[153] Elstner et al.[154] have shown that in the presence of quinones and crocidolite-asbestos, xanthine oxidase produces active oxygen species able to decompose methylthioketobutyric acid. Nishino et al.[155] have shown that superoxide ion is formed during the oxidation of chicken liver xanthine dehydrogenase by dioxygen, apparently as a result of oxidation of flavosemiquinone and the iron-sulfur centers. It has also been shown that oxygen radicals are produced in the reactions catalyzed by pyruvate: ferredoxin oxidoreductase from *Tritrichomonas foetus* hydrogenosomes,[156] NADPH-ferricyanide or NADPH-cytochrome c oxidoreductases and NADPH oxidase of radish plasmalemma vesicles,[157] cysteamine oxygenase,[158] and lactoperoxidase.[159]

Rashba et al.[160] measured the rates of superoxide production by submitochondrial membranes from rat and mouse livers, using 1-hydroxy-2,2,6,6-tetramethyl-4-oxopiperidine as a free radical scavenger. Ksenzenko et al.[161] have studied superoxide production by bc_1 complex and NADH: ubiquinone reductase segment in the mitochondrial respiratory chain.

Terelius and Ingelman-Sundberg[162] have shown that oxygen radicals are produced by cytochrome P-450 of reconstituted microsomal electron transport chain in the presence of a small amount of Fe(EDTA) in a negatively charged membranous system, whereas in a reconstituted micellar system or in the presence of a high amount of Fe(EDTA), the superoxide ion is generated by "uncoupling" cytochrome P-450 reductase. Ischiropoulos et al.[163] applied the lucigenin-dependent chemiluminescence for studying superoxide production by rat liver microsomes. Kukielka and Cederbaum[164] have shown that NADH is also able to promote the formation of hydroxyl and hydroxyl-like radicals in rat liver microsomes in the presence of iron chelates. Vartanyan and Furevich[165] have studied superoxide production by murine liver nuclei. DelRio et al.[166] have shown that NADH stimulates superoxide production in peroxisomes isolated from pea leaves. Klebanoff et al.[167] recently showed that desferrioxamine accelerates the formation of hydroxyl radicals by the Fenton reaction, forming an active Fe^{2+}-desferrioxamine complex.

REFERENCES

1. **Misra, H. P.,** Generation of superoxide free radical during the autoxidation of thiols, *J. Biol. Chem.*, 249, 2151, 1974.
2. **Saez, G., Thornalley, P. J., Hill, H. A. O., Hems, R., and Bannister, J. V.,** The production of free radicals during the autoxidation of cysteine, *Biochim. Biophys. Acta*, 719, 24, 1982.
3. **Kumar, K. S., Rowse, C., and Hochstein, P.,** Copper-induced generation of superoxide in human red cell membranes, *Biochem. Biophys. Res. Commun.*, 83, 587, 1978.
4. **Jones, G. J., Waite, T. D., and Smith, J. D.,** Light-dependent reduction of copper(II) and its effect on cell-mediated, thiol-dependent superoxide production, *Biochem. Biophys. Res. Commun.*, 128, 1031, 1985.
5. **Sevilla, M. D., Becker, D., Swarts, S., and Herrington, J.,** Sulfinyl radical formation from the reaction of cysteine and glutathione thiyl radicals with molecular oxygen, *Biochem. Biophys. Res. Commun.*, 144, 1037, 1987.
6. **Harman, L. S., Mottley, C., and Mason, R. P.,** Free radical metabolites of L-cysteine oxidation, *J. Biol. Chem.*, 259, 5606, 1985.
7. **Harman, L. S., Carver, D. K., Schreiber, J., and Mason, R. P.,** One- and two-electron oxidation of reduced glutathione by peroxides, *J. Biol. Chem.*, 261, 1642, 1986.
8. **Nakamura, M., Yamazaki, I., Otaki, S., and Nakamura, S.,** Characterization of one- and two-electron oxidations of glutathione coupled with lactoperoxidase and thyroid peroxidase reactions, *J. Biol. Chem.*, 261, 13923, 1986.
9. **Eling, T. E., Curtis, J. F., Harman, L. S., and Mason, R. P.,** Oxidation of glutathione to its thiyl free radical metabolite by prostaglandin H synthase. A potential endogenous substrate for the hydroperoxidase, *J. Biol. Chem.*, 261, 5023, 1986.
10. **Land, E. J. and Swallow, A. J.,** One-electron reactions in biochemical systems as studied by pulse radiolysis. IV. Oxidation of dihydronicotinamide-adenine dinucleotide, *Biochim. Biophys. Acta*, 234, 34, 1971.
11. **Nishikimi, M., Appaji, R. N., and Yagi, K.,** The occurrence of superoxide anion in the reaction of reduced phenazine metosulfate and molecular oxygen, *Biochem. Biophys. Res. Commun.*, 46, 849, 1972.
12. **Picker, S. D. and Fridovich, I.,** On the mechanism of production of superoxide radical by reaction mixtures containing NADH, phenazine methosulfate, and nitroblue tetrazolium, *Arch. Biochem. Biophys.*, 228, 155, 1984.
13. **Bielski, B. H. J., Shiue, G. G., and Bajuk, S.,** Reduction of nitroblue tetrazolium by CO_2^- and O_2^- radicals, *J. Phys. Chem.*, 84, 830, 1980.
14. **Halliwell, B.,** Generation of the superoxide radical during the peroxidative oxidation of NADH by catalase at acid pH values, *FEBS Lett.*, 80, 291, 1977.
15. **Nelson, N., Nelson, H., and Racker, E.,** Photoreaction of FMN-tripcine and its participation in photophosphorylation, *Photochem. Photobiol.*, 16, 481, 1972.
16. **Cunningham, M. L., Johnson, J. S., and Giovanazzi, S. M.,** Photosensitized production of superoxide anion by monochromatic (290—405 nm) ultraviolet irradiation of NADH and NADPH coenzymes, *Photochem. Photobiol.*, 42, 125, 1985.

17. **Bielski, B. H. J. and Chan, P. C.,** Enzyme-catalyzed free radical reactions with nicotine-adenine nucleotides. I. Lactate dehydrogenase-catalyzed chain oxidation of bound NADH by superoxide radicals, *Arch. Biochem. Biophys.*, 159, 873, 1973.
18. **Sakurai, H., Ishizu, K., and Okada, K.,** Superoxide generation by an iron-tetraphenylporphyrin-thiolate-oxygen system and its significance in relation to the coordination site of cytochrome P-450, *Inorg. Chim. Acta*, 91, L9, 1984.
19. **Misra, H. P. and Fridovich, I.,** The generation of superoxide radical during the autoxidation of hemoglobin, *J. Biol. Chem.*, 247, 6960, 1972.
20. **Wever, R., Oudega, B., and VanGelder, B. F.,** Generation of superoxide radicals during the autoxidation of mammalian oxyhemoglobin, *Biochim. Biophys. Acta*, 302, 475, 1973.
21. **Brunoni, M., Falcioni, G., Fioretti, E., Giardina, B., and Rotilio, G.,** Formation of superoxide in the autoxidation of the isolated α and β chains of human hemoglobin and its involvement in hemichrome precipitation, *Eur. J. Biochem.*, 53, 99, 1975.
22. **Gotoh, T. and Shikama, K.,** Generation of the superoxide radical during autoxidation of oxymyoglobin, *J. Biochem.*, 80, 397, 1976.
23. **Berman, M. C., Adnams, C. M., Ivanetich, K. M., and Kench, J. E.,** Autoxidation of soluble trypsin-cleaved microsomal ferrocytochrome b_5 and formation of superoxide radicals, *Biochem. J.*, 157, 237, 1976.
24. **Demma, L. S. and Salhany, J. M.,** Direct generation of superoxide anions by flash photolysis of human oxyhemoglobin, *J. Biol. Chem.*, 252, 1367, 1977.
25. **Demma, L. S. and Salhany, J. M.,** Subunit inequivalence in superoxide anion formation during photooxidation of human oxyhemoglobin, *J. Biol. Chem.*, 254, 4532, 1979.
26. **Goldberg, B. and Stern, A.,** The generation of O_2^- by the interaction of the hemolytic agent, phenylhydrazine, *J. Biol. Chem.*, 250, 2401, 1975.
27. **Goldberg, B., Stern, A., Peisach, J., and Blumberg, W. E.,** The detection of superoxide anion from the reaction of oxyhemoglobin and phenylhydrazine using EPR spectroscopy, *Experientia*, 35, 488, 1979.
28. **Goldberg, B., Stern, A., and Peisach, J.,** The mechanism of superoxide anion generation by the interaction of phenylhydrazine with hemoglobin, *J. Biol. Chem.*, 251, 3045, 1976.
29. **Wallace, W. J., Houtchens, R. A., Maxwell, J. C., and Caughey, W. S.,** Mechanism of autoxidation for hemoglobins and myoglobins. Promotion of superoxide production by protons and anions, *J. Biol. Chem.*, 257, 4966, 1982.
30. **Monteiro, H. P., Abdalla, D. S. P., Faljoni-Alario, A., and Bechara, E. J. H.,** Generation of active oxygen species during coupled autoxidation of oxyhemoglobin and δ-aminolevulinic acid, *Biochim. Biophys. Acta*, 881, 100, 1986.
31. **Sampath, V. and Caughey, W. S.,** Prooxidant effects of glutathione in aerobic hemoglobin solutions. Superoxide generation from uncoordinated dioxygen, *J. Am. Chem. Soc.*, 107, 4076, 1985.
32. **Nilsson, R., Pick, F. M., and Bray, R. C.,** ESR studies of reduction of oxygen to superoxide by some biochemical systems, *Biochim. Biophys. Acta*, 192, 145, 1969.
33. **Orme-Johnson, W. H. and Beinert, H.,** Formation of the superoxide anion radical during the reaction of reduced iron-sulfur proteins with oxygen, *Biochem. Biophys. Res. Commun.*, 36, 906, 1969.
34. **Misra, H. P. and Fridovich, I.,** The generation of superoxide radical during the autoxidation of ferredoxins, *J. Biol. Chem.*, 246, 6886, 1971.
35. **Nakamura, S. and Kimura, T.,** Studies on aggregated multienzyme systems. Stimulation of oxygen uptake of ferredoxin-nicotinamide adenine dinucleotide phosphate reductase-ferredoxin complex by cytochrome c, *J. Biol. Chem.*, 247, 6462, 1972.
36. **Allen, J. F.,** A two-step mechanism for the photosynthetic reduction of oxygen by ferredoxin, *Biochem. Biophys. Res. Commun.*, 66, 36, 1975.
37. **Hosein, B. and Palmer, G.,** The kinetics and mechanism of oxidation of reduced spinach ferredoxin by molecular oxygen and its reduced products, *Biochim. Biophys. Acta*, 723, 383, 1983.
38. **Bernhardt, F.-H. and Kuthan, H.,** Dioxygen activation by putidamonooxin. The oxygen species formed and released under uncoupling conditions, *Eur. J. Biochem.*, 120, 547, 1981.
39. **Jones, A. F., Winkles, J. W., Thornalley, P. J., Lunec, J., Jennings, P. E., and Barnett, A. H.,** Inhibitory effect of superoxide dismutase on fructosamine assay, *Clin. Chem.*, 33, 147, 1987.
40. **Sakurai, T. and Tsuchia, S.,** Superoxide production from nonenzymatically glycated proteins, *FEBS Lett.*, 236, 406, 1988.
41. **McCord, J. M. and Fridovich, I.,** The reduction of cytochrome c by milk xanthine oxidase, *J. Biol. Chem.*, 243, 5753, 1968.
42. **Fridovich, I. and Handler, P.,** Detection of free radicals generated during enzymatic oxidations by the initiation of sulfite oxidation, *J. Biol. Chem.*, 236, 1836, 1961.
43. **Knowles, P. F., Gibson, J. F., Pick, F. M., and Bray, R. C.,** Electron-spin-resonance evidence for enzyme reduction of oxygen to a free radical the superoxide ion, *Biochem. J.*, 111, 53, 1969.
44. **Bray, R. C., Pick, F. M., and Samuel, D.,** Oxygen-17 hyperfine splitting in the electron paramagnetic resonance spectrum of enzymatically generated superoxide, *Eur. J. Biochem.*, 15, 352, 1970.

45. **Arneson, R. M.,** Substrate-induced chemiluminescence of xanthine oxidase and aldehyde oxidase, *Arch. Biochem. Biophys.*, 136, 352, 1970.
46. **Buettner, G. R. and Oberley, L. W.,** Considerations in the spin trapping of superoxide and hydroxyl radical in aqueous systems using 5,5-dimethyl-1-pyrroline-1-oxide, *Biochem. Biophys. Res. Commun.*, 83, 69, 1978.
47. **Ueno, I., Kohno, M., Yoshihira, K., and Hirono, I.,** Quantitative determination of the superoxide radicals in the xanthine oxidase reaction by measurement of the electron spin resonance signal of the superoxide radical spin adduct of 5,5-dimethyl-1-pyrroline-1-oxide, *J. Pharmaco-Din.*, 7, 563, 1984.
48. **Kahl, R., Weiman, A., Weinke, S., and Hildebrandt, A. G.,** Detection of oxygen activation and determination of the activity of antioxidants toward reactive oxygen species by use of the chemiluminigenic probes luminol and lucigenin, *Arch. Toxicol.*, 60, 158, 1987.
49. **Hodgson, E. K. and Fridovich, I.,** The accumulation of superoxide radical during the aerobic action of xanthine oxidase, *Biochim. Biophys. Acta*, 430, 182, 1976.
50. **Olson, J. S., Ballou, D. P., Palmer, G., and Massey, V.,** The reaction of xanthine oxidase with molecular oxygen, *J. Biol. Chem.*, 249, 4350, 1974.
51. **Anderson, R. F., Hille, R., and Massey, V.,** The radical chemistry of milk xanthine oxidase as studied by radiation chemistry technique, *J. Biol. Chem.*, 261, 15870, 1986.
52. **Porras, A. G., Olson, J. S., and Palmer, G.,** The reaction of reduced xanthine oxidase with oxygen. Kinetics of peroxide and superoxide production, *J. Biol. Chem.*, 256, 9096, 1981.
53. **Hille, R. and Massey, V.,** Studies on the oxidative half-reaction of xanthine oxidase, *J. Biol. Chem.*, 256, 9090, 1981.
54. **Fridovich, I.,** Quantitative aspects of the production of superoxide anion radical by xanthine oxidase, *J. Biol. Chem.*, 245, 4053, 1970.
55. **Nagano, T. and Fridovich, I.,** Superoxide radical from xanthine oxidase acting upon lumazine, *J. Free Rad. Biol. Med.*, 1, 39, 1985.
56. **Halliwell, B.,** Generation of hydrogen peroxide, superoxide and hydroxyl radicals during the oxidation of dihydroxyfumaric acid and by peroxidase, *Biochem. J.*, 163, 441, 1977.
57. **Rotilio, G, Falcioni, G., Fioretti, E., and Brunoni, M.,** Decay of oxyperoxidase and oxygen radicals: a possible role for myeloperoxidase, *Biochem. J.*, 145, 405, 1975.
58. **Medeiros, M. H. G., Wefers, H., and Sies, H.,** Generation of excited species catalyzed by horseradish peroxidase or hemin in the presence of reduced glutathione and H_2O_2, *Free Rad. Biol. Med.*, 3, 107, 1987.
59. **Yokota, K. and Yamazaki, I.,** Analysis and computer stimulation of aerobic oxidation of reduced nicotinamide adenine dinucleotide catalyzed by horseradish peroxidase, *Biochemistry*, 16, 1913, 1973.
60. **Halliwell, B.,** Lignin synthesis: the generation of hydrogen peroxide and superoxide by horseradish peroxidase and its stimulation by manganese(II) and phenols, *Planta*, 140, 81, 1978.
61. **Avigliano, L., Carelli, V., Casini, A., Finazzi-Agro, A., and Liberatore, F.,** Oxidation of NAD dimers by horseradish peroxidase, *Biochem. J.*, 226, 391, 1985.
62. **Kukreja, R. C., Kontos, H. A., Hess, M. L., and Ellis, E. F.,** PGH synthase and lipoxygenase generate superoxide in the presence of NADH or NADPH, *Circ. Res.*, 59, 612, 1986.
63. **Rotilio, G., Calabrese, L., Agro, A. F., and Mondovi, B.,** Indirect evidence for the production of superoxide anion radicals by pig kidney diamine oxidase, *Biochim. Biophys. Acta*, 198, 618, 1970.
64. **Younes, M. and Weser, U.,** Involvement of superoxide in the catalytic cycle of diamine oxidase, *Biochim. Biophys. Acta*, 526, 644, 1978.
65. **Nakamura, S.,** Initiation of sulfite oxidation by spinach ferredoxin-NADP reductase and ferredoxin system: a model experiment on the superoxide anion radical production by metalloflavoproteins, *Biochem. Biophys. Res. Commun.*, 41, 177, 1970.
66. **Kido, T., Soda, K., and Asada, K.,** Properties of 2-nitropropane dioxygenase of *Hansenula mraki*. Formation and participation of superoxide, *J. Biol. Chem.*, 253, 226, 1978.
67. **Bhagwat, A. S. and Saue, P. V.,** Evidence for the involvement of superoxide anions in the oxygenase reaction of ribulose-1,5-diphosphate carboxylase, *Biochem. Biophys. Res. Commun.*, 84, 865, 1978.
68. **Fontecave, M., Graslund, A., and Reichard, P.,** The function of superoxide dismutase during the enzymatic formation of the free radical of ribonucleotide reductase, *J. Biol. Chem.*, 262, 12332, 1987.
69. **Hodgson, E. K. and Fridovich, I.,** Reversal of the superoxide dismutase reaction, *Biochem. Biophys. Res. Commun.*, 54, 270, 1973.
70. **Loschen, G., Azzi, A., Richter, C., and Flohe, L.,** Superoxide radicals as precursors of mitochondrial hydrogen peroxide, *FEBS Lett.*, 42, 68, 1974.
71. **Boveris, A. and Cadenas, E.,** Mitochondrial production of superoxide anions and its relationship to the antimycin insensitive respiration, *FEBS Lett.*, 54, 311, 1975.
72. **Nohl, H. and Hegner, D.,** Do mitochondria produce oxygen radicals *in vivo*?, *Eur. J. Biochem.*, 82, 563, 1978.
73. **Takeshige, K. and Minakami, S.,** NADH- and NADPH-dependent formation of superoxide anions by bovine heart submitochondrial particles and NADH-ubiquinone reductase preparation, *Biochem. J.*, 180, 129, 1979.

74. **Grigolava, I. V., Ksenzenko, M. Yu., Konstantinov, A. A., Tichonov, A. N., Kerimov, T. M., and Ruuge, E. K.**, Tyron as a spin-trap for superoxide radicals produced by the respiratory chain of submitochondrial particles, *Biokhimia*, 45, 75, 1980.
75. **Turrens, J. F. and Boveris, A.**, Generation of superoxide anion by the NADH dehydrogenase of bovine heart mitochondria, *Biochem. J.*, 191, 421, 1980.
76. **Forman, H. J. and Kennedy, J.**, Dihydroorotate-dependent superoxide production in rat brain and liver. A function of the primary dehydrogenase, *Arch. Biochem. Biophys.*, 173, 219, 1976.
77. **Dionisi, O., Galeotti, T., Terranova, T., and Azzi, A.**, Superoxide radicals and hydrogen peroxide formation in mitochondria from normal and neoplasmic tissues, *Biochim. Biophys. Acta*, 403, 292, 1975.
78. **Rich, P. R. and Bonner, W. D.**, The sites of superoxide anion generation in higher plant mitochondria, *Arch. Biochem. Biophys.*, 188, 206, 1978.
79. **Ledenev, A., Popova, E. Yu., Konstantinov, A. A., and Ruuge, E. K.**, Detection of superoxide radicals produced by intact heart mitochondria by spin trapping, *Biofizika*, 30, 708, 1985.
80. **Koshkin, V. V.**, Superoxide radical formation in skeletal muscle mitochondria, *Biokhimia*, 48, 1965, 1983.
81. **Tsirigotis, M. and Rydstroem, J.**, Generation of superoxide anion and lipid peroxidation in different cell types and subcellular fractions from rat testis, *Toxicol. Appl. Pharmacol.*, 94, 362, 1988.
82. **Oyanagui, Y. and Hagihara, B.**, Active oxygen production of mitochondria from tumors and liver of adrenalectomized rodents, in *Superoxide and Superoxide Dismutase in Chemistry, Biology, and Medicine*, Rotilio, G., Ed., Elsevier, Amsterdam, 1986, 429.
83. **Konstantinov, A. A., Peskin, A. V., Popova, E. Yu., Khomutov, G. B., and Ruuge, E.**, Superoxide generation by the respiratory chain of tumor mitochondria, *Biochim. Biophys. Acta*, 894, 1, 1987.
84. **Boveris, A. and Cadenas, E.**, Production of superoxide radicals and hydrogen peroxide in mitochondria, in *Superoxide Dismutase*, Vol. 2, Oberley, L. W., Ed., CRC Press, Boca Raton, FL, 1982, 15.
85. **Nohl, H.**, A novel superoxide radical generator in heart mitochondria, *FEBS Lett.*, 214, 269, 1987.
86. **Ksenzenko, M., Konstantinov, A. A., Khomutov, G. B., Tikhonov, A. N., and Ruuge, E. K.**, Effect of electron transfer inhibitors on superoxide generation in the cytochrome bc_1 site of the mitochondrial respiratory chain, *FEBS Lett.*, 155, 19, 1983.
87. **Cadenas, E., Boveris, A., Ragau, C. I., and Stoppani, A. O. M.**, Production of superoxide radicals and hydrogen peroxide by NADH-ubiquinone reductase and ubiquinol-cytochrome c reductase from beef-heart mitochondria, *Arch. Biochem. Biophys.*, 180, 248, 1977.
88. **Trumpower, B. L. and Simmons, Z.**, Diminished inhibition of mitochondrial electron transfer from succinate to cytochrome c by theonoyltrifluoroacetone induced by antimycin, *J. Biol. Chem.*, 254, 4608, 1979.
89. **Turrens, J. F., Alexandre, A., and Lehninger, A. L.**, Ubisemiquinone is the electron donor for superoxide formation by complex III of heart mitochondria, *Arch. Biochem. Biophys.*, 237, 408, 1985.
90. **Nohl, H. and Jordan, W.**, Investigation of the identity of the autoxidizable component of the mitochondrial respiratory chain, in *Superoxide and Superoxide Dismutase in Chemistry, Biology, and Medicine*, Rotilio, G., Ed., Elsevier, Amsterdam, 1986, 125.
91. **Nohl, H. and Jordan, W.**, The superoxide mitochondrial site of superoxide formation, *Biochem. Biophys. Res. Commun.*, 138, 533, 1986.
92. **Afanas'ev, I. B. and Polozova, N. I.**, Equilibrium constants for the reaction of radical anion $O_2^{\cdot-}$ with natural quinones and their model compounds, *Zh. Org. Khim.*, 15, 1802, 1979.
93. **Forman, H. J. and Kennedy, J.**, Superoxide production and electron transport in mitochondrial oxidation of dihydroorotic acid, *J. Biol. Chem.*, 250, 4322, 1975.
94. **Henry, M.-F. and Vignais, P. M.**, Production of superoxide anions in *Paracoccus denitrificans, Arch. Biochem. Biophys.*, 203, 365, 1980.
95. **Aust, S. D., Roerig, D. L., and Pederson, T. C.**, Evidence for superoxide generation by NADPH-cytochrome c reductase of rat liver microsomes, *Biochem. Biophys. Res. Commun.*, 47, 1133, 1972.
96. **Nelson, D. H. and Ruhmann-Wennhold, A.**, Corticosteroids increase superoxide anion production by rat liver microsomes, *J. Clin. Invest.*, 56, 1062, 1975.
97. **Auclair, C., DeProst, D., and Hakim, J.**, Superoxide production by liver microsomes from phenobarbital treated rat, *Biochem. Pharmacol.*, 27, 355, 1978.
98. **Deley, P. and Balny, C.**, Production of superoxide ions in rat microsomes, *Biochimie*, 55, 329, 1973.
99. **Lai, C.-S., Grover, T. A., and Piette, L. H.**, Hydroxyl radical production in a purified NADPH-cytochrome c (P-450) reductase system, *Arch. Biochem. Biophys.*, 193, 373, 1979.
100. **Lyakhovich, V. V., Mishin, V. M., and Pokrovsky, A. G.**, Interrelationship between the generation of oxygen anion-radicals and the reduction of artificial acceptors and cytochrome P-450 by NADPH-cytochrome c reductase, *Biokhimia*, 42, 1323, 1977.
101. **Ishiguro, I., Shinohara, R., Ishikara, A., and Naito, J.**, Formation of the red neotetrazolium formazan by reduced nicotinamide adenine dinucleotide phosphate-cytochrome c reductase in the presence of triton X-100, *Chem. Pharm. Bull.*, 22, 2935, 1974.
102. **Bosterling, B. and Trudell, J. R.**, Spin trap evidence for production of superoxide radical anions by purified NADPH-cytochrome P-450 reductase, *Biochem. Biophys. Res. Commun.*, 98, 569, 1981.

103. **Grover, T. A. and Piette, L. H.**, Influence of flavin addition and removal on the formation of superoxide by NADPH-cytochrome P-450 reductase: a spin-trap study, *Arch. Biochem. Biophys.*, 212, 105, 1981.
104. **Auclair, C., Torres, M., and Hakim, J.**, Superoxide anion involvement in NBT reduction catalyzed by NADPH-cytochrome P-450 reductase: a pitfall, *FEBS Lett.*, 89, 26, 1978.
105. **Sligar, S. G., Lipscomb, J. D., Debrunner, P. G., and Gunsalus, I. C.**, Superoxide anion production by the autoxidation of cytochrome P-450_{cam}, *Biochem. Biophys. Res. Commun.*, 61, 290, 1974.
106. **Sligar, S. and Debrunner, P.**, Superoxide production from the autoxidation of cytochrome P-450_{cam}, *Proc. Fed. Am. Soc. Exp. Biol.*, 33, 1256, 1974.
107. **Bartoli, G. M., Galeotti, T., Palombini, G., Parisi, G., and Azzi, A.**, Different contribution of rat liver microsomal pigments in the formation of superoxide anions and hydrogen peroxide during development, *Arch. Biochem. Biophys.*, 184, 276, 1977.
108. **Kuthan, H., Tsuji, H., Graf, H., Ullrich, V., Werringloer, J., and Estabrook, R. W.**, Generation of superoxide anion as a source of hydrogen peroxide in a reconstituted monooxygenase system, *FEBS Lett.*, 91, 343, 1978.
109. **Kuthan, H. and Ullrich, V.**, Oxidase and oxygenase function of the microsomal cytochrome P-450 monooxygenase system, *Eur. J. Biochem.*, 126, 583, 1982.
110. **Ingelman-Sundberg, M. and Johansson, I.**, Cytochrome b_5 as electron donor to rabbit liver cytochrome P-450_{LM_2} in reconstituted phospholipid vesicles, *Biochem. Biophys. Res. Commun.*, 97, 582, 1980.
111. **Zhukov, A. A. and Archakov, A. I.**, Complete stoichiometry of free NADP oxidation in liver microsomes, *Dokl. Akad. Nauk SSSR*, 269, 1235, 1983.
112. **Morehouse, L. A., Thomas, C. E., and Aust, S. D.**, Superoxide generation by NADPH-cytochrome P-450 reductase: the effect of iron chelators and the role of superoxide in microsomal lipid peroxidation, *Arch. Biochem. Biophys.*, 232, 366, 1984.
113. **Karasch, E. D. and Novak, R. E.**, Bis(alkylamino)anthracenedione antineoplastic agent metabolic activation by NADPH-cytochrome P-450 reductase and NADH dehydrogenase: diminished activity relative to anthracycline, *Arch. Biochem. Biophys.*, 224, 682, 1983.
114. **Bartoli, G. M., Galeotti, T., and Azzi, A.**, Production of superoxide anions and hydrogen peroxide in Ehrlich ascites tumor cell nuclei, *Biochim. Biophys. Acta*, 497, 622, 1977.
115. **Patton, S. E., Rosen, G. M., and Rauckman, E. J.**, Superoxide production by purified hamster hepatic nuclei, *Mol. Pharmacol.*, 18, 588, 1980.
116. **Peskin, A. V., Konstantinov, A. A., and Zbarskii, I. B.**, An unusual NAD(P)H-dependent superoxide radical anion-generating redox system in hepatoma 22a nuclei, *Free Rad. Res. Commun.*, 3, 47, 1987.
117. **Peskin, A. V., Zbarsky, I. B., and Konstantinov, A. A.**, A novel type of superoxide generating system in nuclear membranes from hepatoma 22a ascites cells, *FEBS Lett.*, 117, 44, 1980.
118. **Yusa, T., Crapo, J. D., and Freeman, B. A.**, Hyperoxia enhances lung and liver nuclear superoxide generation, *Biochim. Biophys. Acta*, 798, 167, 1984.
119. **Peskin, A. V. and Shlyakova, L.**, Cell nuclei generate DNA-nicking superoxide radicals, *FEBS Lett.*, 194, 317, 1986.
120. **Beachamp, C. and Fridovich, I.**, A mechanism for the production of ethylene from methional. The generation of the hydroxyl radical by xanthine oxidase, *J. Biol. Chem.*, 245, 4641, 1970.
121. **Fong, K.-L., McCay, P. B., Poyer, J., Misra, H. P., and Keele, B. B.**, Evidence for superoxide-dependent reduction of Fe^{3+} and its role in enzyme-generated hydroxyl radical formation, *Chem. Biol. Interact.*, 15, 77, 1976.
122. **Richmond, R., Halliwell, B., Chauhan, J., and Darbre, A.**, Superoxide-dependent formation of hydroxyl radicals: detection of hydroxyl radicals by hydroxylation of aromatic compounds, *Anal. Biochem.*, 118, 328, 1981.
123. **Motohashi, N. and Mori, I.**, Thiol-induced hydroxyl radical formation and scavenger effect of thiocarbamides on hydroxyl radicals, *J. Inorg. Biochem.*, 26, 205, 1986.
124. **Winterbourn, C. C. and Sutton, H. C.**, Iron and xanthine oxidase catalyze formation of an oxidant species distinguishable from OH·: comparison with the Haber-Weiss reaction, *Arch. Biochem. Biophys.*, 244, 27, 1986.
125. **Aruoma, O. I., Grootveld, M., and Halliwell, B.**, The role of iron in ascorbate-dependent deoxyribose degradation. Evidence consistent with a site-specific hydroxyl radical generation caused by iron ions bound to the deoxyribose molecule, *J. Inorg. Biochem.*, 29, 289, 1987.
126. **Morehouse, K. M. and Mason, R. P.**, The transition metal-mediated formation of the hydroxyl free radicals during the reduction of molecular oxygen by ferredoxin-ferredoxin:$NADP^+$ oxidoreductase, *J. Biol. Chem.*, 263, 1204, 1988.
127. **Lai, C. and Piette, L. H.**, Hydroxyl radical production involved in lipid peroxidation of rat liver microsomes, *Biochem. Biophys. Res. Commun.*, 78, 51, 1977.
128. **Lai, C.-S. and Piette, L. H.**, Spin-trapping studies of hydroxyl radical production involved in lipid peroxidation, *Arch. Biochem. Biophys.*, 190, 27, 1978.

129. **Cohen, G. and Cederbaum, A. I.,** Microsomal metabolism of hydroxyl radical scavenging agents: relationship to the microsomal oxidation of alcohols, *Arch. Biochem. Biophys.*, 199, 438, 1980.
130. **Puntarulo, S. and Cederbaum, A.,** Chemiluminescence studies on the generation of oxygen radicals from the interaction of NADPH-cytochrome P-450 reductase with iron, *Arch. Biochem. Biophys.*, 258, 510, 1987.
131. **Ingelman-Sundberg, M. and Ekstrom, G.,** Aniline is hydroxylated by the cytochrome P-450-dependent hydroxyl radical-mediated mechanism, *Biochem. Biophys. Res. Commun.*, 106, 625, 1982.
132. **Ingelman-Sundberg, M. and Hagbjork, A.-L.,** On the significance of the cytochrome P-450-dependent hydroxyl radical-mediated oxygenation mechanism, *Xenobiotica*, 12, 673, 1982.
133. **van Steveninck, J., Boegheim, J. P. J., Dubbelman, T. M. A. R., and van der Zee, J.,** The influence of porphyrins on iron-catalysed generation of hydroxyl radicals, *Biochem. J.*, 250, 197, 1988.
134. **von Steveninck, J., van der Zee, J., and Dubbelman, T. M. A. R.,** Site-specific and bulk-phase generation of hydroxyl radicals in the presence of cupric ions and thiol compounds, *Biochem. J.*, 232, 309, 1985.
135. **Fong, K. L., McCay, P. B., Poyer, J. L., Keele, B. B., and Misra, H.,** Evidence that peroxidation of lysosomal membranes is initiated by hydroxyl free radicals produced during flavin enzyme activity, *J. Biol. Chem.*, 248, 7792, 1973.
136. **Moorhouse, C. P., Halliwell, B., Grootveld, M., and Gutteridge, J. M. C.,** Cobalt(II) ion as a promoter of hydroxyl radical and possible "crypto-hydroxyl" radical formation under physiological conditions. Differential effects of hydroxyl radical scavengers, *Biochim. Biophys. Acta*, 843, 261, 1985.
137. **Sadrzadeh, S. M. H., Graf, E., Panter, S. S., Hallaway, P. E., and Eaton, J. W.,** Hemoglobin. A biologic Fenton reagent, *J. Biol. Chem.*, 259, 14354, 1984.
138. **Benatti, U., Morelli, A., Guida, L., and DeFlora, A.,** The production of activated oxygen species by an interaction of methemoglobin with ascorbate, *Biochem. Biophys. Res. Commun.*, 111, 980, 1983.
139. **Gutteridge, J. M. C.,** Iron promoters of the Fenton reaction and lipid peroxidation can be released from haemoglobin by peroxides, *FEBS Lett.*, 201, 291, 1986.
140. **Whitburn, K. D.,** The interaction of oxymyoglobin with hydrogen peroxide: the formation of ferrylmyoglobin at moderate excesses of hydrogen peroxide, *Arch. Biochem. Biophys.*, 253, 419, 1987.
141. **Puppo, A. and Halliwell, B.,** Formation of hydroxyl radicals from hydrogen peroxide in the presence of iron. Is haemoglobin a biological Fenton reagent?, *Biochem. J.*, 249, 185, 1988.
142. **Puppo, A. and Halliwell, B.,** Formation of hydroxyl radicals in biological systems. Does myoglobin stimulate hydroxyl radical formation from hydrogen peroxide?, *Free Rad. Res. Commun.*, 4, 415, 1988.
143. **Harel, S. and Kanner, J.,** The generation of ferryl or hydroxyl radicals during interaction of haemoproteins with hydrogen peroxide, *Free Rad. Res. Commun.*, 5, 21, 1988.
144. **Biemond, P., Eijk, H. G., van Swaak, J. G., and Koster, J. F.,** Iron mobilization from ferritin by superoxide derived from stimulated polymorphonuclear leukocytes. Possible mechanism in inflammation disease, *J. Clin. Invest.*, 73, 1576, 1984.
145. **O'Connell, M., Halliwell, B., Moorhouse, C. P., Aruoma, O. I., Baum, H., and Peters, I. J.,** Formation of hydroxyl radicals in the presence of ferritin and hemosiderin. Is hemosiderin formation a biological protective mechanism?, *Biochem. J.*, 234, 727, 1986.
146. **Ozaki, M., Kawabata, T., and Awai, M.,** Iron release from hemosiderin and production of iron-catalyzed hydroxyl radicals in vitro, *Biochem. J.*, 250, 589, 1988.
147. **Ambruso, D. R. and Johnston, R. B.,** Lactoferrin enhances hydroxyl radical production by human neutrophils, neutrophil particulate fractions and an enzymatic generating system, *J. Clin. Invest.*, 67, 352, 1984.
148. **Aruoma, O. I. and Halliwell, B.,** Superoxide-dependent and ascorbate-dependent formation of hydroxyl radicals from hydrogen peroxide in the presence of iron. Are lactoferrin and transferrin promoters of hydroxyl-radical generation?, *Biochem. J.*, 241, 273, 1987.
149. **Sibille, J. C., Doi, K., and Aisen, P.,** Hydroxyl radical formation and iron-binding proteins. Stimulation by the purple acid phosphatases, *J. Biol. Chem.*, 262, 59, 1987.
150. **Tajima, G. and Shikama, K.,** Autoxidation of oxymyoglobin. An overall stoichiometry including subsequent side reactions, *J. Biol. Chem.*, 262, 12603, 1987.
151. **Storch, J. and Ferber, E.,** Detergent-amplified chemiluminescence of lucigenin for determination of superoxide anion production by NADPH oxidase and xanthine oxidase, *Anal. Chem.*, 169, 262, 1988.
152. **Radi, R. A., Rubbo, H., and Prodanov, E.,** Comparison of the effects of superoxide dismutase and cytochrome c on luminol chemiluminescence produced by xanthine oxidase-catalyzed reactions, *Biochim. Biophys. Acta*, 994, 89, 1989.
153. **Kuppusamy, P. and Zweier, J. L.,** Characterization of free radical generation by xanthine oxidase. Evidence for hydroxyl radical generation, *J. Biol. Chem.*, 264, 9880, 1989.
154. **Elstner, E. F., Schnetz, W., and Vogl, G.,** Cooperative stimulation by sulfite and crocidolite asbestos fibers of enzyme catalyzed production of reactive oxygen species, *Arch. Toxicol.*, 62, 424, 1989.
155. **Nishino, T., Nishino, T., Schopfer, L. M., and Massey, V.,** The reactivity of chicken liver xanthine dehydrogenase with molecular oxygen, *J. Biol. Chem.*, 264, 2518, 1989.

156. **Docampo, R., Moreno, S. N. J., and Mason, R. P.,** Free radical intermediates in the reaction of pyruvate: ferredoxin oxidoreductase in *Tritrichomonas foetus* hydrogenosomes, *J. Biol. Chem.*, 262, 12417, 1987.
157. **Vianelli, A. and Macri, F.,** NAD(P)H oxidation elicits anion superoxide formation in radish plasmalemma vesicles, *Biochim. Biophys. Acta*, 980, 202, 1989.
158. **Ricei, G., Dupre, S., Frederici, G., Nardini, M., Spoto, G., and Cavallini, D.,** Cysteamine oxygenase: possible involvement of superoxide ion in the catalytic mechanism, *Free Rad. Res. Commun.*, 3, 365, 1987.
159. **Kohler, H. and Jenzer, H.,** Interaction of lactoperoxidase with hydrogen peroxide: formation of enzyme intermediates and generation of free radicals, *Free Rad. Biol. Med.*, 6, 323, 1989.
160. **Rashba, Yu. E., Vartanyan, L. S., Baider, L. M., and Krinitskaya, L. A.,** Quantitative estimate of the rate of superoxide radical generation in mitochondrial membranes by ESR techniques, *Biofizika*, 34, 57, 1989.
161. **Ksenzenko, M. Yu., Konstantinov, A. A., Khomutov, G. B., and Ruuge, E. K.,** Studies of superoxide radical generation in the NADH: ubiquinone reductase segment of the respiratory chain with the aid of the spin probe 2,2,5,5-tetramethyl-4-oxopiperidine-*N*-oxyl, *Biol. Membr.*, 6, 840, 1989.
162. **Terelius, Y. and Ingelman-Sundberg, M.,** Cytochrome P-450-dependent oxidase activity and hydroxyl radical production in micellar and membranous types of reconstituted systems, *Biochem. Pharmacol.*, 37, 1383, 1988.
163. **Ischiropoulos, N., Kumae, T., and Kikkawa, Y.,** Effect of interferon inducers on superoxide anion generation from rat liver microsomes detected by lucigenin chemiluminescence, *Biochem. Biophys. Res. Commun.*, 161, 1042, 1989.
164. **Kukielka, E. and Cederbaum, A. I.,** NADH-dependent microsomal interaction with ferric complexes and production of reactive oxygen intermediates, *Arch. Biochem. Biophys.*, 275, 540, 1989.
165. **Vartanyan, L. S. and Gurevich, S. M.,** NADH- and NADPH-dependent superoxide radical generation in liver nuclei, *Biokhimiya*, 54, 1020, 1989.
166. **DelRio, L. A., Fernandez, V. M., Ruperez, F. L., Sandalio, L. M., and Palma, J. M.,** NADH induces the generation of superoxide radicals in leaf peroxisomes, *Plant Physiol.*, 89, 728, 1989.
167. **Klebanoff, S. J., Waltersdorph, A. M., Bryce, B. R., and Rosen, H.,** Oxygen-based free radical generation by ferrous ions and deferoxamine, *J. Biol. Chem.*, 264, 19765, 1989.

Chapter 3

PRODUCTION OF OXYGEN RADICALS BY XENOBIOTICS

I. INTRODUCTION

In Chapter 2 the generation of oxygen radicals in enzymatic and nonenzymatic processes occurring in biological systems was discussed. Another important pathway to oxygen radicals is the stimulation of their production by xenobiotics. Such stimulation may be a consequence of various phenomena: the redox transformation of drugs, environmental or professional poisoning, ethanol intoxication, or even nutrition of some components of food. All these phenomena (excluding the first one) may apparently be considered as events stimulating pathological changes in living organisms. In contrast to this, the drug-stimulated generation of oxygen radicals may have both favorable and unfavorable effects: it may be responsible for the therapeutic activity of drugs or be a cause of their side cytotoxic action.

We already discussed the mechanism of generation of the superoxide ion by electron donors (pyrogallol, 1,8-dihydro-9-anthrone, some pyridinium compounds, etc. [Volume I, Chapter 3]). There are a great many compounds which are able to stimulate production of the superoxide ion and hydroxyl radicals in cells. These compounds may be divided into two groups, of reducible and oxidizable species. In the case of reducible compounds or oxidants (quinones, pyridinium compounds, nitro compounds, etc.), the first step is the "activation" of xenobiotics into the free radical state, that is, the formation of a free radical by one-electron reduction. The subsequent step frequently is the reduction of dioxygen by a free radical to the superoxide ion. However, this route is by no means a single one; it is now known that the free radical formed can directly interact with cell components or induce the generation of hydroxyl radicals without the participation of the superoxide ion.

Oxidizable compounds (for example, hydroquinones) may directly reduce oxygen to form superoxide ion. However, a more important route is apparently the enzymatic one-electron oxidation catalyzed by peroxidases. Below, we will discuss the mechanisms of free radical activation of drugs and other toxic compounds and the generation of oxygen radicals as well as the significance of free radical mechanisms in the therapeutic and cytotoxic action of drugs.

II. PRODUCTION OF OXYGEN RADICALS BY REDUCIBLE COMPOUNDS

A. ANTHRACYCLINES

Anthracycline antibiotics are widely used in modern cancer therapy for treatment of leukemias and solid tumors. They possibly belong to the most powerful up-to-date anticancer drugs, and, therefore, the mechanism of their action is the subject of a great number of experimental works. Many anthracycline derivatives are now known, but all of them contain the anthraquinone moiety, shown here for two important antibiotics: doxorubicin (adriamycin) (I) and daunorubicin (daunomycin) (II).

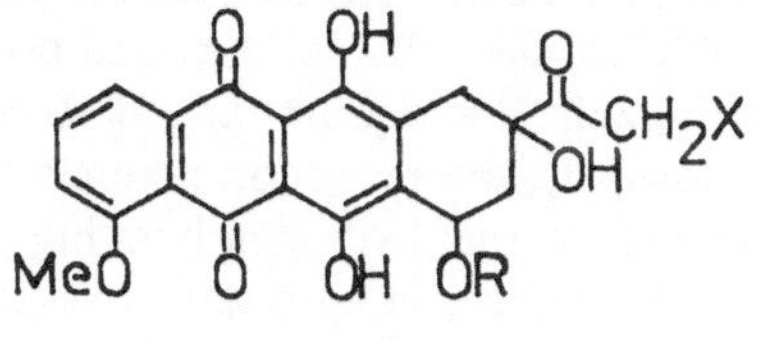

I. X = OH
II. X = H
R = daunosamine

It should be expected that the occurrence of the quinone moiety makes anthracyclines capable of participating in various redox processes including the reactions with dioxygen and the superoxide ion (direct and reverse reactions of Equation 1).

$$\text{Anth}^{\bar{\cdot}} + O_2 \rightleftarrows \text{Anth} + O_2^{\bar{\cdot}} \tag{1}$$

The first experimental evidence of the participation of anthracyclines in Reaction 1 was obtained by Handa and Sato,[1] who showed that doxorubicin and daunorubicin stimulated sulfite oxidation by rat liver microsomes. Free radical transformations of anthracyclines in mitochondria and microsomes and in the reaction catalyzed by microsomal NADPH-cytochrome P-450 reductase resulting in the formation of anthracycline semiquinones, consumption of dioxygen, and the generation of superoxide ion were later studied in detail.[2-6] It might be concluded that production of the superoxide ion and other oxygen radicals (via the superoxide-driven Fenton reaction) is a main mechanism of free radical cytotoxic activity of anthracyclines in normal and cancer cells.

However, there is another pathway from anthracyclines to free radical metabolites in biological systems in which iron-anthracycline complexes participate. It has been shown that anthracyclines are strong chelating agents and readily bind iron ions.[7-10] The iron-anthracycline complexes formed may generate oxygen radicals without involving the quinone moiety in a redox process:

$$Fe^{3+}\text{-Anth} \xrightarrow{e} Fe^{2+}\text{-Anth} \tag{2}$$

$$Fe^{2+}\text{-Anth} + O_2 \rightleftarrows Fe^{3+}\text{-Anth} + O_2^{\bar{\cdot}} \tag{3}$$

$$2\,O_2^{\bar{\cdot}} + 2\,H^+ \rightarrow H_2O_2 + O_2 \tag{4}$$

$$Fe^{2+}\text{-Anth} + H_2O_2 \rightarrow Fe^{3+}\text{-Anth} + HO\cdot + HO^- \tag{5}$$

Therefore, to understand the possible pathways of generating oxygen radicals by anthracyclines in cells, it is necessary to consider the mechanisms of the interaction of these antibiotics with dioxygen, enzymatic and nonenzymatic reductants, and metal ions. After this, free radical hypotheses of anticancer and toxic action of anthracyclines may be discussed.

1. The Interaction of Anthracycline Antibiotics with Dioxygen and the Superoxide Ion

It was shown in Volume I (Chapter 3) that the ability of reductants to generate superoxide ion by the one-electron reduction of dioxygen depends on their reduction potentials. Therefore, kinetics and thermodynamics of redox processes with participation of anthracyclines and dioxygen were widely studied. In aqueous solution, owing to the rapid completion of one-electron transfer reactions and the instability of anthracycline semiquinones and the superoxide ion, these reactions have been studied by the pulse-radiolysis method. The rate constants for one-electron transfer reactions of anthracyclines and the calculated one-electron reduction potentials of these antibiotics are cited in Tables 1 and 2.

It is seen that in aqueous solution the one-electron reduction potential of doxorubicin is about 0.15 V smaller than that of dioxygen, that is, the equilibrium constant for Reaction 1 is about 850.[13,15] Thus for doxorubicin, the equilibrium of Reaction 1 is indeed shifted to the right in aqueous solution. However, the reduction potentials of anthracyclines increase substantially in aprotic media due to internal hydrogen bonding[16] and become more positive than that of dioxygen (Table 2). Therefore, in aprotic media (including possibly lipid medium), the equilibrium of Reaction 1 must be shifted to the left.

These data allow us to explain the discrepancy in the results obtained by studying the interaction of the superoxide ion with doxorubicin in aqueous and aprotic solutions. It has

TABLE 1
Rate Constants for Free Radical Reactions of Quinone Drugs in Aqueous Solution

Reaction	pH	k ($M^{-1}s^{-1}$)	Ref.
e_{aq}^- + $^+$HDoxorubicinH$_2$ → $^+$HDoxorubicinH$_2^{\cdot -}$ + H_2O	6.5	$(2.5 \pm 0.3) \cdot 10^9$	11
e_{aq}^- + DoxorubicinH$^-$ → DoxorubicinH$_2^{\cdot -}$ + HO$^-$	11.5	$(1.5 \pm 0.2) \cdot 10^{10}$	11
·COOH + $^+$HDoxorubicinH$_2$ → $^+$HDoxorubicinH$_3$· + CO_2	1.1	$(3.5 \pm 0.4) \cdot 10^9$	11
$CO_2^{\cdot -}$ + $^+$HDoxorubicinH$_2$ → $^+$HDoxorubicinH$_2^{\cdot -}$ + CO_2	6.5	$(3.4 \pm 0.4) \cdot 10^9$	11
$O_2^{\cdot -}$ + Doxorubicin → O_2 + Doxorubicin$^{\cdot -}$		$2.67 \cdot 10^8$ (?)	12
		$(3.5 \pm 0.2) \cdot 10^5$	104
Doxorubicin$^{\cdot -}$ + O_2 → Doxorubicin + $O_2^{\cdot -}$		$4.4 \cdot 10^7$	12
	7.0	$(3.0 \pm 0.2) \cdot 10^8$	13
$^+$HDoxorubicinH$_2^{\cdot -}$ + O_2 → $^+$HDoxorubicinH$_2$ + $O_2^{\cdot -}$	6.0	$(3.5 \pm 0.4) \cdot 10^8$	11
Doxorubicin$^{\cdot -}$			
+ Fe(III)Desferrioxamine	7.0	$<4 \cdot 10^4$	13
+ Fe(III)DTPA	7.0	$(7.0 \pm 0.3) \cdot 10^8$	13
+ Fe(III)EDTA	7.0	$(2.8 \pm 0.2) \cdot 10^8$	13
+ Fe(III)ATP	7.0	$(8.0 \pm 1.4) \cdot 10^6$	13
e_{aq}^- + Daunorubicin → Daunorubicin$^{\cdot -}$ + H_2O	7.0	$(1.63 \pm 0.03) \cdot 10^{10}$	14
$CO_2^{\cdot -}$ + Daunorubicin → Daunorubicin$^{\cdot -}$ + CO_2	7.0	$(2.0 \pm 0.2) \cdot 10^9$	14
$O_2^{\cdot -}$ + Mitomycin C → O_2 + (Mitomycin C)$^{\cdot -}$		$(5.2 \pm 0.5) \cdot 10^5$	104
(Mitomycin C)$^{\cdot -}$			
+ O_2 → Mitomycin C + $O_2^{\cdot -}$	7.0	$(2.2 \pm 0.2) \cdot 10^8$	13
+ Fe(III)Desferrioxamine	7.0	$<6 \cdot 10^4$	13
+ Fe(III)DTPA	7.0	$(2.4 \pm 0.2) \cdot 10^7$	13
+ Fe(III)EDTA	7.0	$(9.0 \pm 0.8) \cdot 10^6$	13
+ Fe(III)ATP	7.0	$<6 \cdot 10^4$	13
(CI 941)$^{\cdot -}$ + O_2 → CI 941 + $O_2^{\cdot -}$		$(1.8 \pm 0.6) \cdot 10^9$	151
Mitoxantrone$^{\cdot -}$ + O_2 → Mitoxantrone + $O_2^{\cdot -}$		$(5.1 \pm 0.5) \cdot 10^8$	104
$O_2^{\cdot -}$ + Mitoxantrone → O_2 + Mitoxantrone$^{\cdot -}$		200	104
AZQ$^{\cdot -}$ + O_2 → AZQ + $O_2^{\cdot -}$	7.0	$(1.1 \pm 0.1) \cdot 10^7$	174
$O_2^{\cdot -}$ + AZQ → O_2 + AZQ$^{\cdot -}$	7.0	$(2.7 \pm 0.2) \cdot 10^8$	174
BZQ$^{\cdot -}$ + O_2 → BZQ + $O_2^{\cdot -}$	7.0	$(8.2 \pm 0.5) \cdot 10^8$	174
$O_2^{\cdot -}$ + BZQ → O_2 + BZQ$^{\cdot -}$	7.0	$(1.5 \pm 0.1) \cdot 10^5$	174

been shown[20] that contrary to the data obtained in the *in vitro* experiments in aqueous solution (see above), in aprotic medium (DMF), the superoxide ion reacts with doxorubicin practically irreversibly. In reality, however, this difference is apparently not explained only by the medium effect. First of all, Reaction −1 is not the only direction of the interaction of the superoxide ion with an anthracycline molecule (although the direct Reaction 1 does not have alternating pathways). It is a consequence of the presence in anthracyclines of phenolic hydroxyls which, as has been proposed by some authors,[21,22] can react with the superoxide ion via disproportionation and not one-electron transfer.

$$O_2^{\cdot -} + AnthH_2 \rightleftarrows HO_2\cdot + AnthH^- \qquad (6)$$

Recently, the mechanism of the interaction of the superoxide ion with anthracyclines was reinvestigated.[23,24] It was shown that the same product X was formed in the reactions of doxorubicin with the superoxide ion, benzosemiquinone, and ascorbate ion as well as during the electrochemical reduction of doxorubicin in aprotic media. As it is impossible for benzosemiquinone and especially ascorbate anion to deprotonize doxorubicin,[23] these reactions must proceed by the one-electron reduction mechanism.

$$\text{Red} + \text{Anth} \rightarrow \text{Ox} + \text{product X} \qquad (7)$$

TABLE 2
One-Electron Reduction Potentials of Quinone Drugs

Compound	Medium	E^1_7 (V) (vs. NHE)	$E_{1/2}$ (V) (vs. SCE)	Ref.
O_2	H_2O	−0.16		Vol. I
O_2	DMF		−0.84 ± 0.03	Vol. I
Doxorubicin	Isopropanol	−0.289		12
	t-Butanol	−0.295		12
	H_2O, pH 7	−0.328		15
	DMF		−0.665	16
	H_2O		−0.64[a]	17
Daunorubicin	t-Butanol	−0.305		12
	H_2O, pH 7	−0.43		14
	DMF		−0.77	18
	H_2O		−0.64[a]	17
5-Iminodaunorubicin	H_2O, pH 7.1		−0.675[a]	19
Aclacinomycin A	30% aq. MeCN		−0.55, −0.65[a]	17
Mitomycin C	t-Butanol	−0.271		12
	Isopropanol	−0.238		12
	H_2O, pH 7	−0.310		13
	DMF		−0.937[b]	159
Ametantrone	Isopropanol	−0.348		12
Mitoxantrone	H_2O, pH 7	−0.527		49
Anthrapyrazole CI 941	H_2O	−0.538		151
	H_2O		−1.040	19
Anthraquinonyl glucosaminosides				
XIX	DMF		−0.97	18
XX	DMF		−0.73	18
XXI	DMF		−0.72	18
XXII	DMF		−0.68	18
XXIII	DMF		−0.70	18
Diaziquone AZQ	t-Butanol	−0.168 (?)		12
	H_2O, pH 7	−0.070		174
	DMF		−0.77[b]	170
BZQ	H_2O, pH 7	−0.336		174

[a] Two-electron potential.
[b] vs. Ag/AgCl.

Similar results were obtained in studying the reactions of the superoxide ion and ascorbate with another anthracycline antibiotic, aclacinomycin A (III).[24]

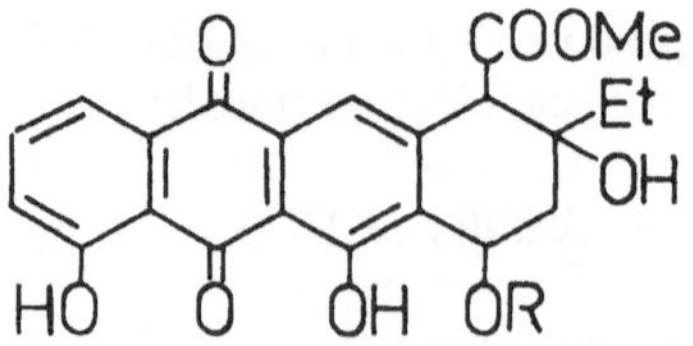

III. R = aminotrisaccharide

There is an important difference between Reactions −1 and 7 (for Red = $O_2^{\cdot-}$). Reaction 7 is an irreversible process, and X is a diamagnetic compound, whereas the product of one-electron transfer Reaction −1 is paramagnetic semiquinone. Furthermore, it was found[23] that Reaction 7 also proceeds in water-acetonitrile solutions up to a water content of 90%. Therefore, it may be concluded that the irreversible transformation of the semiquinone formed by Reaction −1 may shift the equilibrium of Reaction 1 to the left even in the case of the unfavorable reduction potential difference of reagents (as in the case of the reactions of doxorubicin with superoxide ion in aqueous solution or with benzosemiquinone and ascorbate anion in aprotic media).

An irreversible process which shifts the equilibrium of Reaction 1 to the left probably is deglycosidation,[23] which is a typical mode of the transformation of anthracycline semiquinones.[25] The transformation of anthracycline semiquinones formed during the reduction of parent antibiotics by the superoxide ion may be presented as follows:

$$O_2^{-} + \text{Anth} \rightleftharpoons O_2 + \text{Anth}^{-}$$

IV

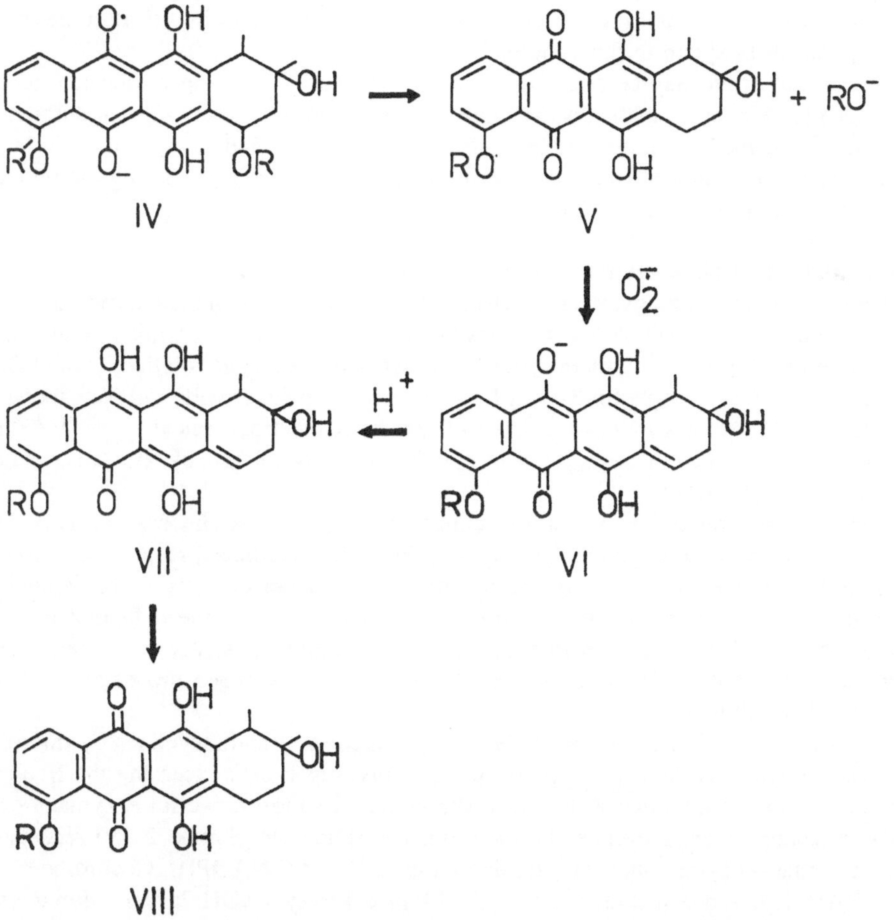

In accord with this scheme, the cleavage of the semiquinone glycoside bond results in the formation of a neutral radical of deoxyaglycone (V), which is further reduced by the superoxide ion to anion (VI). The protonation of VI gives a tautomer of deoxyaglycone (VII), which is transformed into deoxyaglycone (VIII). It was proposed that the observed "product X" is actually anion VI. This scheme is a modification of the one proposed by Kleyer and Koch[26] for the reduction of daunorubicin by a free radical, 3,5,5-trimethyl-2-oxomorpholin-3-yl. However, these authors assumed that deoxyaglycone is formed during the decomposition of the hydroquinone form of anthracycline. Such a process possibly also takes place when doxorubicin is reduced by $NaBH_4$;[23] however, it is of little consequence for the reaction of the superoxide ion which is unable to reduce anthracyclines to hydroquinones.

How can these findings be interpreted in connection with the indisputable fact of the generation of superoxide ion by anthracyclines in cells and cellular components?

It seems that a main factor here is the competition between the rates of reversible one-electron transfer processes (Equation 1) and the irreversible transformation of anthracycline semiquinones. A rate constant for the direct Reaction 1 in aqueous solution is about 10^8 M^{-1} s^{-1} (Table 1). On the other hand, the rate constant for the heterolytic cleavage of the glycoside bond must be significantly smaller. (Land et al.[11] determined the rate constant for the elimination of daunosamine from the reduced doxorubicin as about 5 s^{-1}, but they related it to the decomposition of doxorubicin hydroquinone.) Therefore, anthracycline semiquinones generated in cells will be instantly converted to parent anthracyclines in the presence of dioxygen. Indeed, it has been shown that the formation of deoxyaglycone VIII during enzymatic reduction of anthracyclines occurs under anaerobic conditions or after the complete consumption of dioxygen in the system.[17,27]

Another situation may be realized in lipid membranes if the superoxide ion generated enzymatically or nonenzymatically will have an opportunity to react with an anthracycline molecule. Lipid medium must be closer to aprotic solvents than to aqueous solution, and therefore the equilibrium of Reaction 1 will be shifted to the left, increasing the probability of deglycosidation of anthracycline semiquinones.

2. Enzymatic and Nonenzymatic Reduction of Anthracyclines

The reduction of anthracyclines in cells is a first and key stage of their activation resulting in the generation of oxygen radicals. It has been shown that many chemical reductants can reduce anthracyclines, among them sodium borohydride in aqueous solution[28-30] and DMF,[23] 3,5,5-trimethyl-2-oxomorpholin-3-yl (a free radical),[26] the superoxide ion (see above), and ascorbic acid.[23,24] Anthracyclines are also readily reduced electrochemically.[16,25,31,32] Kinetics of anthracycline reduction by e_{aq} and $CO_2^{\cdot-}$(·COOH) was studied by the pulse-radiolysis technique[11,14,33] (Table 1).

Nonenzymatic reduction of anthracyclines may proceed by the one- or two-electron pathways to form the semiquinone and hydroquinone intermediates, respectively. Semiquinones are formed electrochemically and by chemical reductants in aprotic media and pulse-radiolysis experiments in aqueous solutions. It is of interest that the reduction by sodium borohydride may lead to both semiquinones and hydroquinones. As in the reaction with superoxide ion, semiquinones formed in the absence of dioxygen are deglycosidated or reduced to hydroquinones.

It was already pointed out that study of the reduction of anthracyclines by microsomes and mitochondria was the starting point of many investigations concerning the free radical properties of these antibiotics. It has been shown that the most powerful enzymatic catalyst for the reduction of anthracyclines is microsomal NADPH cytochrome P-450 reductase.[34-39] However, other enzymes such as xanthine oxidase,[27,35,39,40] NADPH cytochrome c reductase,[41] NADPH ferredoxin oxidoreductase,[42-44] *Vibrio harvey* NADH flavin oxidoreductase,[45] and NADH cytochrome c reductase[35] are also capable of reducing anthracyclines. Important studies of the reduction of anthracyclines were carried out using microsomes, mitochondria, submitochondrial preparations, cardiac sarcosomes, and erythrocyte membranes.[6,17,46-52]

Enzymatic reduction of anthracyclines was studied by monitoring indirect characteristics such as substrate oxidation, dioxygen consumption, and superoxide ion generation, as well as by the direct determination of anthracycline semiquinones and deoxyaglycones. Usually, broad, unresolved ESR signals of semiquinones were observed in many enzymatic systems. However, recently, the highly resolved ESR spectra of semiquinones of daunorubicin and its aglycones (daunomycinone and 7-deoxydaunomycinone) were obtained during the reduction of anthracyclines by the xanthine-xanthine oxidase system in water-DMSO solutions.[53]

In spite of numerous investigations, many questions concerning the mechanism of enzymatic reduction of anthracyclines remain unanswered. It was already pointed out that under aerobic conditions, the formation of deoxyaglycones occurred only after all dioxygen was consumed. However, this conclusion is challenged by the finding that the change of the doxorubicin spectrum during its reduction by NADPH-cytochrome c-(ferredoxin)-oxidoreductase under aerobic conditions is accompanied by the appearance of a maximum at 568 to 571 nm,[54] which is very similar to that of a tautomer of deoxyaglycone.

It is possible that deoxyaglycones are formed during the decomposition of enzymatically generated anthracycline semiquinones. Such a mechanism was proposed by Pan and Bachur,[27] Pan et al.,[35] Gutierrez et al.,[41] and Bachur.[55]

$$\text{Anth(OR)} \xrightarrow{e} \text{Anth(OR)}^{\overline{\cdot}} \rightarrow \text{Anth}\cdot + \text{RO}^- \xrightarrow[\text{H}^+]{\text{Red}} \text{AnthH} \qquad (8)$$

There are data supporting this mechanism. Thus, it has been shown[36] that the reduction of doxorubicin by NADPH-cytochrome P-450 reductase consists of two phases characterized by different ESR spectra. An unstable ESR spectrum observed during the first stage is believed to correspond to doxorubicin semiquinone, whereas a stable and markedly asymmetrical ESR spectrum observed during the second stage apparently belongs to deoxyaglycone semiquinone. (Aglycones are, as a rule, insoluble in aqueous solution, and therefore yield "power" asymmetrical ESR spectra.) Therefore, one may propose that the doxorubicin semiquinone is transformed during the reaction into the deoxyaglycone semiquinone.

Other important findings supporting the above mechanism were obtained by Pan et al.[35] These authors showed that the reductive glycoside cleavage of daunorubicin is catalyzed only by single-electron transport and mixed-electron transport flavoenzymes (NADPH-cytochrome P-450 reductase, xanthine oxidase, and NADH-cytochrome c reductase), whereas two-electron transport and predominantly two-electron transport flavoenzymes (L-amino acid oxidase and glutathione reductase) were unable to catalyze this reaction. These data also indicate that semiquinones must be precursors of deoxyaglycones.

Another mechanism of the reductive glycoside cleavage of anthracyclines is based on the results obtained in studying the reduction of daunorubicin by 3,5,5-trimethyl-2-oxomorpholin-3-yl (a free radical).[26] In this reaction daunorubicin hydroquinone is formed, and on these grounds it was assumed that deglycosidation took place during the transformation of hydroquinone into a tautomer of deoxyaglycone, which was finally converted to deoxyaglycone.

$$\text{Anth(OR)} \xrightarrow[2\,\text{H}^+]{2e} \text{H}_2\text{Anth(OR)} \rightarrow \text{``tautomer''} + \text{RO}^- \xrightarrow{\text{H}^+} \text{AnthH} \qquad (9)$$

It was proposed that this mechanism is realized in the reduction of anthracyclines by spinach ferredoxin-NADPH reductase and ferredoxin.[44] Indeed, it was found that one of the anthracycline antibiotics studied, nogalamycin, was converted to the intermediate with a maximum at 428 nm typical for hydroquinone.

At the present time, it is difficult to determine the real significance of both mechanisms in the *in vivo* and *in vitro* reductive glycoside cleavage of anthracyclines. The problem is further complicated by the fact that the reduction mechanisms may be different for various anthracyclines, which differ significantly by their reductive ability characterized by dioxygen consumption and superoxide production. The same may be said about the structure of the products of glycoside cleavage. For example, in contrast to doxorubicin and daunorubicin, which are converted to deoxyaglycones, aclacinomycin A is reduced to a deglycosidation dimer, a typical product of semiquinone recombination.

It was already shown that depending on reductants, anthracyclines may be reduced to semiquinones or hydroquinones. Recently, it was found that a reductant may be responsible for the mechanism of the reductive cleavage. Thus, 3,5,5-trimethyl-2-oxomorpholin-3-yl reduced 5-iminodaunorubicin to deoxyaglycone with no hydroquinone and quinone methide (a tautomer of deoxyaglycone) formed, whereas both these intermediates were observed during reduction with dithionite.[32] Therefore, it may be expected that the mechanism of the reductive cleavage of anthracyclines *in vivo* will depend on both the reductant and the anthracycline. Nonetheless, the importance of one-electron transport enzymes as catalysts of the glycoside cleavage and the impossibility of two-electron reduction of anthracyclines by superoxide ion possibly indicate the importance of the one-electron reduction mechanism (Reaction 8).

The reduction of anthracyclines in microsomes is catalyzed by NADPH-cytochrome P-450 reductase, whereas in mitochondria, this reaction is catalyzed by some component of Complex I. Firstly, it was shown by Nohl and Jordan[47] who found that rotenone, an inhibitor of electron transfer from Complex I to the rest of the respiratory chain, did not affect the reduction of doxorubicin in isolated rat heart mitochondria. These findings were confirmed in the experiments with beef heart submitochondrial preparations[48] where it was shown that various inhibitors blocking electron transport from Complex I to ubiquinone (rotenone, amytal, and piericidin) also have no effect on the formation of doxorubicin and daunorubicin semiquinones. On these grounds it was concluded that the anthracycline reduction is catalyzed by NADH dehydrogenase or one of the iron-sulfur proteins belonging to Complex I and that the reduction of anthracyclines by succinic dehydrogenase or ubiquinone is impossible. The ability of iron-sulfur proteins to reduce anthracyclines was also shown in the experiments with Fe_2S_2-containing protein, ferredoxin, which was an even more efficient reductant of anthracyclines (daunorubicin, aclacinomycin A, nogalamycin, etc.) than spinach ferredoxin-NADPH reductase.[44]

3. "Oxidation" of Anthracyclines

Doxorubicin, daunorubicin, and other anthracyclines contain both the quinone and hydroquinone moieties; therefore, these compounds can participate in both the reduction and oxidation process. However, until now there were only fragmentary data concerning the oxidation of anthracyclines. Chinami et al.[56] have shown that doxorubicin forms a free radical (which was detected by ESR spectroscopy) at basic pH values and even at pH 7.5 in Tris®-HCl buffer. It was assumed that this free radical is phenoxyl, formed during the oxidation of phenolic hydroxyl by dioxygen:

$$\text{Anth–OH} \rightleftarrows \text{Anth–O}^- + \text{H}^+ \xrightarrow{O_2} \text{Anth–O}\cdot + O_2^{\overline{\cdot}} \qquad (10)$$

The same mechanism was recently accepted for the oxidation of a new antibiotic, fredericamycin A (IX),[57] which, similarly to anthracyclines, has the hydroquinone-quinone moiety. This compound is easily oxidized by dioxygen in DMSO, supposedly to form a phenoxyl radical. Trifluoroacetic acid quantitatively converted the radical formed to the parent antibiotic.

Although in both cases it was logical to accept the oxidation mechanism of the reactions, the identification of free radicals as phenoxyls is doubtful. Phenoxyls are neutral free radicals; therefore, they cannot be very stable in DMSO or aqueous solution due to the disproportionation into quinone and hydroquinone. However, the lifetime of the fredericamycin free radical was about several days.[57] The most surprising fact about this radical is its conversion back to the parent antibiotic after acidification with trifluoroacetic acid. It is impossible to understand how a neutral free radical can be reduced by a proton donor. In our opinion, this phenomenon is possible only for the semiquinone anion radical which will disproportionate after protonation to form hydroquinone and quinone. Therefore, it may be suggested that the formation of free radicals of antibiotics in both discussed works is a result of some reduction process, for example, the reduction of antibiotics by hydroxyl anion or superoxide ion formed during the saturation of DMSO or basic aqueous solution with dioxygen in the presence of traces of metal ions (Volume I, Chapter 9).

$$HO^- + Ant \rightleftarrows (HOAnt)^- \xrightarrow{Ant} HOAnt\cdot + Ant^{\cdot-} \quad (11)$$

$$O_2^{\cdot-} + Ant \rightleftarrows O_2 + Ant^{\cdot-} \quad (12)$$

$$Ant^{\cdot-} + H^+ \rightleftarrows HAnt\cdot \quad (13)$$

$$2\ HAnt\cdot \rightarrow Ant + AntH_2 \quad (14)$$

4. Metal Complexes of Anthracyclines and their Participation in Redox Processes

Anthracyclines are good chelators and therefore easily form complexes with metal ions. The most important metal complexes of anthracyclines are iron complexes, which apparently play an important role in the production of oxygen radicals and cytotoxic action of anthracyclines. The structure of ferric ion-doxorubicin complexes was first determined by May et al.[7] These authors showed that doxorubicin binds ferric ions into complexes of 1:1, 2:1, and 3:1 stoichiometry with step association constants of 10^{10}, 10^{11}, and $10^{4.4}$, respectively. Therefore, an over-all association constant of 3:1 doxorubicin-iron complex is equal to $10^{33.4}$. The 3:1 stoichiometry of ferric ion-doxorubicin complex was confirmed in subsequent works.[8,10] This complex can be obtained by the oxidation of ferrous ion-doxorubicin complex by dioxygen, the exchange reaction between the ferric ion-acetohydroxamic acid complex and doxorubicin, and the interaction of ferric salt with doxorubicin at pH 7.4 in sodium cacodylate, Hepes, and Tris®-HCl buffers. The formation of a complex is accompanied by a decrease in the absorption of doxorubicin at 479 nm and the appearance of a band at 600 nm.[8,10,58] Surprisingly, Bachur et al.[9] concluded that the ferric ion-doxorubicin complex has uncertain stoichiometry, but their results are apparently distorted by the use of an excess of ferric salt.

In contrast to the ferric ion-doxorubicin complex, the ferrous ion-doxorubicin complex was not as carefully studied. It was found that this complex can be obtained only under anaerobic conditions in sodium cacodylate and Tris®-HCl buffers.[8,58] The complex has a maximum at 500 to 515 nm and was readily oxidized by dioxygen to the ferric ion-doxorubicin complex. Iron-doxorubicin complexes may be prepared in both aqueous and aprotic solutions, but not in a phosphate buffer where the phosphate anion successfully competes with anthracyclines for iron ions.[58]

Doxorubicin is not the only anthracycline antibiotic able to bind iron ions. The 1:3 ferric ion-anthracycline complexes having absorption spectra very similar to that of Fe^{3+}-doxorubicin were obtained with daunorubicin,[10] carminomycin,[59] and 5-iminodaunorubicin.[60,61] It is of interest that 5-iminodaunorubicin is able to extract quantitatively ferric ion from the ferric ion-doxorubicin complex.[60] As 11-deoxydoxorubicin does not form a complex with ferric ion, it was concluded[62] that the presence of the C_{11}-hydroxy group in anthracycline is an obligatory condition for the complexation with metal ions. However, it may be true for 1,1-deoxydoxorubicin only, because aclacinomycin A, which also lacks the C_{11}-hydroxy group, nonetheless, does form Fe^{3+} and Fe^{2+} complexes.[58] However, these complexes possibly have a different structure.

Anthracyclines also are able to form complexes with metal ions other than iron ions. For example, the complexes of doxorubicin with Cu^{2+} and doxorubicin and daunorubicin with Pd^{2+} were obtained.[63,64]

We already discussed the ability of anthracyclines to be reduced by mitochondrial and microsomal enzymes to semiquinones with the subsequent generation of superoxide ion. It seems that binding metal ions may drastically change the ability of anthracyclines to interact with these enzymes. Thus, it has been shown that the iron- and palladium-doxorubicin complexes completely lost the ability to transport electrons from NADH to dioxygen through mitochondrial NADH dehydrogenase.[10,64] This fact may be of importance for understanding the mechanism of cytotoxic activity of anthracyclines (see below).

The formation of metal-anthracycline complexes and, first of all, iron-anthracycline complexes *in vivo* may have various important biological consequences. It is proposed by many workers that iron-anthracycline complexes are efficient generators of oxygen radicals by Reactions 3 and 5 and that these reactions are an important and maybe even a main mode of cytotoxic action of anthracyclines. The occurrence of Reaction 3 is supported by the fact that the Fe^{2+}-doxorubicin complex exists only under anaerobic conditions and that its oxidation by dioxygen results in the appearance of an absorption spectrum similar to that of Fe^{3+}-doxorubicin.[8] Direct evidence of the oxidation of ferrous ion bound to doxorubicin by dioxygen was obtained in the ESR experiments where it was shown that on exposure to dioxygen, the ESR-silent Fe^{2+}-Dox complex is transformed into Fe^{3+}-Dox with its well-known ESR spectrum ($g = 2.0$ and 4.2).[65-67]

In some cases, Reaction 3 is accompanied by the formation of products other than the Fe^{3+}-Anth complex. For example, the ferrous ion-aclacinomycin A complex (Fe^{2+}-Acl) is easily oxidized by dioxygen into Fe^{3+}-Acl, but the latter then dissociates to form free aclacinomycin A, probably due to the small association constant of Fe^{3+}-Acl and the absence of excess Fe^{3+} ions in solution.[58]

$$Fe^{2+}\text{–Acl} + O_2 \rightleftarrows Fe^{3+}\text{–Acl} + O_2^{\overline{\cdot}} \quad (15)$$

$$Fe^{3+}\text{–Acl} \rightleftarrows Fe^{3+} + \text{Acl} \quad (16)$$

Fe^{3+}-Dox is also not the only product of the oxidation of Fe^{2+}-Dox. It has been shown[58] that in addition to maxima at 496 and 600 nm, which correspond to the Fe^{3+}-doxorubicin spectrum, the spectrum of reaction product(s) contains a maximum at 547 nm. This maximum possibly indicates the formation of the deoxyaglycone tautomer (see above), which can be formed as a result of the reaction of the superoxide ion formed in Reaction 3 with the anthracycline moiety of iron-doxorubicin complexes. However, another interpretation of the spectrum of reaction products is also possible (see below).

Fe^{3+}-doxorubicin catalyzes the reduction of dioxygen by glutathione and cysteine (but not by NADPH or NADH).[8] This process can be described by Reactions 17 and 3.

$$\text{Red} + Fe^{3+}\text{–Dox} \rightleftarrows \text{Ox} + Fe^{2+}\text{–Dox} \quad (17)$$

Another reductant which is able to reduce Fe^{3+}-Anth complexes is the ascorbate anion.[58] Unexpectedly, the reduction of Fe^{3+}-Dox by ascorbate turned out to be a complicated process consisting of two stages. At the first stage (at pH 5 to 7.4 in water or Tris®-HCl buffer), free doxorubicin was formed. Apparently, it is explained by the reduction of ferric ion and the subsequent dissociation of the ferrous ion-doxorubicin complex.

$$AH^{-} + Fe^{3+}\text{–Dox} \rightarrow AH\cdot + Fe^{2+}\text{–Dox} \quad (18)$$

$$Fe^{2+}\text{–Dox} \rightleftarrows Fe^{2+} + \text{Dox} \quad (19)$$

Indeed, the Fe^{2+}-Dox complex was quantitatively formed only with a great excess of ferrous ions;[58] therefore, at 0.1 m*M* concentrations, the equilibrium of Reaction 19 must be completely shifted to the right. On the second stage, a new product was formed which was identical to the product formed during the oxidation of Fe^{2+}-Dox by dioxygen. This fact apparently indicates that the same individual compound and not a mixture of deoxyaglycone tautomer and the Fe^{3+}-doxorubicin complex is a product of these two reactions. Its structure remains obscure, but possibly, it is a new Fe^{3+} complex such as Fe^{3+}-deoxyaglycone or Fe^{3+}-$Dox^{\overline{\cdot}}$. Under the same conditions, Fe^{3+}-Acl was easily reduced by ascorbate anion to form a single product, supposedly, a tautomer of aclacinomycin deoxyaglycon. The next mechanism was accepted.[58]

$$AH^- + Fe^{3+}\text{–Acl} \rightarrow AH\cdot + Fe^{2+}\text{–Acl} \tag{20}$$

$$Fe^{2+}\text{–Acl} \rightleftarrows Fe^{2+} + \text{Acl} \tag{21}$$

$$Fe^{2+} + O_2 \rightleftarrows Fe^{3+} + O_2^{\overline{\cdot}} \tag{22}$$

$$O_2^{\overline{\cdot}} + \text{Acl} \rightarrow \text{deoxyaglycone tautomer} \tag{23}$$

A very interesting, unusual reaction was described for the Fe^{3+}-doxorubicin complex: the self-reduction of the ferric ion by the doxorubicin moiety. First, such a mechanism was proposed for the stimulation of the ferric ion-dependent peroxidation of phospholipid liposomes by doxorubicin.[68-70] It was suggested that ferric ion is reduced by the anthracycline ligand via an intramolecular electron transfer:

$$Fe^{3+}\text{–DoxH} \rightleftarrows Fe^{2+}\text{–Dox}\cdot + H^+ \rightleftarrows Fe^{2+} + \text{Dox}\cdot + H^+ \tag{24}$$

It is obvious that this reaction generates a new Fe^{2+}-Dox complex containing the semiquinone ligand. This proposal was supported by the ESR studies of Fe^{3+}-Dox.[65,66] It has been found that under anaerobic conditions, Fe^{3+}-Dox is transformed into an ESR-silent complex which is supposed to be Fe^{2+}-Dox. In addition, the ESR spectrum of organic free radical with $g = 2.004$ appeared when an excess of doxorubicin (Fe^{3+}: Dox = 1:8) was applied.

Although these findings are in agreement with the mechanism proposed (Reaction 24), many data remain unexplained. Thus an ESR spectrum of the semiquinone formed in Reaction 24 is the same as that obtained by one-electron *reduction* of doxorubicin, although in Reaction 24 semiquinone must be formed as a result of one-electron *oxidation* of the antibiotic. In the last work[67] the authors try to explain this fact by proposing that an organic free radical is formed during the oxidation of the side-chain hydroxyl group of doxorubicin:

$$\underset{\text{X.}}{\text{Dox–COCH}_2\text{OH}} \xrightarrow{Fe^{2+}} \underset{\text{XI.}}{\text{Dox–COCH}_2\text{O}\cdot} \xrightarrow{Fe^{2+}} \underset{\text{XII.}}{\text{Dox–COCHO}} \tag{25}$$

If Reaction 25 is possible, then it becomes understandable why self-reduction is observed for doxorubicin and is not observed for daunorubicin, which has no side-chain hydroxyl group. It was also proposed that an unpaired electron of an alkoxy radical (XI) can be transferred to the hydroquinone C-ring to form semiquinone very similar (but not identical, due to the oxidation of the side chain) to the semiquinone formed in the reduction of doxorubicin.

Unfortunately, this explanation lacks chemical support as an electron transfer through five saturated bonds is impossible as well as the one-electron oxidation of aliphatic hydroxyl by ferric ion. It is also impossible to understand another experimental fact: the regeneration of Fe^{3+}-Dox on the exposure of the product(s) of self-reduction to dioxygen.[65-67] Such a process is possible only for the semiquinone formed in the reduction of doxorubicin. Thus, the mechanism of self-reduction of Fe^{3+}-Dox remains unexplained.

5. Generation of Hydroxyl Radicals by Anthracyclines

The generation of hydroxyl radicals by anthracyclines is usually considered as the most important free radical process stimulated by these antibiotics, because these active radicals may be a main cause of their cytotoxic activity. The anthracycline-stimulated formation of hydroxyl radicals is well documented in many biological systems, using various experimental techniques. Lown and Chen[28] used spin-trapping by DMPO for the demonstration of HO· formation during the oxidation of daunorubicin by air. In the presence of sodium formate, an ESR spectrum of DMPO-adduct was substituted by that of DMPO-CO_2^- that ruled out the possibility of the formation of DMPO-OH during the decomposition of DMPO-OOH. Rowley and Halliwell[42] measured the production of hydroxyl radicals during the reduction of doxorubicin by the NADPH-ferredoxin reductase system using hydroxylation of salicylate. The generation of hydroxyl radicals in rat liver microsomes catalyzed by doxorubicin was determined by measuring $^{14}CO_2$ formed in the decarboxylation of ^{14}C benzoate.[71] Spin-trapping was also successfully applied for detection of hydroxyl radicals generated by doxorubicin in the NADPH-cytochrome P-450 reductase system,[72] the NADH-NADH dehydrogenase system,[73] in rat heart sarcosomes,[74] in erythrocytes,[75] and in the sensitive subline of human breast tumor cells,[76] as well as by Fe^{3+}-Dox complex in the presence of DNA.[62] On the other hand, Doroshow[77,79,80] and Doroshow and Davies[78] developed a method of measuring the anthracycline-stimulated production of hydroxyl radicals by cardiac mitochondrial NADH dehydrogenase and beef heart submitochondrial particles on the basis of evolution of methane from DMSO. Komiyama et al.[81] have shown that microsomal NADPH-cytochrome P-450 reductase in the presence of doxorubicin and daunorubicin catalyzes the formation of ethylene from methional. This reaction is also frequently interpreted as an evidence of HO· production, although other active radicals may also be a reason for ethylene production.

In contrast to well-documented data on the anthracycline-stimulated generation of hydroxyl radicals in various systems, the mechanism of their formation is not clear. At the beginning, it was proposed[47,81-84] that the anthracycline semiquinones formed during enzymatic reduction of anthracyclines react directly with hydrogen peroxide to form hydroxyl radicals.

$$Anth^{\cdot -} + H_2O_2 \rightarrow Anth + HO\cdot + HO^- \qquad (26)$$

We already discussed this mechanism in Volume I (Chapter 9); it was shown there that even the paraquat radical cation is unable to reduce hydrogen peroxide, although the reduction potential of paraquat was as low as -0.43 V. Therefore, for most organic free radicals, including anthracycline semiquinones with reduction potentials of (-0.3) to (-0.4) V, Reaction 26 may proceed only in the presence of metal catalysts.

This conclusion was confirmed experimentally. Tero-Kubota et al.[38] have shown that the only product of the reduction of aclacinomycin A by cytochrome P-450 reductase under aerobic conditions was the superoxide ion which was identified as the PBN-OOH spin adduct. The PBN-OH spin adduct was formed only in the presence of ferric ions. Thus, aclacinomycin semiquinone was not able to reduce hydrogen peroxide without iron ions. Similar results were obtained with doxorubicin and daunorubicin. Sushkov et al.[85,86] have studied the generation of hydroxyl radicals during the reduction of 9,10-anthraquinone-2-sulfonate (AQS) and 2-(dimethylamino)-3-chloro-1,4-naphthoquinone by microsomal NADPH cytochrome

P-450 reductase. It was found that hydrogen peroxide did not affect the rate of decay of AQS semiquinone; it proves that their direct interaction is insignificant. On the other hand, ferric ions greatly stimulated the formation of the DMPO-Me spin adducts (in the presence of DMSO) by both quinones. It was concluded that the production of hydroxyl radicals in the presence of these quinones (which are considered as the model compounds of anticancer antibiotics) is always an iron-dependent reaction.

Now, the possibility of noncatalyzed reduction of hydrogen peroxide by anthracycline semiquinones (Reaction 26) can be completely excluded.[87] However, the substitution of Reaction 26 by Reaction 5 is insufficient for explaining the ability of anthracycline antibiotics to generate hydroxyl radicals in cells. First of all, there is no evidence that Reaction 5 is a more effective route to hydroxyl radicals than the classical Fenton reaction:

$$Fe^{2+} + H_2O_2 \rightarrow Fe^{3+} + HO\cdot + HO^- \qquad (27)$$

Moreover, it was shown that the reduction of Fe^{3+}-Dox by ascorbate leads to immediate dissociation of Fe^{2+}-Dox complex (Reaction 19). This fact was recently supported by study of the effect of doxorubicin on the ascorbate-dependent lipid peroxidation in lecithin liposomes.[88] It was found that at equal iron ion concentrations, the level of lipid peroxidation was practically the same in the ferrous ion-dependent system and in the ascorbate-Fe^{3+}-Dox system. Therefore, one may suppose that in both cases, lipid peroxidation was initiated by free ferrous ions.

Thus, anthracyclines are in no way true "pro-oxidants", i.e., the hydroxyl radical promoters by Reactions 5 or 26. What is more, they may even be effective free radical scavengers, as is seen from their ability to scavenge superoxide ion (see above). For example, it was recently shown that doxorubicin inhibits the luminol- and lucigenin-sensitized chemiluminescence by activated polymorphonuclear leukocytes[89,90] and macrophages.[91] These antioxidative (antiradical) properties of anthracyclines may be well explained by their ability, as quinones, to oxidize the superoxide ion via a one-electron transfer mechanism. Thus, the stimulation of hydroxyl radical production by anthracycline antibiotics must be a consequence of other reasons.

Two phenomena apparently may be the origin of indirect pro-oxidant activity of anthracyclines: the anthracycline-induced superoxide production and chelating properties of anthracyclines. We have already seen that anthracyclines are readily reduced by microsomal and mitochondrial reductases with subsequent generation of superoxide ion. For example, from 31 antibiotics and their derivatives, one half, including the most important representatives such as doxorubicin, 11-deoxydoxorubicin, etc., sharply increased superoxide production in rat liver microsomes.[17] It is known that the superoxide ion is further converted to hydrogen peroxide as a result of catalytic and noncatalytic dismutation. Thus, the contact of anthracycline antibiotics with cellular membranes must result in a sharp increase in hydrogen peroxide concentration.

Another consequence of this contact probably is an increase in ferrous ion concentration. The origin of "free" iron ions in cells is uncertain. It is possible that most iron ions chelating by anthracyclines come from ferritin and transferrin.[92] It is interesting that two types of interaction between anthracyclines and these proteins which lead to chelating iron ions by antibiotics are possible. It has been shown[93] that doxorubicin chelates ferric ions released from iron-loaded serum transferrin at acidic pH. As the release of iron ions from transferrin in acidic cell compartments takes place during the transfer of iron by transferrin to the physiological targets (ferritin or cytochromes), doxorubicin may act as a "false" acceptor of ferric ions.

The release of iron from ferritin by anthracyclines apparently proceeds via another mechanism, namely, by the reduction of ferric ions stored in this protein. Thus, it was found[94] that ferric ions in ferritin are reduced by the superoxide ion produced during the aerobic reduction of daunorubicin and doxorubicin by NADPH-cytochrome P-450 reductase. It is of interest that under anaerobic conditions, anthracyclines retain the ability to release iron ions from ferritin probably by a direct electron transfer from semiquinones to ferric ions. Thus, it may be concluded that the ability of anthracyclines to stimulate superoxide generation and to release ferrous ions at the same site (after the reduction of chelating ferric ions) is possibly a main reason for stimulation by these antibiotics of the formation of hydroxyl radicals.

6. Free Radical Mechanisms of Cytotoxicity and Anticancer Activity of Anthracyclines

a. Stimulation of Lipid Peroxidation

As oxygen radicals play an important role in the initiation of lipid peroxidation (Chapter 5), one may expect that anthracyclines will stimulate this process. Indeed, in 1977, Goodman and Hochstein[3] have shown that doxorubicin and daunorubicin enhanced the lipid peroxidation initiated by microsomal NADPH-cytochrome P-450 reductase. Later on, it was shown that anthracyclines stimulated NADPH- and NADH-dependent lipid peroxidation in rat heart, liver, and kidney microsomes,[95-97] mouse heart, liver, lung, and kidney microsomes,[97,98] rat and mouse liver and kidney mitochondria,[97,99] and rabbit liver microsomes and heart mitochondria;[100] SOD inhibited (in part) the anthracycline-stimulated lipid peroxidation that confirmed the participation in this process of the superoxide ion. Lipid peroxidation was also inhibited by hydroxyl radical scavengers (for example, 1,3-dimethylurea) and iron chelators (EDTA and DTPA).

Phenolic antioxidant, VP-16 (a semisynthetic derivative of podophyllotoxin), manifested a strong inhibitory effect on the daunorubicin-stimulated NADPH-dependent lipid peroxidation in mouse liver microsomes.[101] Endogenous free radical scavengers, α-tocopherol and ubiquinone, were also effective inhibitors. Thus, Mimnaugh and co-workers[96,98] found that an increase in the anthracycline-stimulated lipid peroxidation in mouse heart microsomes in comparison with rat heart microsomes is explained by a smaller tocopherol level in mouse heart microsomes (0.25 ± 0.04 and 0.51 ± 0.09 μg of tocopherol per milligram of protein for mouse and rat heart microsomes, respectively). Similarly, lipid peroxidation in beef heart ubiquinone-depleted mitochondria stimulated by Fe^{3+}-Dox increased about two times in comparison with lipid peroxidation in reconstituted mitochondria (after reincorporating ubiquinone Q_{10}).[102] It has also been shown[103] that the daunorubicin-stimulated lipid peroxidation in rat heart and liver homogenates was inhibited by magnesium ascorbate and cystamine. Effects of endogenous scavengers (antioxidants) are further considered in animal studies.

To a considerable degree, lipid peroxidation depended on the nature of the antibiotics: doxorubicin, daunorubicin, carminomycin, and aclacinomycin A caused a significant increase in NADPH- and NADH-dependent lipid peroxidation, whereas 5-iminodaunorubicin had little or no effect. Another important feature of this process is its dependence on molecular oxygen. It was found[95,97] that the increase in rat liver microsomal peroxidation in the presence of doxorubicin is considerably smaller under air than under dioxygen. It is also of interest that significant quantities of aglycone metabolites were formed under air.

Stimulation of lipid peroxidation by anthracyclines has also been shown in the *in vivo* experiments.[100,105-112] Doxorubicin and daunorubicin were usually administered intraperitonally. It has been shown that anthracyclines significantly stimulated lipid peroxidation in rat cardiac mitochondria,[105] mouse heart and liver microsomes,[110] and mouse heart and liver.[111,112] Lipid peroxidation also sharply increased in the lens of rat with doxorubicin-induced cataract.[107] Antioxidant enzymes (SOD and catalase) and free radical scavengers (vitamin E, ubiquinone, glutathione, cysteamine, cysteine, and mannitol) decreased the MDA formation.[105,112] On the other hand, there was no increase in MDA production in the isolated, vitamin E-deficient rat hearts after perfusion with doxorubicin,[108] or in mice injected with doxorubicin and fed Mn-sufficient and vitamin E-sufficient or Mn-deficient and vitamin E-deficient diets.[109]

In contrast to enzymatic anthracycline-dependent lipid peroxidation which was studied in numerous works, the effect of anthracyclines on nonenzymatic peroxidation is uncertain. Mimnaugh and co-workers[98] concluded that doxorubicin potently inhibits the ascorbate-dependent lipid peroxidation in mouse heart and liver microsomes. Recently, it was found[88] that doxorubicin does not affect entirely the ascorbate-iron-dependent lipid peroxidation of lecithin liposomes in phosphate buffer. It was explained by rapid dissociation of the Fe^{2+}-Dox complex formed after the reduction of Fe^{3+}-Dox by ascorbate: owing to that, in the presence and the absence of doxorubicin, lipid peroxidation was initiated by the same reactive species — free ferrous ions.

Anthracycline-stimulated lipid peroxidation is one of the most important lipid peroxidation systems studied. It seems now that several different mechanisms of lipid peroxidation may be realized in the presence of anthracyclines. It is usually accepted that even partial inhibition of lipid peroxidation by SOD indicates an important role of the superoxide ion in the peroxidation mechanism.[97] Really, however, the anthracycline-stimulated generation of the superoxide ion may indirectly affect lipid peroxidation, increasing a pool of hydrogen peroxide. It may explain the different conclusions arrived at by various workers about the effect of SOD on the anthracycline-dependent lipid peroxidation.

Data obtained in studying the anthracycline-dependent lipid peroxidation cast doubt on the role of free hydroxyl radicals in this process.[72,113,114] Inhibitory and other studies[72,113,114] have shown that active oxygen radicals initiating lipid peroxidation in the presence of anthracyclines must be "crypto-hydroxyl" radicals or iron-oxygenated complexes and not free hydroxyl radicals (for further discussion, see Chapter 5 and Volume I, Chapter 9). Therefore, iron ions must play an obligatory role in the peroxidation process. Indeed, it was found[115] that the NADPH-dependent lipid peroxidation in rat liver microsomes stimulated by doxorubicin and measured by ethane and pentane production proceeds only in the presence of iron salt.

Phospholipid peroxidation catalyzed by doxorubicin and iron ions was inhibited by iron chelators (EDTA, DTPA, and desferrioxamine) and was not affected by hydroxyl radical scavengers such as mannitol and thiourea as well as by SOD and catalase.[113] Therefore, it was concluded that the iron-doxorubicin complex must play a decisive role in this process. Sugioka et al.[68] have shown that although the incubation of microsomal NADPH-cytochrome P-450 reductase with doxorubicin does induce the generation of superoxide ion and hydrogen peroxide to a great extent, it does not promote the peroxidation of unsaturated phospholipid micelles; the latter process took place only in the presence of the Fe^{3+}-ADP complex. In the subsequent work,[72] these authors concluded that the formation of the iron-doxorubicin complex is an obligatory condition of stimulation by doxorubicin of lipid peroxidation because lipid peroxidation proceeded only in a Tris®-HCl buffer where the tightly coordinated Fe^{3+}-ADP-doxorubicin complex exists. At the same time, there was no reaction in the phosphate buffer where this complex dissociated.

Another peroxidation mechanism is apparently realized under anaerobic conditions.[87] Contrary to the results obtained by studying lipid peroxidation in rat liver microsomes,[95,97] the doxorubicin-dependent phospholipid peroxidation catalyzed by xanthine oxidase or NADPH-ferredoxin reductase was highest under nitrogen containing traces of dioxygen and decreased with increasing proportion of dioxygen.[87] It was proposed that in this case, lipid peroxidation is initiated by the direct reduction of ferric ions by doxorubicin semiquinones (Reaction 28) and is independent of superoxide generation, which requires the presence of dioxygen.

$$Anth^{\cdot-} + Fe^{3+} \rightleftarrows Anth + Fe^{2+} \qquad (28)$$

The importance of the reduction of Fe^{3+} ions in anthracycline-dependent lipid peroxidation is also supported by the inhibitory effect of ceruloplasmin,[116] probably due to the oxidation of ferrous ions by this protein.

Another possible pathway for the initiation of anthracycline-dependent lipid peroxidation is self-reduction of ferric ions by the doxorubicin moiety of the Fe^{3+}-Dox complex (Reaction 24).[68-70,116]

Summarizing the data concerning the mechanisms of anthracycline-induced lipid peroxidation, one may conclude that it is a metal (iron)-dependent process. In all cases, an important function of anthracyclines is the reduction of ferric ions as a result of superoxide production, the direct reduction of ferric ion by anthracycline semiquinone, or the self-reduction of Fe^{3+} inside the iron-anthracycline complex. Another important function of anthracyclines is the stimulation of lipid peroxidation due to the production of hydrogen peroxide by dismutation of superoxide ion. The results obtained by Sugioka et al.[72] demonstrate an important role of iron-anthracycline complexes in lipid peroxidation. However, their actual role in the peroxidation mechanism remains uncertain due to instability of the ferrous ion-anthracycline complex.[58] It makes doubtful the possibility of this complex being an effective catalyst of the production of active oxygen radicals (hydroxyl and "crypto-hydroxyl" radicals). Therefore, it is possible that the formation of iron-anthracycline complexes may stimulate enzymatic lipid peroxidation via a self-reduction reaction or as a result of transportation of iron ions to the membrane surface where after reduction, these complexes release ferrous ions — genuine catalysts of lipid peroxidation.

b. Deoxyribose Degradation

Another possible target of anthracycline-dependent free radical attack is deoxyribose, the sugar moiety of DNA. It has been shown[69,82,117] that the reduction of doxorubicin by xanthine oxidase or ferredoxin reductase results in the degradation of deoxyribose and the formation of TBA-reactive products. The reaction proceeded under partially anaerobic conditions and decreased with increasing concentration of dioxygen. Under strict anaerobic conditions, little deoxyribose degradation was observed. The most efficient inhibitor of deoxyribose degradation was catalase; hydroxyl radical scavengers (ethanol, mannitol, formate, benzoate, and thiourea) also inhibited the reaction in a concentration-dependent manner. Later on, it was found[118] that ferredoxin reductase well stimulates deoxyribose degradation by doxorubicin, daunorubicin, and epirubicin under air. If in previous works the effect of SOD was uncertain, this reaction was entirely insensitive to SOD. Catalase, desferrioxamine, and ceruloplasmin were effective inhibitors.

The variable effects of different chelators (desferrioxamine completely inhibited deoxyribose degradation and EDTA greatly enhanced it) and an unusual inhibitory activity of various hydroxyl radical scavengers (mannitol and thiourea possessed unexpectedly high inhibitory properties) led to the conclusion that the reactive species formed in the anthracycline-stimulated degradation of deoxyribose are not free hydroxyl radicals.[89] Therefore, one may suggest that in this process, as in the case of anthracycline-dependent lipid peroxidation, crypto-hydroxyl radicals or site-specific hydroxyls play a main role.

c. Free Radical Degradation of DNA

There are several potential pathways for the free radical interaction of anthracycline antibiotics with DNA: the intercalation of anthracycline semiquinones into DNA,[34,119] the DNA scission by oxygen radicals generated during oxidation of anthracycline semiquinones by dioxygen,[36] and the DNA damage by iron-anthracycline complexes.[62,120,121] The participation of oxygen radicals is supported by the inhibition of the anthracycline-induced DNA scission by catalase and (partially) hydroxyl radical scavengers[114,122] and by the formation of HO-spin adducts.[62,121] Cleaving DNA by oxygen radicals is apparently possible only in the presence of free anthracyclines, as anthracyclines bound to DNA are not reduced by reductases.[36,42,62] On these grounds, it was concluded[42] that the inability of doxorubicin to induce DNA fragmentation in the presence of ferredoxin reductase is explained by the intercalation of this antibiotic in DNA.

Iron-anthracycline complexes may apparently interact with DNA without intercalation, forming ternary complexes.[120] In contrast to DNA-intercalated anthracyclines, ternary complexes can be reduced by glutathione and are able to reduce dioxygen to superoxide ion. As a rule, these complexes stimulate hydroxyl radical formation more efficiently than iron-anthracycline complexes.[62,121] Therefore, they cause extensive DNA fragmentation in the presence of glutathione. DNA damage by ternary iron-anthracycline-DNA complexes depends on the anthracycline structure: doxorubicin, daunorubicin, carminomycin, and 4-demethoxydaunorubicin are effective catalysts of DNA degradation in the presence of iron ions, whereas 11-*O*-methyldaunorubicin, 11-deoxydaunorubicin, and 11-deoxydoxorubicin have a smaller activity.[121] It was explained by the inability of these anthracyclines to form strong iron complexes.

Catalase resulted in nearly complete inhibition of DNA cleaving by ternary complexes, while the effect of SOD was uncertain. Hydroxyl radical scavengers (mannitol, histidine, and DMSO) inhibited DNA cleaving partially and only in high concentrations.[120] Thus, hydrogen peroxide formed during dismutation of superoxide ion plays an important role in the DNA damage induced by ternary complexes. The low efficiency of hydroxyl radical scavengers indicates a possible participation in DNA damage of crypto-hydroxyl radicals or "site-specific" hydroxyls (similarly to the anthracycline-dependent lipid peroxidation), but, at the same time, the formation of free hydroxyl radicals is well documented by the formation of their spin adducts. This discrepancy is discussed below.

d. Other Possible Modes of Free Radical Damage Induced by Anthracyclines

One of the toxic effects of anthracyclines which is probably related to their free radical activity is the damage of the respiratory chain activity in mitochondria. It has been shown[123] that the incubation of pig heart submitochondrial particles with doxorubicin in the presence of Fe^{3+} ions resulted in the inactivation of NADH oxidase, NADH-cytochrome c reductase, and cytochrome c oxidase. The inactivation of respiratory chain activity occurred simultaneously with phospholipid peroxidation and was inhibited by EDTA and (partially) by a free radical scavenger, BHT.[124] Therefore, it seems likely that the respiratory chain inhibition is a consequence of lipid peroxidation or is induced by the same reactive oxygen species. There was no inactivation and lipid peroxidation without iron ions. It is of interest that an increase in NADH oxidation and the rate of electron flux sharply diminished enzyme inactivation and lipid peroxidation. It was proposed that this effect may be explained by increasing the reduction of endogenous ubiquinone into ubiquinol possessing antioxidant properties.

It has also been shown that doxorubicin inhibits mitochondrial NADH-cytochrome c oxidoreductase activity in the guinea pig heart[125] and both mitochondrial and cytoplasmic isocitrate dehydrogenase activities in homogenate of rat heart.[126] Recently, it was determined by *in vivo* ^{31}P-NMR that doxorubicin decreases the phosphocreatine/ATP ratio in the rat heart.[127] Muhammed and co-workers[128,129] concluded that doxorubicin is able to inhibit succinate oxidation and phosphorylation in heart and liver mitochondria in the presence of hexokinase. The addition of ubiquinone Q_9 reversed the inhibition to a large extent. However, these findings were recently questioned[48] as in mitochondria, anthracyclines are reduced by the NADH dehydrogenase flavin of Complex I and not by succinate dehydrogenase or ubiquinone. The results obtained by Muhammed and co-workers may be explained by reverse electron transport to Complex I in the presence of ADP or some shortcomings of the manometric technique of dioxygen determination.

On the other hand, Pollakis et al.[130] concluded that doxorubicin (but not 5-iminodaunorubicin) facilitates electron transport between NADH and cytochrome c in the inner mitochondrial membrane as a result of covalent-like binding to cardiolipin. This phenomenon is supposedly accompanied by phospholipid peroxidation stimulated by anthracycline semiquinones and the disturbance of mitochondrial membranes. Thus, the effect of anthracyclines on respiratory chain activity in mitochondria may have a very complicated nature.

It has been shown[131] that doxorubicin inhibits Ca^{2+} uptake by rat heart sarcoplasmic reticulum. Catalase, hydroxyl radical scavengers (*N*-acetylcysteine, glutathione, histidine, and dimethylurea), and an iron chelator, desferrioxamine, substantially reduced the toxic effect of doxorubicin. Although the mechanism of doxorubicin action on Ca^{2+} uptake remains unclear, these authors proposed that it may be related to free radical oxidation of sulfhydryl groups participating in Ca^{2+} uptake.

e. Anthracycline-Dependent Free Radical Damage to Whole Cells

Important results relevant to the mechanism of anticancer action and cardiotoxicity of anthracyclines were obtained by studying the effects of anthracyclines on whole cells. There is strong evidence of a decisive role of oxygen radicals in the killing of tumor cells by doxorubicin, including MCF-7 human breast cancer cells[76,132,133] and Ehrlich ascites carcinoma cells.[79] The generation of hydroxyl radicals was shown by the spin-trapping technique[76,133] and the formation of methane from DMSO.[79] SOD, catalase, *N*-acetylcysteine, DMSO, thiourea, and diethylurea reduced or abolished the killing of tumor cells by doxorubicin, which confirmed the participation of both superoxide ion and hydroxyl (or crypto-hydroxyl) radicals. Effects of chelators seem to depend on their ability to traverse the tumor cell surface: bipyridine and *N,N,N',N'*-tetrakis(2-pyridylmethyl)ethylenediamine, which freely enter the cell, significantly reduced cell killing while desferrioxamine, EDTA, and DTPA (which in most cases are unable to penetrate the cell surface) produced no effects.[132] (However, Sinha et al.[76] have found that desferrioxamine decreased the doxorubicin-stimulated HO formation in cells.)

In contrast to its toxic action on the sensitive subline of tumor cells (MCF-7), doxorubicin did not stimulate the formation of HO spin adducts in the resistant subline.[76,133] The difference between the sensitive and resistant sublines may be mainly relevant to the increased glutathione peroxidase activity in the resistant subline as the activities of other antioxidant enzymes (SOD and catalase) in both systems were identical. It should be noted that selective damage of Se-glutathione peroxidase in human erythrocytes and rat liver hepatocytes by doxorubicin was shown earlier.[134] This fact apparently confirms the importance of increasing glutathione peroxidase activity for cell survival.

There is other evidence of the importance of the glutathione redox cycle, consisting of glutathione peroxidase and glutathione reductase, in the protection of cells against the toxic action of anthracyclines. It has been shown[135,136] that the inactivation of glutathione reductase by *N,N*-bis(2-chloroethyl)-*N*-nitrosourea (BCNU) greatly increased lipid peroxidation and lactate dehydrogenase leakage induced by doxorubicin in hepatocytes and decreased cell viability. The protective action of the glutathione redox cycle is possibly explained by the regulation of the level of hydrogen peroxide in cells.

It was recently shown[137] that in addition to the well-known direct cytotoxic activity, doxorubicin and daunorubicin possess another toxic activity, the enhancement of complement susceptibility in human melanoma cells. It was found that anthracyclines converted the cells to susceptible ones, the Fe^{3+}-doxorubicin complex being about two times more effective than doxorubicin. The effects of doxorubicin and daunorubicin were partially inhibited by SOD, catalase, and DMSO and completely blocked by iron ion chelators: EDTA, DTPA, and bathophenanthroline sulfonate. Based on these findings, it was concluded that anthracyclines generate oxygen radicals in the presence of iron ions and, by this, cause alterations in the melanoma cells, resulting in the enhanced complement susceptibility. In accord with this proposal, 5-iminodaunorubicin, possessing the reduced ability to generate oxygen radicals, did not enhance the complement susceptibility of cells. It should be noted that doxorubicin and daunorubicin were more effective after immobilization onto glycerol-coated glass beads.

f. **In Vivo** ***Investigations of Free Radical Activity of Anthracyclines***

It is understandable that usually we have only indirect and frequently disputable *in vivo* evidence of the participation of free radicals in the anthracycline-dependent processes. We already discussed one such piece of evidence: the anthracycline-dependent lipid peroxidation in animal experiments. Other, and possibly more important, data may be obtained by studying the effects of free radical scavengers and iron chelators on anthracycline-induced cytotoxicity. These investigations are of great practical importance as they are directed to developing new methods of decreasing the cardiotoxicity of anthracyclines without changing their anticancer activity. It is supposed that only cardiac toxicity and not the anticancer action of anthracyclines depends on their free radical activity.

Various free radical scavengers, antioxidants, and metal chelators were tried as the *in vivo* inhibitors of anthracycline-dependent free radical processes. It has been shown[105] that doxorubicin-induced cardiac toxicity in mice can be suppressed by α-tocopherol; however, until now, clinical trials with α-tocopherol, SOD, and *N*-acetylcysteine have been unsuccessful.[138] A great amount of attention is now being attached to bisdioxopiperazine ICRF-187, which is believed to be transformed in cells into putative chelating species.[92]

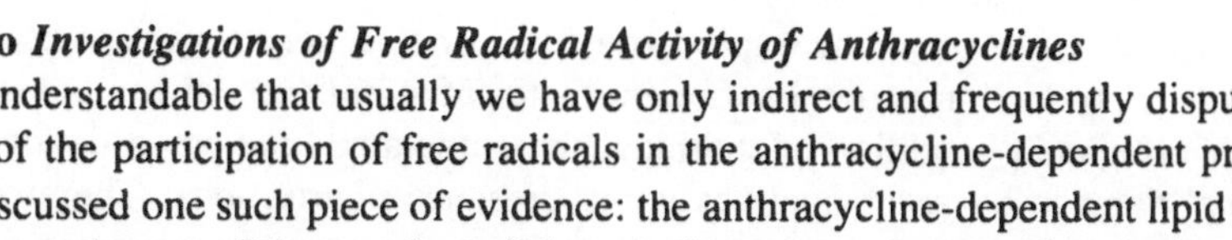

(29)

ICRF-187

It has been shown that ICRF-187 diminished the severity of doxorubicin-induced cardiomyopathy in beagle dogs,[139] rats,[140] and mice.[141] In the last work an analogous effect was observed for the doxorubicin analogs, epirubicin and idarubicin. It was also earlier shown[142] that the reduction of cardiac toxicity of doxorubicin in mice can be achieved by pretreatment with a typical hydroxyl scavenger, DMSO.

Recently, the first successful clinical trials of ICRF-187 as an agent preventing cumulative doxorubicin-induced cardiac toxicity were reported for women having metastatic breast cancer.[138] The authors believe that if their results are confirmed, then the possibility of suppressing doxorubicin-induced cardiomyopathy without interfering with anticancer activity of the drug will be proven. It was proposed that ICRF-187 is able to block all iron-dependent pathways of oxygen radical production stimulated by doxorubicin, but the difference in the protective effects of ICRF-187 on normal and tumor cells may lie in the bioavailability of the drug active form in both types of cells. However, it should be noted that the chelating mechanism of the protective action of ICRF-187 needs additional confirmation.

g. Evolution of Free Radical Hypothesis of Toxicity and Anticancer Activity of Anthracycline Antibiotics

Discovery of anthracycline-stimulated superoxide production in microsomes and mitochondria[1-6] greatly stimulated studies of free radical mechanisms of anticancer activity and cardiotoxicity of these antibiotics. At the beginning, it was proposed that there are two modes of anthracycline activity in cells: the anticancer activity depending on intercalation of antibiotics in DNA and the side cardiotoxicity relating to the generation of oxygen radicals which induce lipid peroxidation in heart mitochondria. Based on this proposal, one may expect that free radical scavengers, chelators, and antioxidants may be useful for suppression of anthracycline cardiotoxicity. If this proposal is valid, then a decrease in cardiotoxicity may also be achieved by the modification of anthracycline structure resulting in diminishing generation of the superoxide ion.

However, as we have seen, the superoxide ion is not a very toxic species, and the enhancement of its production is not a sufficient factor of stimulation of lipid peroxidation. In addition, the obligatory condition of anthracycline-stimulated lipid peroxidation and cellular free radical damage by anthracyclines is the presence of iron ions. Therefore, at the second stage of developing free hypothesis of anthracycline activity, a great amount of attention was paid to the role of iron-anthracycline complexes. It was proposed that iron-anthracycline complexes are genuine catalysts of the Fenton reaction in which very active hydroxyl radicals are formed. Indeed, as we have seen, many convincing examples of hydroxyl radical generation by anthracyclines in cells and its inhibition by metal chelators were obtained.

However, there are three important points which were not explained. The first one is the fact that there is evidence[72] that the anthracycline-dependent lipid peroxidation is really stimulated by oxygenated iron complexes (''crypto-hydroxyl'' radicals) and not by genuine hydroxyl radicals. The second one is the inability of hydroxyl radicals to diffuse from the site of their generation to a substantial distance, due to their high reactivity. The third one is the instability of the ferrous ion-anthracycline complexes that brings into question their importance as real catalysts of the Fenton reaction.

Important findings which greatly transform a traditional point of view on the anticancer activity of anthracyclines were recently obtained in the experiments with immobilized anthracyclines including the immobilization onto glycerol-coated glass beads,[137] the incorporation into biodegradable albumin microspheres,[143] and covalent binding to agarose polymers.[144] It was found that in all cases, immobilized anthracyclines possessed a stronger anticancer activity and a diminished cardiac toxicity. As immobilized anthracyclines are not able to penetrate lipid membranes, it follows that the anticancer activity of anthracyclines cannot be explained by intercalation in DNA. Obviously, this conclusion completely changes a generally adopted attitude that anthracyclines are the DNA-intercalating agents. However, until now, there is no other explanation of retaining anticancer activity by immobilized anthracyclines.

We now venture to propose the next mechanism of cytotoxic action of anthracyclines. The primary effect of anthracyclines is an increase in superoxide generation as a result of the formation of anthracycline semiquinones during the reduction of anthracyclines by NADPH- and NADH-reductases. As was noted above, the superoxide ion itself probably does not manifest toxic action, and a main consequence of its generation is the following increase in hydrogen peroxide concentration due to superoxide dismutation. However, it is possible that the superoxide ion may trigger other important processes including the release of iron ions from ferritin. By this or another way, anthracyclines which remain unchanged during dioxygen reduction acquire iron ions (at present, the most probable sources seem to be ferritin and transferrin) and form strong ferric ion-anthracycline complexes. The reduction of these complexes is apparently a first step leading to active toxic species. There is evidence that this process, which is accompanied by hydroxyl radical production, takes place at the lipid membrane surface,[76,79,133] as the formation of hydroxyl radicals is inhibited by exogenous SOD and catalase which are known to be unable to traverse lipid membranes. There are many physiological reductants (glutathione, cysteine, ascorbate, superoxide ion, etc.) which are able to reduce Fe^{3+}-Anth to Fe^{2+}-Anth. The last complex is unstable[58] and dissociates to form ''free'' ferrous ions. The interaction of Fe^{2+}, H_2O_2, and possibly anthracyclines results in the formation of the main villains: hydroxyl and ''crypto-hydroxyl'' radicals.

We already discussed the nature of "crypto-hydroxyl" radicals (iron-hydroperoxy complexes or iron-dioxygen complexes; Volume I, Chapter 9) which may be precursors of genuine hydroxyl radicals. In contrast to hydroxyl radicals, however, they must be sufficiently long-lived species and may penetrate lipid membranes similarly to hydrogen peroxide. Therefore, it may be proposed that the anticancer activity of anthracyclines is due to the destruction of the DNA of tumor cells by crypto-hydroxyl radicals.

Thus, we may assume that there are two types of free radical activity of anthracyclines: the destruction of lipid membranes by hydroxyl radicals generated during the reduction of iron-anthracycline complexes at the membrane surface, and the intracellular damage of DNA and other cell components by crypto-hydroxyl radicals. The difference in structure of the oxygen radicals involved in both processes may be a reason for the different effects of free radical scavengers and chelators on the cardiac toxicity and anticancer action of anthracyclines.

In this connection some proposals may be made. SOD and catalase must apparently affect both cardiotoxicity and anticancer activity of anthracyclines as they decrease the pools of the main precursors of active oxygen radicals, superoxide ion and hydrogen peroxide. Typical hydroxyl scavengers (for example, mannitol, DMSO) should be more effective in the prevention of cardiac toxic effects. It is difficult to foresee the difference in the action of chelators on both types of free radical activity of anthracyclines, especially if this difference may, in the first place, be determined by the ability of a chelator to penetrate lipid membranes.

Some additional experimental data probably support the participation of free radicals in anticancer activity of anthracyclines. Thus, it has been shown[145] that ICRF-187 and its racemate ICRF-159 enhance the anticancer action of doxorubicin in experimental tumor cell systems. On the other hand, vitamin E increased the toxic effect of doxorubicin on human prostate carcinoma cells and had a protective effect on normal cells.[146] Possibly, these effects of a chelator and an antioxidant have really no connection with free radical processes. However, if they have such a connection, then they indicate the importance of free radical participation in the anticancer activity of anthracyclines.

B. SYNTHETIC ANALOGS OF ANTHRACYCLINES

It is reasonably thought that modifications of the anthracycline structure may be designed which might enhance the anticancer activity of drugs and decrease their cardiotoxicity. This proposal led to developing new groups of anthraquinone drugs which lost many of the features of anthracycline molecules (anthracyclines with minor structural changes such as 5-iminodaunorubicin were considered by us in the preceding section). The modifications were directed to make the drugs more resistant to enzyme reduction (diminishing by this the generation of superoxide ion by the drug semiquinones) while retaining planar, electronic, and other characteristics of parent anthracyclines which are thought to be necessary for the intercalation into DNA.[147] Now, several classes of such "chromophore-modified anthracyclines" were synthesized.

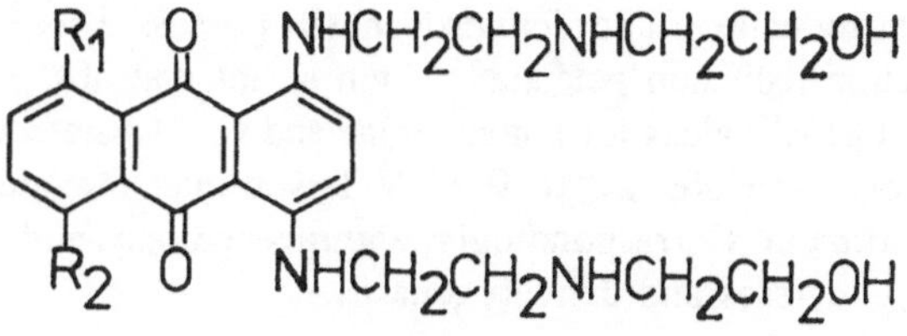

Anthracenediones

XIII. Mitoxantrone, mitozantrone, DHAQ; $R_1 = R_2 = OH$

XIV. MHAQ, (1-hydroxy-5,8-bis(2-[(2-hydroxyethyl)-amino] ethylamino)-9,10-anthracenedione; $R_1 = OH$, $R_2 = H$

XV. Ametantrone, HAQ; $R_1 = R_2 = H$

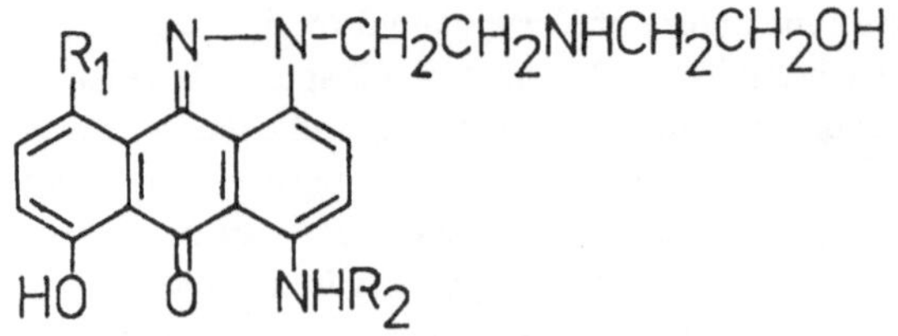

Anthrapyrazoles

XVI.	CI 973; $R_1 = OH$, $R_2 = (CH_2)_2NHCH_3$
XVII.	CI 941; $R_1 = H$, $R_2 = (CH_2)_2NH(CH_2)_2OH$
XVIII.	CI 942; $R_1 = OH$, $R_2 = (CH_2)_3NH_2$

Anthracenediones and anthrapyrazoles indeed turned out to be significantly less effective catalysts of electron transfer (in comparison with anthracyclines) from microsomal and mitochondrial reductases to dioxygen. Thus, mitoxantrone (XIII) and ametantrone (XV) generated $O_2^{\cdot-}$ 5 to 20 times less than doxorubicin and daunorubicin in the presence of NADPH-cytochrome P-450 reductase, NADH dehydrogenase, and rabbit hepatic microsomes.[148] In rat hepatic microsomes, ametantrone was about ten times less effective as a catalyst of superoxide production in comparison with daunorubicin,[31] and mitoxantrone was a much less effective catalyst of superoxide production than doxorubicin in human liver microsomes.[49] Similarly, anthrapyrazoles induced far less (20- to 200-fold) SOD-sensitive oxygen consumption than doxorubicin in rat hepatic microsomes.[19]

Another important distinction between anthracenediones or anthrapyrazoles and anthracyclines is their inability to stimulate lipid peroxidation. Actually, these compounds inhibit lipid peroxidation: ametantrone inhibited lipid peroxidation in mouse heart;[110] mitoxantrone, MHAQ, and ametantrone inhibited rabbit liver microsomal peroxidation;[100] mitoxantrone and ametantrone inhibited the NADPH-cytochrome P-450 reductase-catalyzed oxidation of linoleic acid;[149] and anthrapyrazoles inhibited lipid peroxidation in rabbit and rat liver microsomes.[150,151] In addition, anthracenediones and anthrapyrazoles inhibited the anthracycline-stimulated lipid peroxidation.[100,110,148,150] Both diminished superoxide production, and an inability to stimulate lipid peroxidation is believed to be the reason for a decrease in the cardiotoxicity of anthracenediones and anthrapyrazoles.

The poor ability of these compounds to generate superoxide ion is well explained by diminishing their reduction potentials (Table 2). (It should be stressed that in this case, only one-electron reduction potentials must be considered: E_7^1 values obtained by a pulse-radiolytic technique in water and $E_{1/2}$ values in aprotic media (DMF). In contrast to this, $E_{1/2}$ values obtained in water are two-electron reduction potentials and therefore are not supposed to be a correct estimate of the reduction rates of substrates by one-electron transfer reductases.) It was supposed[152] that reductases and dehydrogenases possess a maximal activity toward quinones with one-electron reduction potentials within an interval of (-0.20) to (-0.06) V. It is seen from Table 2 that E_7^1 values for mitoxantrone and CI 941 are equal to -0.527 V and -0.538 V, respectively, and are 0.2 to 0.25 V lower than the reduction potentials of doxorubicin and daunorubicin. Correspondingly, anthracenediones and anthrapyrazoles must be hard to reduce by reductases and dehydrogenases.

Nonetheless, it was found that under anaerobic conditions, ametantrone and mitoxantrone are reduced by rat liver microsomes and human liver microsomes to semiquinone radicals.[31,49] It is difficult to understand why there is no increase in superoxide production in the presence of these compounds, although their semiquinones must be oxidized by dioxygen with diffusion-controlled rates. (For example, a rate constant for the oxidation of the CI 941 semiquinone by dioxygen is $1.8 \cdot 10^9\ M^{-1}\ s^{-1}$ [151].) Therefore, it seems that diminishing of superoxide production by anthracenediones and anthrapyrazoles cannot be explained only by decreasing their reduction potentials.

Some workers (see, for example, Reference 151) explain the inhibition of lipid peroxidation (including the anthracycline-dependent lipid peroxidation) in the presence of anthracenediones and anthrapyrazoles by the competition of these compounds with anthracyclines for the flavin component of reductases. However, Kharasch and Novak[100] have shown that while doxorubicin-stimulated lipid peroxidation is inhibited half-maximally by 4 μM mitoxantrone, a comparable decrease in superoxide generation is observed only with 2.5 mM mitoxantrone. Ametantrone was nearly equal to mitoxantrone in its ability to inhibit lipid peroxidation, but did not change entirely the doxorubicin-stimulated superoxide production. Similarly, anthrapyrazoles inhibited the doxorubicin-induced lipid peroxidation and did not affect NADPH oxidation and superoxide production in rabbit liver microsomes.[150] Based on these data, it was concluded that anthracenediones and anthrapyrazoles inhibit the subsequent stages of lipid peroxidation, reacting probably with lipid peroxy radicals.

Recently,[18] new anthraquinonyl glucosaminosides (XIX to XXIII) were synthesized which, contrary to anthracenediones and anthrapyrazoles, were significantly more effective producers of the superoxide ion in rat heart sarcosomes even in comparison with daunomycin.

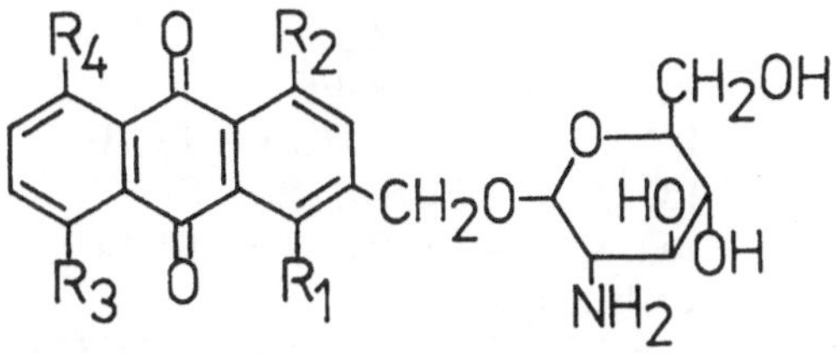

XIX.	$R_1 = R_2 = R_3 = R_4 = H$
XX.	$R_1 = OH, R_2 = R_3 = R_4 = H$
XXI.	$R_1 = R_2 = OH, R_3 = R_4 = H$
XXII.	$R_1 = R_3 = OH, R_2 = R_4 = H$
XXIII.	$R_1 = R_3 = H, R_2 = R_4 = OH$

It seems that there is no correlation between one-electron reduction potentials of these compounds measured in DMF and their ability to generate superoxide ion.

C. OTHER QUINONE ANTICANCER DRUGS: MITOMYCINS, ACTINOMYCINS, STREPTONIGRIN, AND AZIRIDINYLQUINONES

All these compounds are anticancer drugs containing the quinone moiety. Together with anthracyclines, they undoubtedly are the most important quinone anticancer drugs.

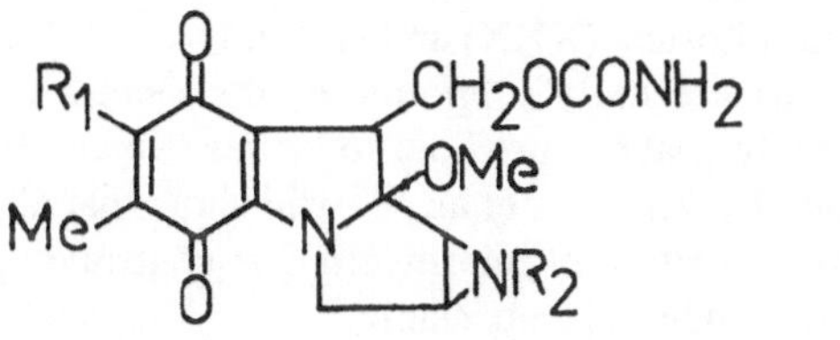

XXIV. Mitomycin C, $R_1 = NH_2$, $R_2 = H$
Mitomycin B, $R_1 = NH_2$, $R_2 = Me$

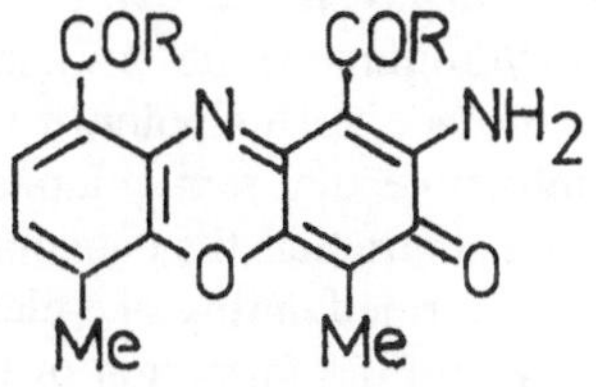

XXV. Actinomycin D
R = -Thr-D-Val-Pro-Sar-MeVal
└——O——┘

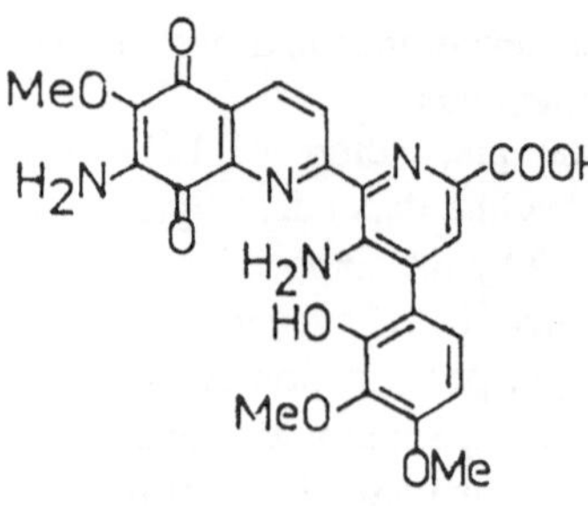

XXVI. Streptonigrin

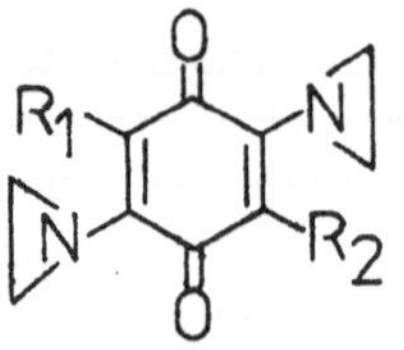

XXVII. Aziridinylquinones:
Diaziquone or AZQ
$R_1 = R_2 = NHCOOEt$
BZQ, $R_1 = R_2 = NHCH_2CH_2OH$
RQ2, $R_1 = N(CH_2)_2$, $R_2 = Cl$
RQ14, $R_1 = Cl$, $R_2 = NHCOOEt$
Carbazilquinone
$R_1 = Me$, $R_2 = CH(OMe)CH_2OCONH_2$

Mitomycin C is active against many animal and human tumors and has a preferential cytotoxicity toward hypoxic tumor cells. It is thought that the mechanism of its action includes enzymatic reduction to intermediates which alkylate DNA. It has been shown that mitomycin C is reduced by NADPH-cytochrome P-450 reductase and xanthine oxidase under anaerobic conditions[153] and NADPH-cytochrome c reductase,[154] rat liver microsomes,[6,46] and rat sarcosomes[155] under aerobic conditions. In the last case, superoxide ion and hydroxyl radicals were produced. Mitomycin C and its analogs, mitomycin B, BMY-25282 (R_1 = N=CHNMe$_2$, R_2 = H), and BL-6783 (R_1 = N=CHNMe$_2$, R_2 = Me), are also reduced by tumor cells.[156] The mitomycin semiquinone was identified by ESR and optical spectra,[157,158] and the superoxide ion formed as a result of the one-electron oxidation of mitomycin semiquinone by dioxygen was identified as the DMPO and PMN spin adducts.[46,157]

It was supposed that the most probable pathway of reductive transformation of mitomycin C into reactive alkylating intermediates is an intramolecular transformation of its semiquinone (XXVIII).[160]

Another pathway is the two-electron reduction of mitomycin C with the subsequent oxidation of hydroquinone into 2,7-diaminomitosene (XXX) and *cis*- and *trans*-2,7-diamino-1-hydroxymitosene.[161] This point of view was recently supported by the kinetic study[158] as the rate of intramolecular semiquinone transformation turns out to be very small. However, this conclusion contradicts the data obtained by Andrews et al.,[159] who show that all reactive metabolites are formed during one-electron reduction of mitomycin C at controlled potential (−0.950 V) where the formation of hydroquinone is impossible.

Actinomycin D (XXV) is an antibiotic used for the treatment of Wilms' tumor, gestational choriocarcinoma, and other tumors. The reduction of actinomycin D by rat liver microsomes and NADPH-cytochrome P-450 reductase to a free radical which was accompanied by dioxygen consumption has been shown by Bachur et al.[162] An ESR spectrum of the actinomycin semiquinone was also obtained by the use of chemical reductants.[163] An important role of oxygen radical production in the cytotoxic action of actinomycin D was shown in the experiments with its N^2-substituted, spin-labeled derivatives,[164] 2-deamino-2-nitroactinomycin D, and 2-deaminoactinomycin D.[165] All these compounds weakly intercalated or did not intercalate at all into DNA, but nonetheless, showed better anticancer activity than actinomycin D, possibly due to their increased capacities to generate superoxide ion. Streptonigrin (XXVI), another anticancer antibiotic, also catalyzed the production of oxygen radicals by microsomes,[6] xanthine oxidase, and bacteria.[166]

In contrast to the above antibiotics, aziridinyl quinones (XXVII) are synthetic compounds possessing activities in many experimental tumors (intracranial ependymoblastoma, interperitoneal L 1210 tumors, etc.) which are now being studied in clinical trials. It has been shown that diaziquone (3,6-diaziridinyl-2,5-bis[carboethoxyamino]-1,4-benzoquinone) and carbazylquinone are reduced to semiquinones by liver microsomes, NADPH-cytochrome P-450 reductase,[39,167,168] and xanthine oxidase.[169] Diaziquone semiquinone, which was characterized by its ESR spectrum, was stable under aerobic conditions and can also be obtained by electrochemical and chemical reduction ($NaBH_4$, NADPH, and L-cysteine).[170-172] Diaziquone was also reduced to semiquinone by whole cells (human leukemic cell lines K 562 and HL 60, murine leukemic cell line L 1210, and HE_p-2 cells).[171,173]

Diaziridyl quinones differ from anthracyclines, mitomycin C, actinomycin D, and streptonigrin by their capacity to be easily reduced by enzymes and chemical reductants. It is explained by the higher reduction potentials of these compounds. Thus, Butler et al.[174] recently reinvestigated the redox properties of diaziquone and found that its one-electron reduction potential is equal to -0.070 V, which is significantly higher than a previous value of -0.168 V.[12] Owing to that, diaziquone is enzymatically reduced under aerobic conditions. As is seen from Table 1, the rate constant for the reduction of dioxygen by $AZQ^{\overline{\cdot}}$ remains sufficiently high, and therefore diaziquone must retain (at least partially) the capacity to produce superoxide ion. Indeed, diaziquone stimulated oxygen consumption in cells.[171] It is interesting that an analog of diaziquone, 2,5-bis(2-hydroxyethylamino)-3,6-diaziridinyl-1,4-benzoquinone has an essentially smaller reduction potential (Table 2) and therefore must behave similar to anthracyclines in redox processes.

Most of the drugs considered above also stimulated hydroxyl radical production. Thus, mitomycin C, mitomycin B, and streptonigrin produced PBN-OH spin adducts when reduced with $NaBH_4$,[28,175] mitomycin C induced the decarboxylation of benzoate in rat liver microsomes supposedly via HO· generation;[71] mitomycin C and carbazylquinone catalyzed the HO· production by NADPH-cytochrome P-450 reductase as measured by DMPO- and BPN-OH adduct formation and ethylene generation from methional.[39,154,176] Hydroxyl radicals were also formed during the incubation of mitomycin C with EMT6 tumor cells[156] and streptonigrin with bacteria *Neisseria gonorrhoeae*,[166] which is believed to be a reason for the cytotoxicity of these antibiotics. Mitomycin C and streptonigrin stimulated the peroxidation of microsomal phospholipids catalyzed by NADPH-cytochrome P-450 reductase[157] and deoxyribose degradation.[177,178] However, all controversies concerning the true role of hydroxyl radicals in the cytotoxic action of quinone drugs, which were discussed above for anthracyclines, remain valid. Studies of mitomycin C- and streptonigrin-dependent lipid peroxidation and deoxyribose degradation[157,177,178] indicate a principal role of hydroxyl radical-like species and not free hydroxyl radicals. For example, as in the case of doxorubicin, the mitomycin-dependent phospholipid peroxidation proceeded in Tris®-HCl buffer and not in phosphate buffer where HO· generation is maximal.[157] It should be noted that deoxyribose degradation stimulated by mitomycin C and streptonigrin proceeded only at low dioxygen concentrations[177,178] that correspond to the enhanced cytotoxic activities of these antibiotics under hypoxic conditions.

D. BENZOQUINONES AND NAPHTHOQUINONES

Extensive studies of redox cycling of quinone antibiotics might make one suppose that the redox cycling in biological systems is confined to these complicated compounds and is not a property of more simple-structured species. However, this is not the case. There are a great many compounds (mostly quinones, pyridinium salts, and heteroaromatic and nitro compounds) which are reduced by mitochondria, microsomes, and whole cells to free radicals which may or may not be the precursors of oxygen radicals. In this section we consider the redox properties and the toxic action of benzo- and naphthoquinones.

Benzo- and naphthoquinones are readily reduced to semiquinone radical anions by various enzymes: microsomal cytochrome b_5 reductase,[179] NADPH-cytochrome c reductase,[179] xanthine oxidase,[180] mitochondrial NADH dehydrogenase,[181] and NADPH-cytochrome P-450 reductase.[152,182] Under aerobic conditions, some semiquinones reduced dioxygen to produce superoxide ion. The efficiency of superoxide production depended on the redox properties of quinones; for example, benzoquinone, which has the most positive one-electron reduction potential, was not able to generate superoxide ion. Hassan and Fridovich[183] used an increase in the cyanide-resistant respiration in cells of *Escherichia coli* as an index of superoxide generation by cells in the presence of oxidants. From comparison of the efficiency of cyanide-resistant respiration, the activities of various compounds as catalysts of superoxide generation were estimated. It was found that menadione (2-methyl-1,4-naphthoquinone, vitamin K_3), plumbagin (5-hydroxy-2-methyl-1,4-naphthoquinone), and juglone (5-hydroxy-1,4-naphthoquinone) strongly stimulated the cyanide-resistant respiration of *E. coli*, while benzoquinone had a double effect: it slightly stimulated cyanide-resistant respiration at low concentrations, but inhibited it when the concentration increased.

Enzymatic redox cycling of menadione and some other naphthoquinones and benzoquinones with subsequent production of superoxide ion was studied by many workers. Menadione was reduced by rat liver microsomes,[184-187] rat liver mitochondria,[188,189] isolated hepatocytes,[186,190-194] enterocytes,[195,196] endothelial cells,[196,197] human erythrocytes,[198] and human blood serum.[199] Among other quinones, 2,3-dichloro-1,4-naphthoquinone was reduced by rat liver microsomes[200] and beef heart mitochondria;[201] duroquinone and ubiquinone Q_1 were reduced by rat liver microsomes;[184] 2,3-dimethyl-, 2,3-diethyl-, and 2,3-dimethoxy-naphthoquinones participated in redox cycling to form superoxide ion in hepatocytes;[192,194] and *N*-acetyl-*p*-benzoquinone was reduced to semiquinone by rat liver microsomes without catalyzing superoxide generation.[187] β-Lapachone (3,4-dihydro-2,2-dimethyl-2H-naphtho[1,2-b]-pyran-5,6-dione) (XXXII), an antimicrobial agent, was reduced by *Trypanosoma cruzi* epimastigotes[202,203] and Sarcoma 180 cells[204] to form semiquinone under anaerobic conditions. Under aerobic conditions, β-lapachone stimulated formation of superoxide ion by mitochondrial and microsomal fractions of these cells as well as by whole *T. cruzi* cells as it was measured by the SOD-inhibitable oxidation of epinephrine into adrenochrome. Pyrroloquinoline quinone (XXXIII) (a coenzyme of bacterial methanol dehydrogenase and of a bovine serum amine oxidase) was reduced by NADPH to semiquinone under anaerobic conditions and catalyzed superoxide production under aerobic conditions.[205]

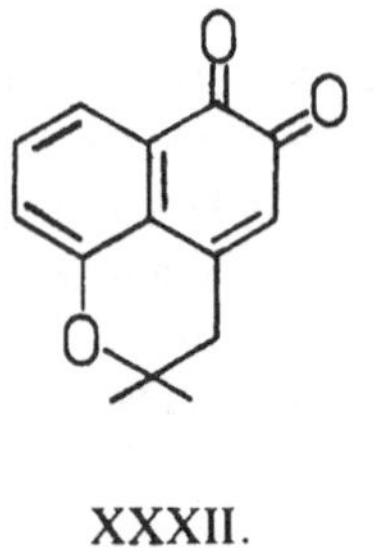

XXXII.

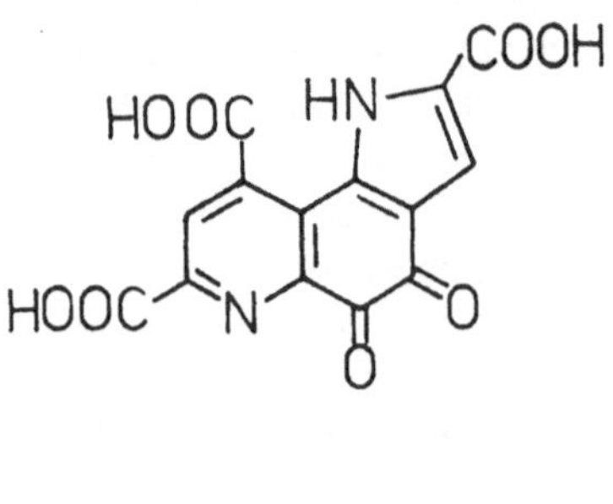

XXXIII.

The redox cycling of quinones was well documented by the SOD-inhibitable reduction of acetylated cytochrome c,[185,190,192] ESR spectra of semiquinones (as a rule, under anaerobic conditions),[182,194,205] and ESR spectra of superoxide spin adducts.[194,197]

In addition to microsomal and mitochondrial one-electron transfer reductases, quinones may be reduced by two-electron reductases.[184,189,195] In this case, methylhydronaphthoquinone formed intracellularly may diffuse back across the cell membrane and generate superoxide ion extracellularly. However, at least for enterocytes,[197] such a process seems to be less important than one-electron reduction of menadione. This conclusion is supported by the data obtained by Orrenius et al.,[186] who did not find any indications of menadione redox cycling by DT-diaphorase in hepatocytes.

In contrast to anthracyclines, naphtho- and benzoquinones are not good promoters of lipid peroxidation. There are several examples of stimulation of this process by naphthoquinones: menadione stimulated lipid peroxidation in endothelial cells,[197] 2,3-dichloro-1,4-naphthoquinone induced beef heart mitochondrial lipid peroxidation,[200] and β-lapachone induced the tocopherol-inhibitable lipid peroxidation in *T. cruzi* cells.[202,203] At the same time, menadione and naphthoquinone completely inhibited NADPH-dependent lipid peroxidation in rat liver microsomes.[187,206] Naphtho- and benzoquinones are apparently also incapable of catalyzing the production of hydroxyl radicals. The formation of a DMPO-OH spin adduct during the incubation of menadione with endothelial cells or enterocytes was explained by the decomposition of DMPO-OOH.[195,197] (However, Komiyama et al.[81] have shown that menadione was a more active catalyst of ethylene generation than anthracyclines in the oxidation of methional by NADPH-cytochrome P-450 reductase in the presence of methemoglobin. The ethylene formation is frequently considered as an index of hydroxyl radical formation; indeed, this reaction was inhibited by tiron, thiourea, and catalase.) Therefore, it is possible that the toxicity of benzo- and naphthoquinones is explained by other reasons than a traditional route:

$$Q \xrightarrow{e} Q^{\overline{\cdot}} \rightarrow O_2^{\overline{\cdot}} \rightarrow HO\cdot$$

The toxic action of quinones was shown in many biological systems. Menadione resulted in thiol depletion, NADPH oxidation, the disruption of calcium homeostasis, and loss in cell viability in hepatocytes[186,190,192,193] and endothelial cells.[197] Mitochondrial calcium release and NADPH oxidation were observed in the presence of menadione,[188,189] benzoquinone,[207] and substituted benzoquinones.[207] 2,3-Dichloro-1,4-naphthoquinone stimulated thiol oxidation and swelling of beef heart mitochondria.[201] Plumbagin exerted an oxygen-dependent lethality on free-living nematodes.[208] 2,5-bis-(Dimethylamino)benzoquinone and 2,5-bis-(di(2′-chloroethylamino)-benzoquinone (benzoquinone dimustard) were toxic to lymphoma cells.[209] Now, the disturbance by quinones of the glutathione redox cycle is recognized as a principal reason for their toxicity. Indeed, the depletion of glutathione after the incubation of cells with menadione was shown in many works. This phenomenon is explained by the coupling of menadione metabolism with the glutathione redox cycle via hydrogen peroxide generated by superoxide dismutation (Reactions 30 to 32):

$$2\,O_2^{\overline{\cdot}} + 2\,H^+ \xrightarrow{SOD} H_2O_2 + O_2 \qquad (30)$$

$$H_2O_2 + 2\,GSH \xrightarrow{GSH\text{-peroxidase}} GSSG + 2\,H_2O \qquad (31)$$

$$GSSG + 2\,NADPH \xrightarrow{GSSG\text{-reductase}} 2\,GSH + 2\,NADP \qquad (32)$$

The reduction of hydrogen peroxide to water by glutathione-peroxide results in an increase in GSSG which leads to increasing the NADPH consumption by Reaction 32. In this way, the menadione-stimulated superoxide production leads to the decrease in protective antioxidant systems and the reduction of cell resistance to oxidative stress via GSH and NADPH depletion. It should be noted that NADPH depletion results not only because of Reaction 32, but also owing to the reduction of menadione to semiquinone by NADPH-cytochrome P-450 reductase.

In addition to the free radical pathway, menadione may also affect the glutathione redox cycle by forming GSH-menadione conjugates[186,191] and by inhibiting glutathione-reductase.[210] Recently, attempts were made to distinguish the effects of redox cycling of quinones and the arylation of intracellular nucleophils. It has been shown that the substituted naphthoquinones (2,3-dimethyl-, 2,3-diethyl-, and 2,3-dimethoxynaphthoquinones),[192,194] which are incapable of arylating thiols, nonetheless resulted in glutathione depletion and were toxic to hepatocytes. In the case of menadione, both redox cycling and thiol arylation as well as the inhibition of glutathione reductase may be the reasons for its toxicity, although it was concluded that about 80% of GSH consumption may be the result of an oxidative mechanism.[185]

The same is true for other quinones. Thus, the benzoquinone toxicity is apparently determined only by the formation of glutathione conjugates while 2,6-dimethoxy-, 2,6-dimethyl-, 2,5-dimethyl-, 2,3,5-trimethyl-, and tetramethylbenzoquinones deplete glutathione by both arylation and redox cycling.[207] It should be noted that one more mode of quinone toxicity was recently found: the inhibition of SOD by superoxide-generating quinones.[211]

In conclusion, it should be pointed out that anthracyclines, benzoquinones, and naphthoquinones may also manifest antioxidative properties in cells. It was already mentioned that naphthoquinones inhibit NADPH-dependent lipid peroxidation.[187,206] However, a more surprising and possibly important fact is that menadione is able to both stimulate and inhibit superoxide production by tumor cells[212] and neutrophils.[213,214] These complicated effects of quinone compounds will be further discussed in Chapter 4.

E. PYRIDINIUM COMPOUNDS

Among pyridinium compounds are strong herbicides, lung toxins, and hepatotoxins. Paraquat (XXXIV) (1,1′-dimethyl-4,4′-bipyridinium dichloride) has been shown to stimulate production of superoxide ion in illuminated chloroplasts,[215] homogenates of lung, liver, and kidney,[216] and rat liver microsomes.[187] Many other bipyridyl herbicides possess the same activity; among them diquat (XXXV) is the most potent catalyst of superoxide production in rat liver mitochondria.[217]

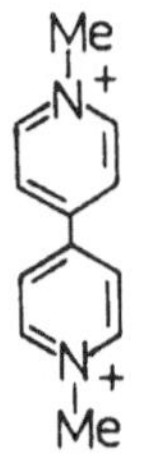

XXXIV.

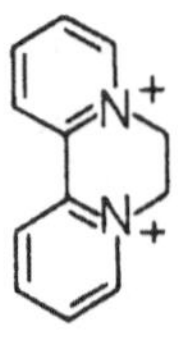

XXXV.

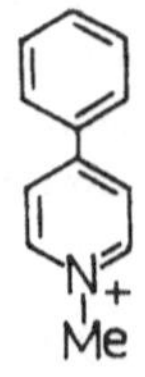

XXXVI.

As in the case of menadione, the toxic action of pyridinium compounds in cells is well documented. It was concluded[166,218-221] that paraquat toxicity in *E. coli, Salmonella typhimurium*, and *N. gonorrhoeae* is relevant to the production of superoxide ion, as paraquat catalyzed a SOD-inhibitable reduction of cytochrome c and stimulated cyanide-insensitive respiration by cell suspensions. It was found that paraquat is reduced inside the cells by cytosolic NADPH:paraquat reductase. (The ESR spectrum of paraquat radical was recorded during the anaerobic incubation of paraquat with intact alveolar type II Clara cells.[222]) The paraquat radical formed was mostly (≥95%) oxidized inside the cell to form superoxide ion intracellularly. In addition, some amount of paraquat radical recrossed the cytoplasmic membrane and generated about 5% of $O_2^{\cdot-}$ extracellularly. Participation of the superoxide ion in the toxic action of paraquat and diquat on Chinese hamster ovary cells,[223] rat hepatocytes,[217] the green alga *Dunaliella salina*,[224] and *N. gonorrhoeae*[166] has also been shown.

3-Methyl-4-phenyl-1,2,3,6-tetrahydropyridine (MPTP), a neurotoxin involved in the development of irreversible Parkinson's-like disease, is not a pyridinium compound; however, it is rapidly metabolized to 1-methyl-4-phenylpyridinium ion (MPP^+ [XXXVI]), which is supposed to be the ultimate neurotoxin. The latter is able to stimulate the SOD-inhibitable formation of superoxide ion and hydroxyl radical (detected as the DMPO-OOH and DMPO-OH spin adducts) by NADPH-cytochrome P-450 reductase;[225] however, MPP^+ was a far less effective catalyst of the one-electron reduction of dioxygen than paraquat. Another drug which apparently should be considered together with pyridinium compounds is an antibiotic, cephaloridine. Cephaloridine contains a pyridinium substituent in the β-lactam ring, which is supposed to participate in redox cycling. Indeed, cephaloridine stimulated superoxide production in rat renal microsomes,[226] although this compound again was a less efficient catalyst than paraquat.

In contrast to menadione, pyridinium compounds readily stimulate the production of hydroxyl and crypto-hydroxyl radicals.[227-230] The mechanism of generation of these radicals is probably very close to that of anthracycline stimulation. As in many other cases, competition between the production of hydroxyl and crypto-hydroxyl radicals or the site-specific mechanism of hydroxyl radical formation depends on the presence of chelators. It is supposed that the generation of these radicals in the presence of dipyridyls and cephaloridine is an initiation step of lipid peroxidation[187,217,226] and deoxyribose degradation,[229] as well as an origin of dipyridyl cytotoxicity.[231] Paraquat- and cephaloridine-dependent microsomal lipid peroxidation was inhibited by SOD+ catalase and hydroxyl radical scavengers (mannitol and [+]-cyanidanol-3),[226] while dipyridyl-induced lipid peroxidation in hepatocytes was inhibited by Trolox C and other antioxidants.[217] As doxorubicin, paraquat, diquat, and benzyl viologen, after reduction by NADPH-cytochrome P-450 reductase, induced the release of iron from ferritin.[232] It was supposed that ferric ions were reduced by both superoxide ion and dipyridyl semiquinones. This process may be one of the mechanisms of the stimulation of dipyridyl-dependent lipid peroxidation.

The depletion of cellular glutathione in the presence of pyridinium compounds is another factor responsible for their toxic action. Recently, it was shown[231] that the diquat toxicity to hepatocytes greatly increased when cells were pretreated with 1,3-bis(2-chloroethyl)-1-nitrosourea (BCNU), an inhibitor of glutathione reductase, that led to a rapid depletion of glutathione. Cytotoxicity of diquat was diminished by catalase and desferrioxamine. On these grounds, it may be assumed that oxygen radicals mediate both main factors of the toxicity of pyridinium compounds: lipid peroxidation and glutathione depletion.

F. NITROAROMATIC AND OTHER REDUCIBLE COMPOUNDS

Nitroaromatic compounds belong to another class of one-electron oxidants which are easily reduced in cells to form radical anions. As semiquinones, these radical anions may take part in redox cycling, producing the superoxide ion under aerobic conditions or (if one-electron transfer to dioxygen is thermodynamically unfavorable) manifesting cytotoxic action by themselves. The class of nitroaromatic compounds includes many important drugs. It is possible that the ability of these compounds to generate oxygen radicals or to form radical anions may be relevant to their therapeutic and cytotoxic action.

Owing to the relative stability of nitroaromatic free radicals, the reduction of nitroaromatic compounds may be studied directly by ESR spectroscopy under anaerobic conditions.[233] Under aerobic conditions, many nitroaromatic radical anions may be oxidized by dioxygen to form superoxide ion. Thus, it has been shown that *p*-nitrobenzoate;[234] nitrofurantoin (*N*-[5-nitro-2-furfurylidene]-1-aminohydantoin) (XXXVII), an antimicrobial agent and urinary antiseptic);[234-236] misonidazole (1-[2-hydroxy-3-methoxypropyl]-2-nitroimidazole) (XXXVIII), a radiosensitizer of hypoxic cells);[236] nifurtimox (4-[(5-nitrofurfurylidene)amino]-3-methylthiomorpholine-1,1-dioxide) (XXXIX), an effective drug against acute Chagas' disease);[237,238] and benznidazole (*N*-benzyl-[2-nitro-1-imidazole]acetamide) (XL), which is also used in the treatment of Chagas' disease,[239,240] were reduced by rat liver and lung microsomes to radical anions which were oxidized by dioxygen to form superoxide ion. Nifurtimox and nitrofurantoin were also reduced by rat liver mitochondria,[241] misonidazole was reduced by xanthine oxidase,[242] and nitrazepam (7-nitro-1,3-dihydro-5-phenyl-2H-1,4-benzodiazepin-2-one) (XLI) was reduced by NADPH-cytochrome c(P-450) reductase.[243] Nitroimidazole derivatives participate in redox cycling in hydrogenosome-enriched fractions from *Tritrichomonas vaginalis,* where ESR spectra of nitroaromatic radical anions and DMPO-OOH spin adduct were obtained.[244]

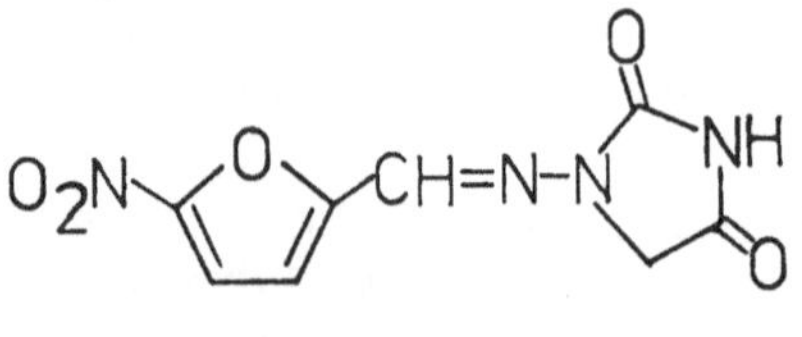

XXXVII. Nitrofurantoin

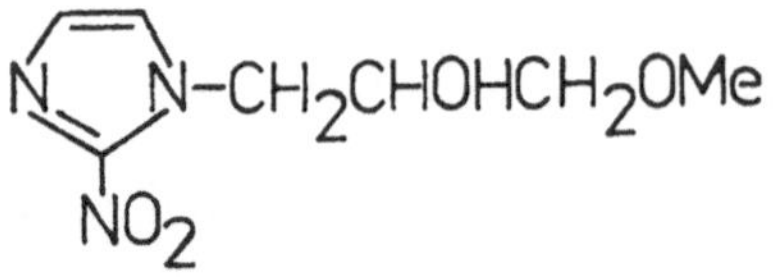

XXXVIII. Misonidazole

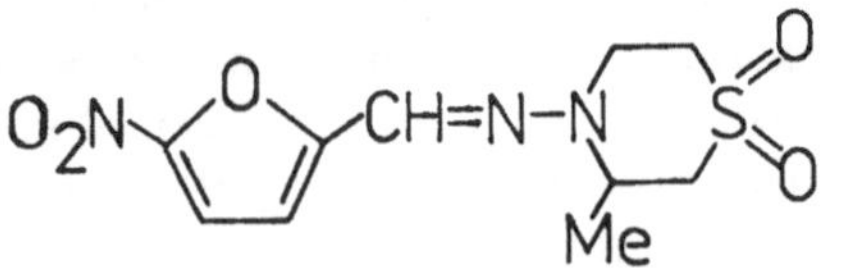

XXXIX. Nifurtimox

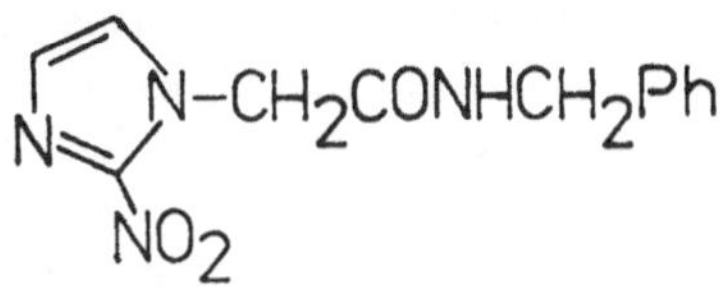

XL. Benznidazole

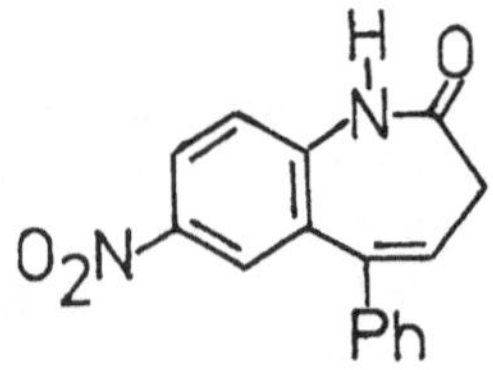

XLI. Nitrazepam

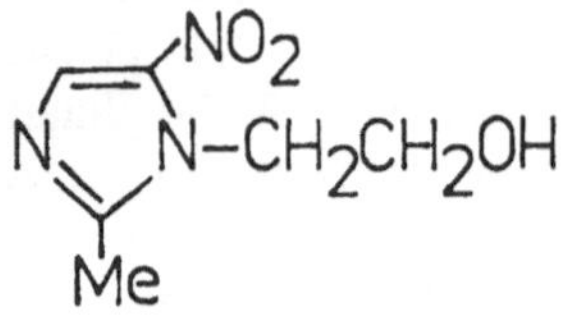

XLII. Metronidazole

All these and some other nitroaromatic compounds are cytotoxic. Their cytotoxicity may be a consequence of the production of both nitroaromatic radical anions and oxygen radicals. Therefore, the reduction of nitroaromatic drugs by whole cells was studied and compared with their cytotoxic action. Nifurtimox inhibited growth of *Trypanosoma cruzi* cells, inducing an increase in the respiratory rate and the release of hydrogen peroxide.[245] At the same time, nifurtimox stimulated superoxide production by the *T. cruzi* mitochondrial and microsomal fractions. Nitrofurantoin injured pulmonary endothelial cells, increasing superoxide production.[246] It was assumed that stimulation of the formation of oxygen radicals by endothelial cells may represent one mechanism of pulmonary toxicity by nitrofurantoin. Metronidazole (1-[2-hydroxyethyl]-2-methyl-5-nitroimidazole [XLII]) was reduced by *Tritrichomonas foetus* and *T. vaginalis* cells to a radical anion under anaerobic conditions.[247,248] Dioxygen inhibited an ESR spectrum of radical anion, evidently due to its oxidation and the formation of a superoxide ion. Nitrazepam induced dysfunction into endothelial cells and enterocytes,[195,196] probably also due to the stimulation of superoxide production. It is of interest that the toxicity of misonidazole to hypoxic Chinese hamster ovary cells was increased by Fe^{3+}EDTA and decreased by desferrioxamine; it shows that iron plays an important role in the redox cycling of this drug.[249]

As usual, lipid peroxidation may be one mechanism of the cytotoxicity of nitroaromatic drugs. Indeed, the nifurtimox-stimulated lipid peroxidation in rat liver microsomes was documented.[238] However, there are at least two more mechanisms of cytotoxicity related to the free radical activity of nitroaromatic drugs. The first one is the stimulation of the oxidation of some natural substrates. Thus, nitrofurantoin stimulated epinephrine oxidation by rat lung microsomes mediated by superoxide ion.[235] It was supposed that an increase in superoxide production may be (partially) responsible for the oxidation of NADPH in cells in the presence of nitrofurantoin. Later on, it was found that under anaerobic conditions, many catechols (epinephrine, norepinephrine, dopamine, and L-DOPA) may be directly oxidized by nitrofurantoin, nifurtimox, and other nitroaromatic drugs.[250] Nitrofurantoin also oxidized oxyhemoglobin to form superoxide ion,[251] which it was supposed might be the reason for NADPH oxidation in erythrocytes in the presence of this drug.

Another origin of nitroaromatic cytotoxicity is the disturbance by these drugs of glutathione redox cycling. As menadione, some nitroaromatic drugs (for example, nitrofurantoin[252] and nifurtimox[253]) inhibit glutathione reductase and reduce the glutathione level by producing extra hydrogen peroxide via the dismutation of the superoxide ion.

Quinones, pyridinium, and nitroaromatic compounds represent main classes of substances reducible to radical anions by cells and cellular components. However, there are compounds with different structures which nonetheless can also be reduced in biological systems and can participate in redox cycling to produce superoxide ion. 3-Methylindole (the main ruminal fermentation product of L-tryptophan) was reduced by goat lung microsomes to a nitrogen-centered free radical (radical anion [?]) which was detected as a PBN spin adduct.[254] This spin adduct was converted *in vitro* (in microsomes) and *in vivo* (in lungs of goats) into a spin adduct of the carbon-centered free radical.[255,256] This fact is believed to indicate the initiation of lipid peroxidation by the methylindole free radical.

Phenazine methosulfate (XLIII) catalyzed superoxide production in isolated rat hepatocytes and induced a substantial decrease in intracellular glutathione level.[257] Primaquine (XLIV), an antimalarial drug, catalyzed dioxygen consumption by *Trypanosoma cruzi* extracts, but in this case hydroxyl radicals were detected.[258]

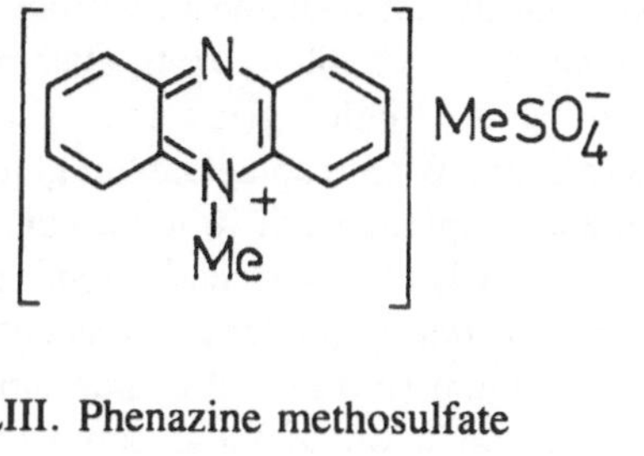

XLIII. Phenazine methosulfate

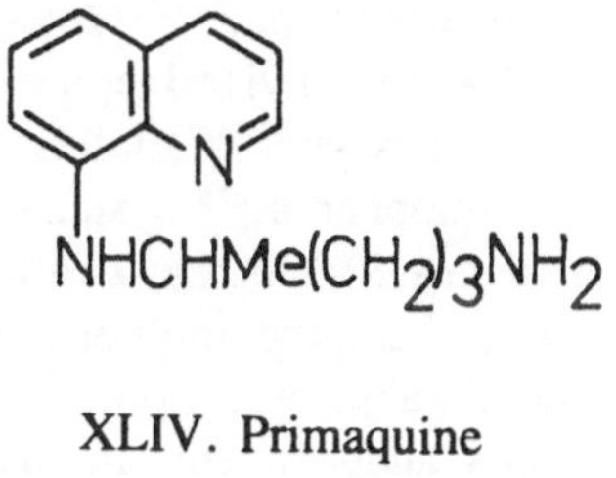

XLIV. Primaquine

In contrast to most compounds able to catalyze oxygen radical production, primaquine has a very negative one-electron reduction potential (−1.25 V vs. SCE, DMF).[259] It indicates that real catalysts of oxygen radical production are probably orthoquinones, the oxidative metabolites of primaquine.[259]

A special group of reducible compounds consists of metallochromic indicators applied for measurement of calcium transport in cells and cell fractions. It turned out that such metallochromic indicators as murexide (ammonium purpurate), tetramethylmurexide, arsenazo III (2,2′-[1,8-dihydroxy-3,6-disulfonaphthalene-2,7-bisazo]bis[benzenearsonic acid]), and antipyrylazo III (bis[4-antipyrylazo]-4,5-dihydroxy-2,7-naphthalenedisulfonic acid) are reduced to radical anions by rat liver microsomes,[260] mitochondrial,[261] cytosolic fraction,[262,263] and xanthine oxidase.[263] Under aerobic conditions, these radical anions were oxidized by dioxygen to form superoxide ion, which was detected as a DMPO-OOH spin adduct. These findings show that the use of metallochromic indicators in biological studies may cause unfavorable side effects due to their redox cycling.

Aerobic reduction of 4-nitroquinoline-1-oxide (a carcinogen) by isolated tumor cells, cultured human and rodent cells, microsomes, or purified enzymes resulted in the production of the nitro radical anion and the superoxide ion.[212] Crystal violet (a triarylmethane dye used for the elimination of transmission of Chagas' disease by transfusion of blood) was enzymatically reduced by *Trypanosoma cruzi* cells and homogenates to form a carbon-centered free radical.[264] It was proposed that the formation of a free radical which was enhanced by light may be the main cause of the cytotoxicity of crystal violet to *T. cruzi*. Amino-chloroamphenicol, a metabolite of the antibiotic chloroamphenicol, catalyzed the production of superoxide ion in rat liver microsomes determined by the SOD-inhibited reduction of succinoylated cytochrome c.[265]

III. PRODUCTION OF OXYGEN RADICALS BY REDUCTANTS

The ability of reductants to generate superoxide ion during one-electron oxidation by dioxygen was analyzed in Volume I (Chapter 3). It seems reasonable to suggest that analogous processes must take place in biological systems in the presence of various reductants, for example, drugs containing the hydroquinone moiety. The superoxide ion may be formed by direct nonenzymatic oxidation of reductants (Reactions 33 and 34), but this process probably slowed down in cells due to high pK_a values of principal oxidizable groups (OH, SH) excluding ascorbic acid with pK_a 4.2.

$$\text{RedH} \rightleftarrows \text{Red}^- + \text{H}^+ \quad (33)$$

$$\text{Red}^- + \text{O}_2 \rightleftarrows \text{Red}\cdot + \text{O}_2^{\overline{\cdot}} \quad (34)$$

Nonetheless, many biologically active compounds are oxidized by dioxygen under physiological conditions. This process is believed to be responsible for their cytotoxicity. Thus, it has been shown that superoxide ion is produced during autoxidation of various phenols (such as 6-hydroxydopamine,[266,267] 6-aminodopamine,[267,271] 6,7-dihydroxytryptamine,[267] dialuric acid,[267-269] 4-dimethylaminophenol,[270] caffeic acid,[272] chlorogenic acid, gallic acid, pyrogallol,[273] and gossypol[274]), thiols (ethyl mercaptan, β-mercaptoethanol, glutathione, dithiothreitol,[275] thiophenol,[276] cysteine,[277] and reduced sporidesmin [a toxin contained in spores of the fungus *Pithomyces chartarum]*[278]), 1,2,4-triaminobenzene,[279] reduced phenazine derivatives,[280] and the antibiotic rifamycin.[281] Superoxide production was determined via the SOD-inhibitable reduction of nitroblue tetrazolium (NBT) and cytochrome c, the formation of spin adducts, etc. In many cases, the oxidation of these compounds was catalyzed by transient metal ions.[261,282] It was supposed that toxic effects may be a consequence of the generation of hydroxyl radicals in the superoxide-driven Fenton reaction.[267,271,280,283] Thus, it was assumed that $O_2^{\overline{\cdot}}$ and H_2O_2 are true agents of the destruction of axonal membranes during the 6-hydroxydopamine-induced damage of catecholamine-secreting neurons.[284]

It has been shown[285] that the cyanide-catalyzed oxidation of α-hydroxycarbonyl compounds proceeds to form superoxide ion, supposedly by the mechanism:

$$\mathrm{RCH(OH)CHO} \rightleftarrows \mathrm{RC(OH){=}CHOH} \tag{35}$$

$$\mathrm{RC(OH){=}CHOH} + \mathrm{CN^-} \rightarrow \mathrm{RC(OH)(CN)\bar{C}HOH} \tag{36}$$

$$\mathrm{RC(OH)(CN)\bar{C}HOH} + \mathrm{O_2} \rightleftarrows \mathrm{RC(OH)(CN)\dot{C}HOH} + \mathrm{O_2^{\overline{\cdot}}} \tag{37}$$

This process must be taken into account in the experiments with cyanide as an enzyme inhibitor.

However, in most cases enzymatic oxidation is probably a more important route to oxygen radicals. Two enzymes, horseradish peroxidase (HRP) and prostaglandin H synthase (PGH) (an enzyme catalyzing the cyclooxygenation of arachidonic acid to hydroxyperoxy-endoperoxide), are widely studied as the catalysts of one-electron oxidation of various drugs. HRP catalyzes the oxidation of phenol, dopamine, catechol, DOPA, pyrogallol, 3,4-dihydroxybenzoic acid, naphthol, and their derivatives to phenoxyl radicals.[286,290] HRP and PGH oxidize aminopyrine (4-[dimethylamino]-1,2-dihydro-1,5-dimethyl-2-phenyl-3H-pyrazol-3-one) (XLV) to a cation radical which then disproportionates to aminopyrine and its demethylated metabolite.[287-289]

$$\mathrm{RedH} \xrightarrow{\text{enzyme }(-e)} \mathrm{Red}^{\overset{+}{\cdot}} \text{ or } (\mathrm{Red}\cdot + \mathrm{H^+}) \tag{38}$$

$$\mathrm{RedH}^{\overset{+}{\cdot}} \text{ or } \mathrm{Red}\cdot + \mathrm{O_2} \rightleftarrows \mathrm{O_2^{\overline{\cdot}}} + \mathrm{Red}\,(+\mathrm{H^+}) \tag{39}$$

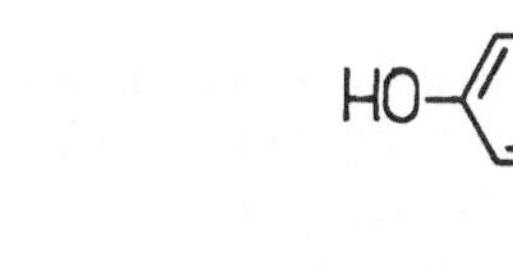

XLV. Aminopyrine

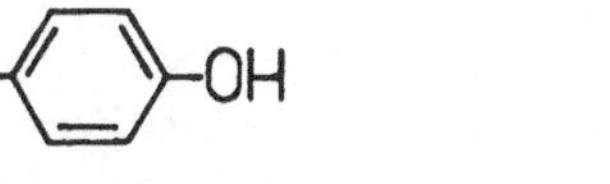

XLVI. Acetaminophen

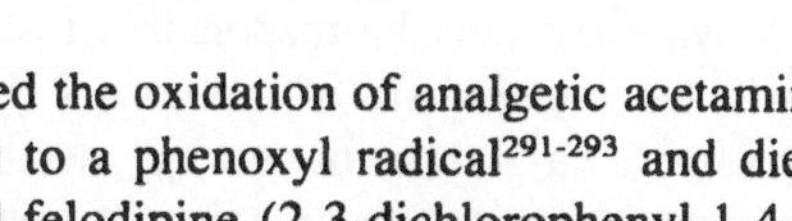

XLVII. Diethylstilbestrol

These enzymes also catalyzed the oxidation of analgetic acetaminophen or paracetamol (4′-hydroxyacetanilide) (XLVI) to a phenoxyl radical[291-293] and diethylstilbestrol (a synthetic estrogen) (XLVII)[294,295] and felodipine (2,3-dichlorophenyl-1,4-dihydropyridine)[296] to free radicals detected as the spin adducts with DMPO and POBN.

HRP and PGH probably play an important role in the peroxidative metabolism of semisynthetic podophyllotoxin derivatives VP-16 and VM-26 (XLVIII) (active anticancer drugs).

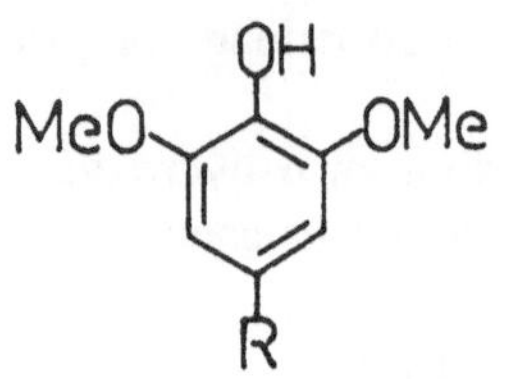

Where R =

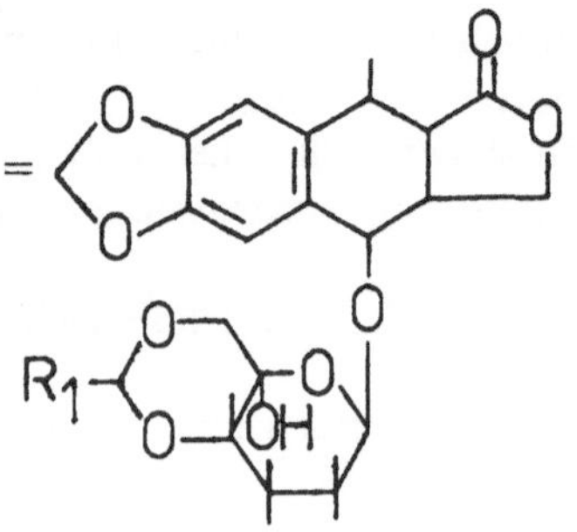

XLVIII.

VP-16, R_1 = Me

VM-26 R_1 =

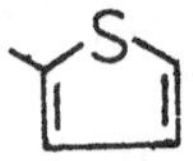

It has been shown[101,297] that both enzymes oxidize these drugs to stable phenoxyl radicals characterized by their ESR spectra. It was assumed that these radicals are intermediates of *O*-demethylation resulting in the formation of real toxic agents, methyl radical and drug *o*-quinone.

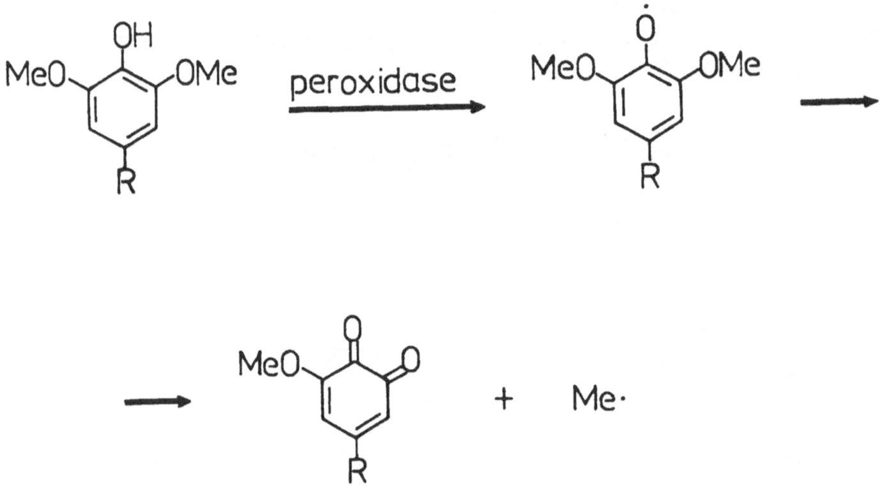

(40)

A similar process apparently proceeds during the incubation of VP-16 with mice liver microsomes where a drug is metabolized by cytochrome P-450.[298] HRP also catalyzed the oxidation of indole-3-acetic acid into a free radical of a benzyl type.[299]

Another important phenolic anticancer drug, 4-hydroxyanisole, is converted to free radical intermediates by a copper-containing enzyme, tyrosinase.[300] It was assumed that free radical metabolites of 4-hydroxyanisole may be responsible for its depigmenting activity.

In many works the free radical metabolism of reductants is considered as an origin of their cytotoxic activity. For example, it is thought that the metabolism of acetaminophen in the kidney by prostaglandin synthase is responsible for its analgetic nephropathy.[292] The oxidation of acetaminophen by cytochrome P-450 in liver microsomes probably stimulates hepatic necrosis in both humans and experimental animals at the administration of high doses of the drug.[301] As always, there may be various mechanisms of free radical cytotoxicity of oxidizable compounds which may catalyze superoxide production by themselves or after converting to quinone forms. For example, gossypol (2,2′-bis[1,6,7-trihydroxy-3-methyl-5-isopropyl-8-naphthaldehyde]), a polyphenolic compound and a potential antiviral drug, stimulated superoxide production by rat liver microsomes.[302] Similarly, the formation of superoxide ion in rat liver microsomes was catalyzed by arylamines,[303] naphthylamines, aminoazo dyes,[304] and possibly hydrazine.[305] Diethylstilbestrol-4′,4″-quinone, a metabolite of diethylstilbestrol, enhanced more than tenfold the production of superoxide ion in hamster kidney microsomes, apparently due to the reduction by NADPH-cytochrome P-450 reductase.[306]

Another possible factor of cytotoxicity is the xenobiotic free radical. These radicals may participate in subsequent reactions inducing undesirable changes in biological processes. Thus, it was proposed[293] that the acetaminophen radical is an intermediate of the enzymatic polymerization of acetaminophen. On the other hand, xenobiotic free radicals may induce glutathione depletion, reducing back to parent compounds. This "futile" metabolism[289] apparently plays a role of cellular protective mechanism against toxic xenobiotic free radicals.

$$\text{Red}\cdot + \text{GSH} \rightarrow \text{RedH} + \text{GS}\cdot \tag{41}$$

$$\text{GS}\cdot + \text{GS}\cdot \rightarrow \text{GSSG} \tag{42}$$

In accord with this proposal, it has been shown that glutathione, cysteine, and *N*-acetylcysteine suppressed the oxidation of acetaminophen into a free radical by prostaglandin synthase,[292] and glutathione reacted with the free radical of diethylstilbestrol.[294] It is interesting that the equilibrium for Reaction 41 is really shifted to the left (k_{41} about (2—3) $\cdot$ 10^4 M^{-1} s^{-1} and k_{-41} = 3 $\cdot$ 10^8 M^{-1} s^{-1} for the acetaminophen free radical).[292] However, owing to a large excess of glutathione relative to Red· and a small stability of GS· in comparison with Red·, the equilibrium is kinetically driven to the right.

A third possible source of toxic action of reductants is the cytotoxicity of their quinone metabolites. For example, benzoquinone is apparently the principal bactericidal agent in the treatment of *E. coli* and other gram-negative bacteria with hydroquinone, as benzosemiquinone obtained during the reduction of benzoquinone by xanthine oxidase did not enhance cytotoxicity.[307] *N*-Acetyl-*p*-benzoquinone imine formed in the oxidation of acetaminophen by cytochrome P-450 also possessed a high toxic activity.[301]

Free radical toxicity of some reductants has been shown in cells and animal studies. It was found[308] that liver injury induced by acetaminophen depended on the metabolism of this drug by mixed-function oxidation. Without the induction of mixed-function oxidation by 3-methylcholanthrene, rat hepatocytes were insensitive to acetaminophen. Killing of hepatocytes by acetaminophen was accelerated by BCNU and was prevented by SOD and catalase. These findings indicate the participation of oxygen radicals in cell killing and the importance of cellular protection by glutathione. It is thought that the production of oxygen radicals is responsible for the induction of DNA strand breaks in Syrian hamster embryo cells by diethylstilbestrol[295] and the toxicity of cysteamine to Chinese hamster ovary cells.[309] It should also be mentioned that the formation of free radicals was demonstrated under physiological conditions in the treatment of pig skin by dithranol (1,8-dihydroxy-9-anthrone), a drug used for treatment of hyperproliferative psoriasis.[310]

IV. OTHER PRODUCERS OF OXYGEN RADICALS

There are xenobiotics which do not possess reducing or oxidizing properties, but nonetheless are able to produce oxygen radicals. These compounds form active metabolites in cells or affect in various ways the reducing activity of other producers of oxygen radicals. The most important drug of that kind is undoubtedly bleomycin (BLM). BLM is a glycopeptide antibiotic used clinically as an antitumor agent. As in the case of anthracyclines, the possible participation of oxygen radicals in the anticancer activity and cytotoxicity of BLM and its analogs has been studied. It has been shown that BLM forms complexes with iron, copper, manganese, and other metal ions which are believed to be real active intermediates inducing DNA degradation. These complexes are also able to generate oxygen radicals.

The Fe(II)BLM complex generated hydroxyl radicals (characterized by the ESR spectrum of a DMPO spin adduct) after saturation of aqueous solution with dioxygen.[311,312] Oxidation of Fe(II)BLM also resulted in the formation of phenoxyl radicals from 2,6-di-tert-butyl-*p*-cresol and α-tocopherol.[312] Under analogous conditions, the Fe(II) complex of tallysomycin (a third-generation BLM analog) did not stimulate hydroxyl radical formation;[313] however, copper-tallysomycin complexes were active HO producers.[313,314] In contrast to this, Cu(II)BLM was unable to produce hydroxyl radicals.[314]

It was established that the activation mechanism of BLM depends on dioxygen. A one-electron reduction potential of BLM is $+0.129 \pm 0.012$ V;[315] therefore, the equilibrium for Reaction 43 is shifted to the left:

$$\text{Fe(II)BLM} + O_2 \rightleftarrows \text{Fe(III)BLM} + O_2^{\overline{\cdot}} \tag{43}$$

Nonetheless, it was shown that Fe(II)BLM reacts with dioxygen via an inner-sphere mechanism. At the first reaction stage, a BLMFe(II)O_2 complex is formed, which is next transformed into "activated Fe(III) complex".[316] The last is believed to be responsible for DNA degradation by BLM. Now, it is accepted that this transformation is really the reduction of BLMFe(II)O_2 by Fe(II)BLM to a compound which may be formally described as BLMFe(III)O_2^{2-}.[317] Correspondingly, the interaction of Fe(II)BLM with dioxygen may be presented as follows:

$$\text{BLMFe(II)} + O_2 \rightleftarrows \text{BLMFe(II)}O_2 \tag{44}$$

$$\text{BLMFe(II)}O_2 + \text{Fe(II)BLM} \rightarrow \text{``activated Fe(III)BLM''} + \text{Fe(III)BLM} \tag{45}$$

The formation of "activated BLM" in the reaction of Fe(II)BLM with dioxygen or in the reaction of Fe(III)BLM with hydrogen peroxide has been studied by many workers.[312,318-320] The interaction of superoxide ion with Fe(III)BLM was also studied.[321] It was found that "activated BLM" is also formed in the last reaction and therefore it is not back-Reaction −43. In addition, it was concluded that during the formation of "activated BLM" the modification of BLM itself may take place.

The most popular mechanism for the toxic action of BLM includes the selective hydrogen atom abstraction from the C4′ atom of deoxyribose, leading to cleavage of the deoxyribose C3′–C4′ bond to form free bases and base propenals.[322-324] However, in this case the question which was discussed in connection with the mechanism of the toxicity of anthracyclines remains sound: is BLM able to penetrate the cell membranes and to interact with DNA *in vivo*? And if this is the case, is such an interaction more important than the production of oxygen radicals in the presence of BLM? The last question may possibly be answered after studying the activity of immobilized BLM.

The BLM-dependent oxygen radical production was widely studied. As a direct one-electron transfer from BLMFe(II) to dioxygen (Reaction 43) is thermodynamically unfavorable, it may be expected that the superoxide ion must be generated as a result of the decomposition of BLMFe(II)O_2 or "activated BLM" by some activators. It has been shown[319] that the oxidation of Fe(II)BLM by dioxygen in the presence of DNA leads to the appearance of the ESR spectrum of a superoxide like free radical. Oxygen radical production stimulated by BLM in the presence of iron ions during DNA destruction[325] was inhibited by SOD, catalase, cysteine, and glutathione.[318]

Activation of the inactive Fe(III)BLM complex can be achieved by the addition of $O_2^{\overline{\cdot}}$, H_2O_2, or UV irradiation.[326] It was supposed that this process is accompanied by the production of superoxide ion and hydroxyl radicals. A more important activation process is probably the enzymatic reduction of Fe(III)BLM. It has been shown that Fe(III)BLM reduced by liver microsomal NADPH cytochrome P-450 reductase,[327,328] liver nuclear NADPH cytochrome P-450 reductase,[329] and nuclear NADH cytochrome b_5 reductase[330] catalyzed the formation of oxygen radicals. Their production was shown by the conversion of methional to ethylene (this reaction was inhibited by hydroxyl radical scavengers, DMSO, mannitol, etc.) and MDA formation (as a result of the oxidation of the DNA deoxyribose moiety).

The Fe(II)BLM complex also stimulated lipid peroxidation (for example, the peroxidation of bovine brain phospholipids[331] and arachidonic acid[332]) although the BLM-dependent lipid peroxidation is a substantially slower process than the BLM-dependent DNA degradation.[331] It should be noted that many features of these processes remain unexplained. Thus, polyphenolic free radical scavengers inhibited BLM-dependent phospholipid peroxidation and accelerated MDA formation in the destruction of DNA by BLM.[331] In contrast to Fe(II)BLM, Cu(II)BLM inhibited the processes initiated by oxygen radicals (the NBT reduction by xanthine oxidase, Fe^{2+}-dependent lipid peroxidation, and Fe(II)BLM-dependent damage to DNA) evidently due to possession of dismutase activity.[333]

Another important mode of free radical activity of BLM is the stimulation of superoxide production by macrophages. Thus, BLM increases superoxide production by pig alveolar macrophages,[334] which may be a cause of BLM pulmonary toxicity.

Production of oxygen radicals is probably a main mechanism of free radical activity of xenobiotics, but some compounds can induce free radical damage in biological systems by direct metabolism to free radicals. The most important class of such compounds are probably halogenated hydrocarbons (especially carbon tetrachloride), which are metabolized in liver to form halogenated free radicals. These processes are widely discussed in the literature.[335-337] It is interesting that some halogenated hydrocarbons are able to catalyze the production of oxygen radicals; for example, lindane (the γ-isomer of hexachlorocyclohexane), an insecticide, stimulated generation of oxygen radicals in rat and mice liver microsomes.[338,339]

Phenelzine (2-phenylethylhydrazine), an antidepressant,[340] is metabolized in rat liver microsomes to the neutral free radical $PhCH_2CH_2\cdot$, but there is no indication of oxygen radical production. Similarly, neocarzinostatin, an anticancer antibiotic, induced DNA sugar damage via a free radical mechanism without oxygen radical participation.[341] On the other hand, gentamicin, an aminoglycoside antibiotic, induces nephrotoxicity inhibitable by dimethylthiourea, desferrioxamine, and DMSO, which indicates the generation of oxygen radicals. However, the mechanism of oxygen radical production by gentamicin is unknown.[342]

V. PHOTOGENERATION OF OXYGEN RADICALS IN THE PRESENCE OF XENOBIOTICS

In Volume I (Chapter 3), the mechanisms of superoxide photoproduction in the presence of various sensitizers (ketones, aromatic amines, phenols, and so on) were considered. Similarly, some drugs and xenobiotics can be applied as the photosensitizers of superoxide generation; the study of these processes is of importance in connection with potential use of these drugs in photodynamic therapy. Balny and Douzou[343] have shown that the irradiation of fluorescein with light at 480 nm in the presence of lactoperoxidase under aerobic conditions resulted in the SOD-inhibitable formation of Compound III. Fluorescein can be replaced by tryptophan and NADPH. It was concluded that Compound III was formed due to the production of $O_2^{\overline{\cdot}}$ during the interaction of dye triplet with dioxygen.

$$D^T + O_2 \rightarrow D^{\dot{+}} + O_2^{\overline{\cdot}} \quad (46)$$

Similarly, superoxide ion was detected during the illumination of acridine dyes, fluorescein, and lucifer yellow CH with visible light in the presence of NADH and NADPH using the SOD-inhibitable reduction of cytochrome c.[344] It was also found that SOD, catalase, DMSO, sodium benzoate, and thiourea protected *E. coli* cells against photodynamic inactivation by acridine dyes, which indicates the participation of both $O_2^{\overline{\cdot}}$ and HO· in cell damage.

It was supposed that superoxide ion is formed in the irradiation of saponified chlorophils with red light as this process was accompanied by the SOD-inhibitable reduction of nitroblue tetrazolium.[345] Superoxide production was also documented by the SOD-inhibitable reduction of NBT and the formation of the DMPO-OOH spin adduct in the irradiation of a phenoxazine derivative in the presence of tetramethylethylenediamine or GSH[346] and melamin suspension[347] with visible light under aerobic conditions.

Oxygen radicals were detected during the illumination of hematoporphyrin and its derivative, Photofrin® II (HPD), used in photodynamic therapy.[348,349] In the case of hematoporphyrin itself, only the DMPO-OH spin adduct was obtained.[348] The HO· formation was inhibited by SOD; in the presence of ethanol and azide, the hydroxyethyl and azide radical spin adducts were formed. On these grounds, it was concluded that $O_2^{\overline{\cdot}}$ is a precursor of hydroxyl radicals and that the formation of DMPO-OH as a result of DMPO-OOH decomposition may be excluded.

In the case of HPD, the superoxide ion was detected via the SOD-inhibitable formation of DMPO-OOH.[349] A main mechanism of $O_2^{\overline{\cdot}}$ production is probably the reduction of the excited states of a photosensitizer by reductants with subsequent oxidation of the radical anions formed by dioxygen.

$$S \xrightarrow{h\nu} S^* \quad (47)$$

$$S^* + RH \longrightarrow S^{\overline{\cdot}} + R\cdot + H^+ \quad (48)$$

$$S^{\overline{\cdot}} + O_2 \longrightarrow S + O_2^{\overline{\cdot}} \quad (49)$$

This mechanism is thought to be realized during the irradiation of HRD at 365 nm in the presence of cysteine, glutathione, and NADH at pH <6.5. At pH >6.5 the interaction of ^{3}HPD* with O_2 (Reaction 48) apparently proceeds in another direction to form singlet oxygen:

$$^3HPD^* + O_2 \rightarrow HPD + {}^1O_2 \quad (50)$$

Owing to that, $O_2^{\overline{\cdot}}$ and H_2O_2 are not formed at physiological pH, and therefore the cell damage induced by HPD may be mediated by $^1O_2$.

Many other drugs may be effective photosensitizers of production of oxygen radicals. Thus, irradiation of doxorubicin and daunorubicin solutions with visible light produced DMPO and 4-MePyBN spin adducts of superoxide ion and hydroxyl radical.[350] Since the methyl spin adduct was not obtained when DMSO was added during the irradiation, it was concluded that hydroxyl radicals were not formed, and DMPO-OH was a product of the decomposition of DMPO-OOH. It was proposed that the superoxide ion was formed by a direct electron transfer from electronically excited antibiotics or their radical anions to dioxygen. Similarly, the photogeneration of superoxide ion by mitomycin C, streptonigrin, and carboquone in DMSO and by mitomycin C in aqueous solution,[351] as well as by tetracycline derivatives (doxycycline, demeclocycline, tetracycline, and oxytetracycline) in aqueous solution[352] was found. Photogeneration of $O_2^{\bar{\cdot}}$ by tetracyclines is apparently a source of their cytotoxicity to *E. coli*, as SOD, catalase, and hydroxyl radical scavengers protected *E. coli* cells against the phototoxicity of these antibiotics.

Buettner et al.[349] have shown that photolysis of chlorpromazine sulfoxide and promazine sulfoxide (oxidative metabolites of phenothiazine drugs with tranquilizing and antihistaminic properties) leads to the formation of hydroxyl radicals detected as DMPO-OH spin adducts. Cysteine, glutathione, ethanol, and some other substrates converted DMPO-OH to the DMPO-substrate radical spin adducts, but SOD and catalase had no effect on the DMPO-OH formation. On these grounds, it was proposed that hydroxyl radicals were formed as a result of homolytic cleavage of the excited sulfoxide:

$$\text{R–SO} \xrightarrow{h\nu} \text{R–SO}^* \tag{51}$$

$$\text{R–SO}^* + \text{H}^+ \longrightarrow \text{R–}\overset{+}{\text{S}}\text{OH}^* \tag{52}$$

$$\text{R–}\overset{+}{\text{S}}\text{OH}^* \longrightarrow \text{RSO}^{\dot{+}} + \text{HO}\cdot \tag{53}$$

Pathak and Joshi[353,354] have shown that furocoumarins (psoralens), which are potent skin-photosensitizing agents, produce singlet oxygen and superoxide ion on irradiation with UV light. It is thought that both $^1O_2$ and $O_2^{\bar{\cdot}}$ may participate in cellular damage by psoralens.

VI. ADDITIONS

It was recently shown[355] that although doxorubicin enhanced cytochrome c reduction by rat liver microsomes, it inhibited the lucigenin-dependent CL under the same conditions. It was found that from the quinones studied (doxorubicin, benzoquinone, and menadione), only menadione was able to enhance the lucigenin-dependent CL induced by xanthine oxidase. Przybyszewski and Malec[356] have shown that doxorubicin as well as hydroxyurea and methotrexate stimulated the reduction of NBT by NADH, which was mediated by superoxide ion. Nohl[357] has found that doxorubicin is reduced by mitochondrial NADH-oxidoreductase. Using the spin-trapping technique, it has been shown[358] that doxorubicin, daunorubicin, and mitomycin C promoted hydroxyl radical formation in the microsomal NADPH cytochrome P-450 reductase system. It has also been shown[359] that doxorubicin stimulated hydroxyl radical production in isolated rat heart (which was inhibited by SOD and catalase).

Pedersen et al.[360] have shown that after treatment of intact red blood cells with doxorubicin, superoxide ion, the doxorubicin semiquinone, and hydrogen peroxide are formed outside of the erythrocyte. Redox cycling of doxorubicin in microsomes stimulated the release of iron ions from microsomal membranes.[361] Monteiro et al.[362] have studied the effects of various quinones on iron release from ferritin. They found that under anaerobic conditions, iron release occurred with semiquinones, having reduction potentials less than that of ferritin (doxorubicin, daunorubicin, paraquat, nitrofurantoin, etc.), while the semiquinones with more positive reduction potentials (menadione, benzoquinone, etc.) were inactive. However, all semiquinones promoted slight SOD-inhibitable iron release under aerobic conditions.

In several studies, the promotion by doxorubicin of lipid peroxidation in rat liver and heart mitochondria and rat liver microsomes[363-365] as well as in human erythrocytes[366] has been studied. α-Tocopherol and β-carotene inhibited doxorubicin-dependent lipid peroxidation.[364,366] Doxorubicin, paraquat, and anthraquinone-2-sulfonate also enhanced the reduction of iron chelates by rat liver microsomes.[365] The treatment of rats with doxorubicin enhanced the formation of fluorescent products of lipid peroxidation in kidney and heart and of lipid peroxides in serum.[367] *N*-Acyldehydroalanines (which are believed to be free radical scavengers) suppressed the reduction of doxorubicin by beef heart mitochondria.[368] Hasinoff and Davey[369] have shown that doxorubicin inactivated mitochondrial cytochrome c oxidase in the presence of cupric ions. Ganey et al.[370] proposed that the hepatotoxicity of doxorubicin depends on oxygen tension, probably due to its effects on the rate of redox cycling of doxorubicin.

In several works the production of oxygen radicals in doxorubicin-sensitive and doxorubicin-resistant cells has been studied. Keizer et al.[371] concluded that there is no difference in the doxorubicin-induced superoxide production by wild-type and oxygen-resistant sublines of CHO cells. Hydroxyl radical scavengers suppressed doxorubicin toxicity in both multidrug-resistant and -sensitive human ovarian cancer cells, whereas ascorbate, SOD, and catalase were ineffective.[372] Alegria et al.[373] also did not find a difference in the superoxide levels produced by doxorubicin-resistant and doxorubicin-sensitive CHO and human breast cancer cells after exposure to doxorubicin. Contrary to this, Sinha et al.[374] have shown that doxorubicin and paraquat produced less hydroxyl radicals in the drug-resistant human breast tumor cells than in the drug-sensitive ones. It was found[375] that doxorubicin-resistant human breast tumor cells were fourfold more resistant to the action of superoxide ion and hydrogen peroxide, evidently due to an increase in the SOD and glutathione peroxidase activities. Furthermore, it was found that the production of oxygen radicals stimulated by doxorubicin was two times less in mitochondria and nuclei from doxorubicin-resistant human breast tumor cells in comparison with doxorubicin-sensitive ones.[376]

Grankvist et al.[377] have shown that demethoxydaunorubicin and epirubicin produced oxygen radicals in the presence of ferrous ions and diminished the clonogenic survival of fibroblasts. Iron chelators, but not SOD or catalase, protected fibroblasts. Recently, it was found[378] that 5-iminodaunorubicin (but not daunorubicin) is oxidized by the H_2O_2-HRP system into a free radical. SOD affected the aclacinomycin-induced cytological alterations in murine peritoneal macrophages, indicating the mediation of this effect by oxygen radicals.[379]

Napetschnig and Sies[380] have shown that mitomycin C stimulated CL after incubation with isolated rat liver microsomes in the presence of glutathione. These authors proposed that CL was produced in the reaction of the superoxide ion with GSH. Mitomycin C and its analogs enhanced the production of superoxide ion and hydroxyl radicals in rat cardiac microsomes and isolated perfused rat hearts.[381] Prisos et al.[382] have shown that the cytotoxic activity of mitomycin antibiotics under aerobic conditions was affected by diethyldithiocarbamate, a SOD inhibitor.

Mahmutoglu and Kappus[383] have shown that microsomal NADH-cytochrome b_5 reductase catalyzes the redox cycling of Fe^{3+} bleomycin (BLM) that promoted MDA formation in the presence of DNA. Using the spin-trapping technique, Turner et al.[384] found that BLM stimulated the formation of only the superoxide ion and not hydroxyl radicals in guinea pig enterocytes. Moseley[385] recently showed that Fe^{3+}-BLM promotes *in vivo* DNA damage (in viral minichromosomes).

It has been shown[386] that actinomycin D catalyzes the reduction of dioxygen to superoxide ion by NADPH ferredoxin reductase. It was supposed[387] that the quinone anticancer antibiotic lactoquinomycin A catalyzes the production of superoxide ion and hydroxyl radicals in the lysates of doxorubicin-resistant mouse leukemia cells. It is important that Dalal and Shi[388] were unable to confirm the formation of superoxide ion and hydroxyl radicals during the autoxidation of fredericamycin A. It has been shown[389,390] that the dihydroxy metabolite of VP-16 (an antitumor drug) catalyzes the iron-dependent hydroxyl radical formation.

Rat hepatocytes are able to produce benzosemiquinone and 2-*S*-glutathionyl-1,4-benzosemiquinone after treatment with benzoquinone.[391] Moore et al.[392] concluded that the hepatocyte cytotoxicity of benzoquinone and its derivatives is originated from calcium release by mitochondria. Thor et al.[393] have found that redox cycling of naphthoquinones (such as menadione and 2,3-dimethoxy-1,4-naphthoquinone) stimulates actin cross-linking in isolated rat hepatocytes. Similarly, it has been shown[394] that the covalent binding of tetrachlorohydroquinone to microsomal proteins is mainly mediated by superoxide ion. Menadione stimulated the production of superoxide ion and hydroxyl radicals by digestive gland microsomes of the common mussel *Mytilus edulis* L.[395]

Buffinton et al.[396] concluded that naphthohydroquinones formed by the reduction of naphthoquinones by DT-diaphorase produce superoxide ion during autoxidation. Similarly, semiquinone and hydroquinone of naphthazarin (5,8-dihydroxy-1,4-naphthoquinone) and its glutathione conjugate formed during reduction of these quinones by rat liver NADPH cytochrome P-450 reductase and DT-diaphorase, respectively, were oxidized by dioxygen to produce superoxide ion.[397] Lea et al.[398] concluded that the rates of superoxide production stimulated by aziridinyl benzoquinones AZQ and BZQ were not correlated with the cytotoxicity of these compounds. In a recent review, Gutierrez[399] considered the mechanisms of reduction of diaziquone by cells, rat liver microsomes, and enzymes as well as the production of superoxide ion during oxidation of its semiquinone and hydroquinone.

Castro et al.[400] have shown that paraquat, but not nifurtimox, stimulates formation of methane from DMSO by liver microsomes in the presence of NADPH, which was apparently mediated by hydroxyl radicals. Ferric complexes such as Fe^{3+}(ATP) and Fe^{3+}(EDTA) increased the paraquat-induced generation of oxygen radicals by NADPH cytochrome P-450 reductase.[401] Sugimoto et al.[402] proposed that the paraquat-dependent induction of *N'*-acetylspermidine and other polyamines in hepatocytes is mediated by the superoxide ion, because SOD suppressed the effect of paraquat. Tsokos-Kuhn[403] concluded that lethal injury of rat liver hepatocytes by diquat is due to its redox cycling, since it took place only when the cells were pretreated with BCNU, an inhibitor of glutathione reductase. It has been shown that 1-methyl-4-phenyl-1,2,3,6-tetrahydropyridine (MPTP) stimulates the formation of oxygen radicals in a combination with dopamine melamine[404] or after incubation with mouse brain mitochondria.[405]

Rao et al.[406] have shown that rat hepatocytes reduce 5-nitrofurans and 2- and 5-nitroimidazoles (nitrofurantoin, misonidazole, etc.) to free radicals. Nitrofurantoin, *p*-nitrobenzoic acid, and *m*-dinitrobenzene stimulated superoxide production by soluble and microsomal fractions from freshwater fish.[407,408] It was proposed that superoxide ion was produced by xanthine oxidase and cytochrome P-450 reductase, respectively. Rossi et al.[409] proposed that the origin of the decrease in the viability of isolated rat hepatocytes in the presence of nitrofurantoin is the inactivation of hepatocyte glutathione reductase activity by hydrogen peroxide, formed as a result of redox cycling of this drug.

Farber et al.[410] have studied the peroxidation-dependent and peroxidation-independent pathways of hepatocyte killing by acetaminophen. Bridge et al.[411] confirmed that the production of reactive oxygen species (oxidative component) may play an important role in the cytotoxic effects of acetaminophen and 2,6-dimethylacetaminophen. Augusto et al.[412] have shown that the HRP + H_2O_2 system oxidizes primaquine into a free radical. Fletcher et al.[413] concluded that the oxidation of oxyhemoglobin and glutathione by 5-hydroxylated metabolites of primaquine is mediated by oxygen radicals. It was also proposed[414] that NBT reduction in mouse cultured cells stimulated by 3-hydroxyamino-1-methyl-5H-pyrido[4,3-b]indole is mediated by superoxide ion.

Isoniazid, iproniazid, and their metabolites (acetylhydrazine and isopropylhydrazine) were reduced to free radicals in isolated hepatocytes.[415] Albano et al.[416] have shown that methyl- and dimethylhydrazines were decomposed after incubation with isolated hepatocytes or liver microsomes to form methyl radicals. The latter were identified as the Me-4POBN spin adducts. Furazolidone stimulated superoxide production in the presence of NADPH by avian cardiac and hepatic microsomes.[417] Similarly, resorufin (7-hydroxyphenoxazone)[418] and 5-(4-nitrophenyl)penta-2,4-dienal[419] catalyzed the production of superoxide ion by rat liver microsomal NADPH-cytochrome P-450 reductase.

It was proposed that superoxide ion is formed during the autoxidation of dialuric acid because this reaction was inhibited by SOD.[420] Winterbourn and Munday[421] have studied the inhibitory effect of SOD on the glutathione-mediated redox cycling of alloxan. It has also been shown[422] that alloxan stimulates the formation of oxygen radicals and the release of calcium in rat liver mitochondria. Nikatsuka et al.[423] have found that rat liver microsomes reduce alloxan to a stable free radical. Sakurai and Miura[424] proposed that alloxan mediated the electron transfer from the sulfhydryl groups of BSA to dioxygen, forming superoxide ion. It has been shown[425,426] that alloxan released iron ions from ferritin in the presence of reduced glutathione. SOD, catalase, DTPA, mannitol, and benzoate inhibited this process.

Pileblad et al.[427] have shown that superoxide ion was formed during the oxidation of dopamine in the presence of ascorbate. Gee and Davison[428] estimated the contribution of the superoxide-dependent pathway in autoxidation of 6-hydroxydopamine under various experimental conditions. Similarly, superoxide ion was formed during autoxidation of the antipsoriatic drug dithranol.[429] Mueller and Kappus[430] concluded that the oxidation of dithranol is accompanied by the formation of hydroxyl radicals detected via the decomposition of KMBA. Hydroxyl radicals are also formed during the autoxidation of δ-aminolevulinic acid, a heme precursor, in the presence of oxyhemoglobin and other iron complexes.[431] Superoxide ion and hydrogen peroxide are formed in autoxidation of *N,N,N',N'*-tetramethyl-*p*-phenylenediamine at neutral pH.[432] Valoti et al.[433] have shown that HRP catalyzed the oxidation of 2-*tert*-butyl-4-methoxyphenol (BHA) and 2,6-di-*tert*-4-methylphenol (BHT) into free phenoxyl radicals.

REFERENCES

1. **Handa, K. and Sato, S.,** Generation of free radicals of quinone-group containing anticancer chemical in NADH-microsome system as evidenced by initiation of sulfite oxidation, *Gann*, 66, 43, 1975.
2. **Bachur, N. R., Gordon, S. L., and Gee, M. V.,** Anthracycline antibiotic augmentation of microsomal electron transport and free radical formation, *Mol. Pharmacol.*, 13, 901, 1977.
3. **Goodman, J. and Hochstein, P.,** Generation of free radicals and lipid peroxidation by redox cycling of Adriamycin and daunomycin, *Biochem. Biophys. Res. Commun.*, 77, 797, 1977.
4. **Sato, S., Iwaizumi, M., Handa, K., and Tamura, Y.,** Electron spin resonance study on the mode of generation of free radicals of daunomycin, Adriamycin, and carboquinone in NAD(P)H-microsome system, *Gann*, 68, 603, 1977.
5. **Thayer, W. S.,** Adriamycin stimulated superoxide formation in submitochondrial particles, *Chem.-Biol. Interact.*, 19, 265, 1977.
6. **Bachur, N. R., Gordon, S. L., and Gee, M. V.,** A general mechanism for microsomal activation of quinone anticancer agents to free radicals, *Cancer Res.*, 38, 1745, 1978.
7. **May, P. M., Williams, G. K., and Williams, D. R.,** Speciation studies of adriamycin, quelamycin and their metal complexes, *Inorg. Chim. Acta*, 46, 221, 1980.
8. **Myers, C. E., Gianni, L., Simone, C. B., Klecker, R. K., and Greene, R.,** Oxidative destruction of erythrocyte ghost membranes catalyzed by the doxorubicin-iron complex, *Biochemistry*, 21, 1707, 1982.
9. **Bachur, N. R., Friedman, R. D., and Hollenbeck, R. G.,** Physicochemical characteristics of ferric adriamycin complexes, *Cancer Chemother. Pharmacol.*, 12, 5, 1984.
10. **Beraldo, H., Garnier-Suillerot, A., Tosi, L., and Lavelle, F.,** Iron(III)-adriamycin and iron(III)-daunorubicin complexes: physicochemical characteristics, interaction with DNA, and antitumor activity, *Biochemistry*, 24, 284, 1985.
11. **Land, E. J., Mukherjee, T., Swallow, A. J., and Bruce, J. M.,** Possible intermediates in the action of adriamycin—a pulse radiolysis study, *Br. J. Cancer*, 51, 515, 1985.

12. **Svingen, B. A. and Powis, G.,** Pulse radiolysis studies of antitumor quinones: radical lifetimes, reactivity with oxygen, and one-electron reduction potentials, *Arch. Biochem. Biophys.*, 209, 119, 1981.
13. **Butler, J., Hoey, B. M., and Swallow, A. J.,** Reactions of the semiquinone free radicals of anti-tumor agents with oxygen and iron, *FEBS Lett.*, 182, 95, 1985.
14. **Houee-Levin, C., Gardes-Albert, M., Ferradini, C., Faraggi, M., and Klapper, M.,** Pulse-radiolysis study of daunorubicin redox cycles: reduction by e_{aq} and COO free radicals, *FEBS Lett.*, 179, 46, 1984.
15. **Land, E. J., Mukherjee, T., Swallow, A. J., and Bruce, J. M.,** One-electron reduction of adriamycin: properties of the semiquinone, *Arch. Biochem. Biophys.*, 225, 116, 1983.
16. **Ashnagor, A., Bruce, J. M., Dutton, P. L., and Prince, R. C.,** One- and two-electron reduction of hydroxy-1,4-naphthoquinones and hydroxy-9,10-anthraquinones. The role of internal hydrogen bonding and its bearing on the redox chemistry of the anthracycline antitumor quinones, *Biochim. Biophys. Acta*, 801, 351, 1984.
17. **Peters, J. H., Gordon, G. R., Kashiwase, D., Lown, J. W., Yen, S.-F., and Plambeck, J. A.,** Redox activities of antitumor anthracyclines determined by microsomal oxygen consumption and assays for superoxide anion and hydroxyl radical generation, *Biochem. Pharmacol.*, 35, 1309, 1986.
18. **Abramson, H. N., Banning, J. W., Nachtman, J. P., Roginski, E. T., Sardessai, M., Wormser, H. C., Wu, J., Nagia, Z., Schroeder, R. R., and Bernardo, M. M.,** Synthesis of anthraquinonyl glucosaminoside and studies of the influence of aglycone hydroxyl substitution on superoxide generation, DNA binding, and antimicrobial properties, *J. Med. Chem.*, 29, 1709, 1986.
19. **Showalter, H. D. H., Fry, D. W., Leopold, W. R., Lown, J. W., Plambeck, J. A., and Reszka, K.,** Design, biochemical pharmacology, electrochemistry and tumor biology of antitumor anthrapyrazoles, *Anti-Cancer Drug Des.*, 1, 73, 1986.
20. **Afanas'ev, I. B., Polozova, N. I., and Samokhvalov, G. I.,** Investigation of the interaction of superoxide ion with adriamycin and the possible origin of cardiotoxicity of the anthracycline anticancer antibiotics, *Bioorgan. Chem.*, 9, 434, 1980.
21. **Anne, A. and Moiroux, J.,** One electron reduction of variously substituted anthraquinones. Reactivity of the radical anions with oxygen in aprotic media, *Nouv. J. Chem.*, 8, 259, 1984.
22. **Nakazawa, H., Andrews, P. A., Callery, P. S., and Bachur, N. R.,** Superoxide radical reactions with anthracycline antibiotics, *Biochem. Pharmacol.*, 34, 481, 1985.
23. **Afanas'ev, I. B. and Polozova, N. I.,** Kinetics and mechanism of the reactions of superoxide ion in solution. Part 6. Interaction of superoxide ion with adriamycin in aprotic and protic media, *J. Chem. Soc. Perkin Trans.*, 2, 835, 1987.
24. **Afanas'ev, I. B., Polozova, N. I., Kuprianova, N. S., and Gunar, V. I.,** Mechanism of the interaction of superoxide ion and ascorbate with anthracycline antibiotics, *Free Rad. Res. Commun.*, 3, 141, 1987.
25. **Malatesta, V., Penco, S., Sacchi, N., Valentini, L., Vigevani, A., and Arcamone, F.,** Electrochemical deglycosation of anthracyclines: stereoelectronic requirements, *Can. J. Chem.*, 62, 2845, 1984.
26. **Kleyer, D. L. and Koch, T. H.,** Mechanistic investigation of reduction of daunomycin and 7-deoxydaunomycinone with bi(3,5,5-trimethyl-2-oxomorpholin-3-yl), *J. Am. Chem. Soc.*, 106, 2380, 1984.
27. **Pan, S.-S. and Bachur, N. R.,** Xanthine oxidase catalyzed reductive cleavage of anthracycline antibiotics and free radical formation, *Mol. Pharmacol.*, 17, 95, 1980.
28. **Lown, J. W. and Chen, H.-H.,** Evidence for the generation of free hydroxyl radicals from certain quinone antitumor antibiotics upon reductive activation in solution, *Can. J. Chem.*, 59, 390, 1981.
29. **Lown, J. W. and Chen, H.-H.,** Electron paramagnetic resonance characterization and conformation of daunorubicin semiquinone intermediate implicated in anthracycline metabolism, cardiotoxicity, and anticancer action, *Can. J. Chem.*, 59, 3212, 1981.
30. **Lown, J. W., Chen, H.-H., Plambeck, J. A., and Acton, E. M.,** Diminished superoxide anion generation by reduced 5-iminodaunorubicin relative to daunorubicin and the relationship to cardiotoxicity of the anthracycline antitumor agents, *Biochem. Pharmacol.*, 28, 2563, 1979.
31. **Sinha, B. K., Motten, A. G., and Hauck, K. W.,** The electrochemical reduction of 1,4-bis-(2-[(2-hydroxyethyl)amino]-ethylamino)-anthracenedione and daunomycin: biochemical significance in superoxide formation, *Chem.-Biol. Interact.*, 43, 371, 1983.
32. **Schreiber, J., Mottley, C., Sinha, B. K., Kalyanaraman, B., and Mason, R. P.,** One-electron reduction of daunomycin, daunomycinone, and 7-deoxydaunomycinone by the xanthine/xanthine oxidase system: detection of semiquinone free radicals by electron spin resonance, *J. Am. Chem. Soc.*, 109, 348, 1987.
33. **Houee-Levin, C., Gardes-Albert, M., and Ferradini, C.,** Reduction of daunorubicin aqueous solutions by $COO^{\bullet}$ free radicals. Reactions of reduced transients with H_2O_2, *FEBS Lett.*, 173, 27, 1984.
34. **Bachur, N. R., Gordon, S. L., Gee, M. V., and Kon, H.,** NADPH cytochrome P-450 reductase activation of quinone anticancer agents to free radicals, *Proc. Natl. Acad. Sci. U.S.A.*, 76, 954, 1979.
35. **Pan, S.-S., Pedersen, L., and Bachur, N. R.,** Comparative flavoprotein catalysis of anthracycline antibiotic. Reductive cleavage and oxygen consumption, *Mol. Pharmacol.*, 19, 184, 1981.
36. **Berlin, V. and Haseltine, W. A.,** Reduction of adriamycin to a semiquinone-free radical by NADPH cytochrome P-450 reductase produces DNA cleavage in a reaction mediated by molecular oxygen, *J. Biol. Chem.*, 256, 4747, 1981.

37. **Komiyama, T., Oki, T., Inui, T., Takeuchi, T., and Umezawa, H.,** Reduction of anthracycline glycoside by NADPH cytochrome P-450 reductase, *Gann,* 70, 403, 1979.
38. **Tero-Kubota, S., Ikegami, Y., Sugioka, K., and Nakano, M.,** Spin trapping study on the generation mechanism of active oxygen radicals in the enzymatic reduction of quinoid antitumor agents, *Chem. Lett.,* 1583, 1984.
39. **Komiyama, T., Kikuchi, T., and Sugiura, Y.,** Interaction of anticancer quinone drugs, aclacinomycin A, adriamycin, carbazilquinone, and mitomycin C with NADPH-cytochrome P-450 reductase, xanthine oxidase and oxygen, *J. Pharmacobiol.-Dyn.,* 9, 651, 1986.
40. **Ledenev, A. N., Peskin, A. V., Konstantinov, A. A., and Ruuge, E. K.,** Free radical generation during the interaction of adriamycin with xanthine oxidase, *Biofizika,* 31, 519, 1986.
41. **Gutierrez, P. L., Gee, M. V., and Bachur, N. R.,** Kinetics of anthracycline antibiotic free radical formation and reductive glycosidase activity, *Arch. Biochem. Biophys.,* 223, 68, 1983.
42. **Rowley, D. A. and Halliwell, B.,** DNA damage by superoxide-generating systems in relation to the mechanism of action of the antitumor antibiotic adriamycin, *Biochim. Biophys. Acta,* 761, 86, 1983.
43. **Paur, E., Youngman, R. J., Lengfelder, E., and Elstner, E. F.,** Mechanisms of adriamycin-dependent oxygen activation catalyzed by NADPH-cytochrome c-(ferredoxin)-oxidoreductase, *Z. Naturforsch.,* 39c, 261, 1984.
44. **Fisher, J., Abdella, B. R. J., and McLane, K. E.,** Anthracycline antibiotic reduction by spinach ferredoxin-$NADP^+$ reductase and ferredoxin, *Biochemistry,* 24, 3562, 1985.
45. **Fisher, J., Ramakrishnan, K., and Becvar, J. E.,** Direct enzyme-catalyzed reduction of anthracyclines by reduced nicotinamide adenine dinucleotide, *Biochemistry,* 22, 1347, 1983.
46. **Kalyanaraman, B., Perez-Reyes, E., and Mason, R. P.,** Spin-trapping and direct electron spin resonance investigations of the redox metabolism of quinone anticancer drugs, *Biochim. Biophys. Acta,* 630, 119, 1980.
47. **Nohl, G. and Jordan, W.,** OH-Generation by adriamycin semiquinone and H_2O_2; an explanation for the cardiotoxicity of anthracycline antibiotics, *Biochem. Biophys. Res. Commun.,* 114, 197, 1983.
48. **Davies, K. J. A. and Doroshow, J. H.,** Redox cycling of anthracycline by cardiac mitochondria. I. Anthracycline radical formation by NADH dehydrogenase, *J. Biol. Chem.,* 261, 3060, 1986.
49. **Basra, J., Wolf, C. R., Brown, J. R., and Patterson, L. H.,** Evidence for human liver mediated free-radical formation by doxorubicin and mitozantrone, *Anti-Cancer Drug Des.,* 1, 45, 1985.
50. **Doroshow, J. H. and Reeves, J.,** Daunorubicin-stimulated reactive oxygen metabolism in cardiac sarcosomes, *Biochem. Pharmacol.,* 30, 259, 1981.
51. **Gervasi, P. G., Agrillo, M. G., Citti, L., Danesi, R., and Del Tacca, M.,** *Anticancer Res.,* 6, 1231, 1986.
52. **Peskin, A. V. and Bartosz, G.,** One-electron reduction of an anthracycline antibiotic carminomycin by a human erythrocyte redox chain, *FEBS Lett.,* 219, 212, 1987.
53. **Bird, D. M., Boldt, M., and Koch, T. H.,** A kinetic rationale for the inefficiency of 5-iminodaunomycin as a redox catalyst, *J. Am. Chem. Soc.,* 109, 4046, 1987.
54. **Youngman, R. J. and Elstner, E. F.,** On the interaction of adriamycin with DNA: investigation of spectral changes, *Arch. Biochem. Biophys.,* 231, 424, 1984.
55. **Bachur, N. R.,** Anthracycline antibiotic pharmacology and metabolism, *Cancer Treat. Rep.,* 63, 817, 1979.
56. **Chinami, M., Kato, T., Ogura, R., and Shingu, M.,** Semiquinone formation of adriamycin by oxidation at para-OH residue, *Biochem. Int.,* 8, 299, 1984.
57. **Hilton, B. D., Misra, R., and Zweier, J. L.,** Magnetic resonance studies of fredericamycin A: evidence for oxygen-dependent free radical formation, *Biochemistry,* 25, 5533, 1986.
58. **Afanas'ev, I. B., Polozova, N. I., and Gunar, V. I.,** Mechanism of the reduction of anthracycline antibiotics and their iron complexes by ascorbate: relationship with their effects on lipid peroxidation, in *Organ Directed Toxicities of Anticancer Drugs,* Hacker, M. P., Lazo, J. S., and Tritton, T. R., Eds., Martinus Nijhoff, Boston, 1988, 223.
59. **Fiallo, M. M. L. and Garnier-Suillerot, A.,** Physicochemical studies of the iron(III)-carminomycin complex and evidence of the lack of stimulated superoxide production by NADH dehydrogenase, *Biochim. Biophys. Acta,* 840, 91, 1985.
60. **Fantine, E. O. and Garnier-Suillerot, A.,** Interaction of 5-iminodaunorubicin with Fe(III) and with cardiolipin-containing vesicles, *Biochim. Biophys. Acta,* 856, 130, 1986.
61. **Myers, C. E., Muindi, J. R. F., Zweier, J., and Sinha, B. K.,** 5-Iminodaunomycin. An anthracycline with unique properties, *J. Biol. Chem.,* 262, 11571, 1987.
62. **Muindi, J. R. F., Sinha, B. K., Gianni, L., and Myers, C. E.,** Hydroxyl production and DNA damage induced by anthracycline-iron complex, *FEBS Lett.,* 172, 226, 1984.
63. **Beraldo, H., Garnier-Suillerot, A., and Tosi, L.,** Copper(II)-adriamycin complexes. A circular dichroism and resonance Raman study, *Inorg. Chem.,* 22, 4117, 1985.

64. **Fiallo, M. M. L. and Garnier-Suillerot, A.,** Metal anthracycline complexes as a new class of anthracycline derivatives. Pd(II)-adriamycin and Pd(II)-daunorubicin complexes: physicochemical characteristics and anti-tumor activity, *Biochemistry,* 25, 924, 1986.
65. **Zweier, J. L.,** Reduction of O_2 by iron-adriamycin, *J. Biol. Chem.,* 259, 6056, 1984.
66. **Zweier, J. L.,** Iron-mediated formation of an oxidized adriamycin free radical, *Biochim. Biophys. Acta,* 839, 209, 1985.
67. **Zweier, J. L., Gianni, L., Muindi, J., and Myers, C. E.,** Differences in O_2 reduction by the iron complexes of adriamycin and daunomycin: the importance of the sidechain hydroxyl group, *Biochim. Biophys. Acta,* 884, 326, 1986.
68. **Sugioka, K., Nakano, H., Noguchi, T., Tsuchiya, J., and Nakano, M.,** Decomposition of unsaturated phospholipids by iron-ADP-doxorubicin coordination complex, *Biochem. Biophys. Res. Commun.,* 100, 1251, 1981.
69. **Gutteridge, J. M. C.,** Lipid peroxidation and possible hydroxyl radical formation stimulated by the self-reduction of a doxorubicin-iron(III) complex, *Biochem. Pharmacol.,* 33, 1725, 1984.
70. **Sugioka, K. and Nakano, M.,** Mechanism of phospholipid peroxidation induced by ferric ion-ADP-adriamycin-co-ordinated complex, *Biochim. Biophys. Acta,* 713, 333, 1982.
71. **Tobia, A. J., Coun, D., and Sagone, A.,** The effects of the quinone type drugs on hydroxyl radical (OH·) production by rat liver microsomes, *J. Toxicol. Environ. Health,* 15, 265, 1985.
72. **Sugioka, K., Nakano, H., Nakano, M., Tero-Kubota, S., and Ikegami, Y.,** Generation of hydroxyl radicals during the enzymatic reductions of the Fe^{3+}-ADP-phosphate-adriamycin and Fe^{3+}-ADP-EDTA systems. Less involvement of hydroxyl radical and a great importance of proposed preferryl ion complexes in lipid peroxidation, *Biochim. Biophys. Acta,* 753, 411, 1983.
73. **Thornalley, P. J., Bannister, W. H., and Bannister, J. V.,** Reduction of oxygen by NADH/NADH dehydrogenase in the presence of adriamycin, *Free Rad. Res. Commun.,* 2, 163, 1986.
74. **Thornalley, P. J. and Dodd, N. J. F.,** Free radical production from normal and adriamycin-treated rat cardiac sarcosomes, *Biochem. Pharmacol.,* 34, 669, 1985.
75. **Bannister, J. V. and Thornalley, P. J.,** The production of hydroxyl radicals by adriamycin in red blood cells, *FEBS Lett.,* 157, 170, 1983.
76. **Sinha, B. K., Katki, A. G., Batist, G., Cowan, K. H., and Myers, C. E.,** Differential formation of hydroxyl radicals by adriamycin in sensitive and resistant MCF-7 human breast tumor cells: implication for the mechanism of action, *Biochemistry,* 26, 3776, 1987.
77. **Doroshow, J. H.,** Anthracycline antibiotic-stimulated superoxide hydrogen peroxide, and hydroxyl radical production by NADH dehydrogenase, *Cancer Res.,* 43, 4543, 1983.
78. **Doroshow, J. H. and Davies, K. J. A.,** Redox cycling of anthracycline by cardiac mitochondria. II. Formation of superoxide anion, hydrogen peroxide, and hydroxyl radical, *J. Biol. Chem.,* 261, 3068, 1986.
79. **Doroshow, J. H.,** Role of hydrogen peroxide and hydroxyl radical formation in the killing of Ehrlich tumor cells by anticancer quinones, *Proc. Natl. Acad. Sci. U.S.A.,* 83, 4514, 1986.
80. **Doroshow, J. H.,** Role of reactive oxygen production in doxorubicin cardiac toxicity, in *Organ Directed Toxicities of Anticancer Drugs,* Hacker, M. P., Lazo, J. S., and Tritton, T. R., Eds., Martinus Nijhoff, Boston, 1988, 31.
81. **Komiyama, T., Sawada, M. T., Kobayashi, K., and Yoshimoto, A.,** Enhanced production of ethylene from methional by iron chelates and heme containing proteins in the system consisting of quinone compounds and NADPH-cytochrome P-450 reductase, *Biochem. Pharmacol.,* 34, 977, 1985.
82. **Bates, D. A. and Winterbourn, C. C.,** Deoxyribose breakdown by the adriamycin semiquinone and H_2O_2: evidence for hydroxyl radical participation, *FEBS Lett.,* 145, 137, 1982.
83. **Winterbourn, J. C.,** Evidence for the production of hydroxyl radicals from the adriamycin semiquinone and H_2O_2, *FEBS Lett.,* 136, 89, 1981.
84. **Kalyanaraman, B., Sealy, R. C., and Sinha, B. K.,** An electron spin resonance study of the reduction of peroxides by anthracycline semiquinones, *Biochim. Biophys. Acta,* 799, 270, 1984.
85. **Sushkov, D. G., Gritsan, N. P., and Weiner, L. M.,** Generation of OH radical during enzymatic reduction of 9,10-anthraquinone-2-sulphonate, *FEBS Lett.,* 225, 139, 1987.
86. **Sushkov, D. G., Rumyantseva, G. V., and Weiner, L. M.,** Substrate reduction and oxygen activation in microsomal quinone metabolism, *Biokhimia,* 52, 1898, 1987.
87. **Winterbourn, C. C., Gutteridge, J. M. C., and Halliwell, B.,** Doxorubicin-dependent lipid peroxidation at low partial pressures of O_2, *J. Free Rad. Biol. Med.,* 1, 43, 1985.
88. **Afanas'ev, I. B., Kuprianova, N. S., and Dorozhko, A. I.,** On the mechanism of doxorubicin-dependent nonenzymatic and enzymatic lipid peroxidation, in press.
89. **Nielson, C. P., Brenner, D., and Olson, R. D.,** Doxorubicin and doxorubicinol-induced alterations in human polymorphonuclear leukocyte oxygen metabolite generation, *J. Pharmacol. Exp. Ther.,* 238, 19, 1986.
90. **Schinetti, M. L., Rossini, D., and Bertelli, A.,** Interaction of anthracycline antibiotics with human neutrophils: superoxide production, free radical formation and intracellular penetration, *J. Cancer Res. Clin. Oncol.,* 113, 15, 1987.

91. **Korkina, L. G., Suslova, T. B., Soodaeva, S. K., Afanas'ev, I. B., and Gunar, V. I.,** Effects of quinones on the superoxide production by enzymes and peritoneal macrophages, in *Abstracts of Fifth Conf. Superoxide and Superoxide Dismutase,* The Israel Academy of Sciences and Humanities, Jerusalem, Israel, 33, 1989.
92. **Myers, C. E.,** Role of iron in anthracycline action, in *Organ Directed Toxicities of Anticancer Drugs,* Hacker, M. P., Lazo, J. C., and Tritton, T. R., Eds., Martinus Nijhoff, Boston, 1988, 17.
93. **Demant, E. J. F. and Norskov-Lauritsen, N.,** Binding of transferrin-iron by adriamycin at acidic pH, *FEBS Lett.,* 196, 321, 1986.
94. **Thomas, G. E. and Aust, S. D.,** Release of iron from ferritin by cardiotoxic anthracycline antibiotics, *Arch. Biochem. Biophys.,* 248, 684, 1986.
95. **Mimnaugh, E. G., Trush, M. A., and Gram, T. E.,** Stimulation by adriamycin of rat heart and liver microsomal NADPH-dependent lipid peroxidation, *Biochem. Pharmacol.,* 30, 2797, 1981.
96. **Mimnaugh, E. G., Trush, M. A., Ginsburg, E., and Gram, T. E.,** Differential effects of anthracycline drugs on rat heart and liver microsomal reduced nicotinamide adenine dinucleotide phosphate-dependent lipid peroxidation, *Cancer Res.,* 42, 3574, 1982.
97. **Mimnaugh, E. G., Trush, M. A., and Gram, T. E.,** A possible role for membrane lipid peroxidation in anthracycline nephrotoxicity, *Biochem. Pharmacol.,* 35, 4327, 1986.
98. **Mimnaugh, E. G., Gram, T. E., and Trush, M. A.,** Stimulation of mouse heart and liver microsomal lipid peroxidation by anthracycline anticancer drugs: characterization and effect of reactive oxygen scavengers, *J. Pharmacol. Exp. Ther.,* 226, 806, 1983.
99. **Mimnaugh, E. G., Trush, M. A., Bhatnagar, M., and Gram, T. E.,** Enhancement of reactive oxygen-dependent mitochondrial membrane lipid peroxidation by the anthracycline drug adriamycin, *Biochem. Pharmacol.,* 34, 847, 1985.
100. **Kharasch, E. D. and Novak, R. F.,** Inhibitory effects of anthracenedione antineoplastic agents on hepatic and cardiac lipid peroxidation, *J. Pharmacol. Exp. Ther.,* 226, 500, 1983.
101. **Sinha, B. K., Trush, M. A., and Kalyanaraman, B.,** Free radical metabolism of VP-16 and inhibition of anthracycline-induced lipid peroxidation, *Biochem. Pharmacol.,* 32, 3495, 1983.
102. **Solaini, G., Landi, L., Pasquali, P., and Rossi, C. A.,** Protective effect of endogenous Q on both lipid peroxidation and respiratory chain inactivation induced by an adriamycin-iron complex, *Biochem. Biophys. Res. Commun.,* 147, 572, 1987.
103. **Olinescu, R., Milcoveanu, D., Nita, S., Pascu, E., and Urseanu, I.,** Inhibitory action of magnesium ascorbate on the formation of lipid peroxides by anthracycline antibiotics, *Rev. Roum. Biochim.,* 23, 127, 1986.
104. **Butler, J. and Hoey, B. M.,** Are reduced quinones necessarily involved in the antitumor activity of quinone drugs?, *Br. J. Cancer,* 55 (Suppl.), 53, 1987.
105. **Myers, C. E., McCuire, W. P., Liss, R. H., Ifrim, I., Grotzinger, K., and Young, R. O.,** Adriamycin: the role of lipid peroxidation in cardiac toxicity and tumor response, *Science,* 197, 165, 1977.
106. **Ogura, R.,** Adriamycin-induced lipid peroxidation and its protection, in *Lipid Peroxides in Biology and Medicine,* Yagi, K., Ed., Academic Press, New York, 1982, 255.
107. **Bhuyan, D. K., Podas, S. M., and Bhuyan, K. C.,** Antioxidants in prevention of oxidative damage to the lens and cataract in vivo, in *Superoxide and Superoxide Dismutase in Chemistry, Biology, and Medicine,* Rotillo, G., Ed., Elsevier Science, Amsterdam, 1986, 657.
108. **Julicher, R. H. M., Sterrenberg, L., Bast, A., Riksen, R. O. W. M., Koomen, J. M., and Noordhock, J.,** The role of lipid peroxidation in acute doxorubicin-induced cardiotoxicity as studied in rat isolated heart, *J. Pharm. Pharmacol.,* 38, 277, 1986.
109. **Zidenberg-Cherr, S. and Keen, C. L.,** Influence of dietary manganese and vitamin E on adriamycin toxicity in mice, *Toxicol. Lett.,* 30, 79, 1986.
110. **Patterson, L. H., Gandecha, B. M., and Brown, J. R.,** 1,4-Bis 2-(2-hydroxyethyl)amino ethylamino-9,10-anthracenedione, an anthraquinone antitumor agent that does not cause lipid peroxidation *in vivo;* comparison with daunomycin, *Biochem. Biophys. Res. Commun.,* 110, 399, 1983.
111. **Sazuka, Y., Yoshikawa, K., Tanizawa, H., and Takino, Y.,** Effect of doxorubicin on lipid peroxide levels in tissues of mice, *Gann,* 78, 1281, 1987.
112. **Lenzhofer, R., Magometschigg, D., Dudczak, R., Cerni, C., Bolebruch, C., and Moser, K.,** Indication of reduced doxorubicin-induced cardiac toxicity by additional treatment with antioxidative substances, *Experientia,* 39, 62, 1983.
113. **Gutteridge, J. M. C.,** Adriamycin-iron catalyzed phospholipid peroxidation: a reaction not involving reduced adriamycin or hydroxyl radicals, *Biochem. Pharmacol.,* 32, 1949, 1983.
114. **Youngman, R. J., Gotz, F., and Elstner, E. F.,** Role of oxygen activation in adriamycin-mediated DNA strand scission and the effect of binding on the redox properties of the drug, in *Oxygen Radicals in Chemistry and Biology,* Bors, W., Saran, M., and Tait, D., Eds., Walter de Gruyter, Berlin, 1984, 131.
115. **Kappus, H., Muliawan, H., and Scheulen, M. E.,** The role of iron in lipid peroxidation induced by adriamycin during redox cycling in liver microsomes, in *Oxygen Radicals in Chemistry and Biology,* Bors, W., Saran, M., and Tait, D., Eds., Walter de Gruyter, Berlin, 1984, 359.

116. **Nakano, H., Ogita, K., Gutteridge, J. M. C., and Nakano, M.,** Inhibition by the protein ceruloplasmin of lipid peroxidation stimulated by an Fe^{3+}-ADP-adriamycin complex, *FEBS Lett.*, 166, 232, 1984.
117. **Gutteridge, J. M. C. and Toeg, D.,** Adriamycin-dependent damage to deoxyribose: a reaction involving iron, hydroxyl and semiquinone free radicals, *FEBS Lett.*, 149, 228, 1982.
118. **Gutteridge, J. M. C. and Quinlan, G. J.,** Free radical damage to deoxyribose by anthracycline, aureolic acid and aminoquinone antitumor antibiotics. An essential requirement for iron, semiquinones and hydrogen peroxide, *Biochem. Pharmacol.*, 34, 4099, 1985.
119. **Sinha, B. K. and Gregory, J. L.,** Role of one-electron and two-electron reduction products of adriamycin and daunomycin in deoxyribonucleic acid binding, *Biochem. Pharmacol.*, 30, 2626, 1981.
120. **Eliot, H., Gianni, L., and Myers, C.,** Oxidative destruction of DNA by the adriamycin-iron complex, *Biochemistry*, 23, 928, 1984.
121. **Muindi, J., Sinha, B. K., Gianni, L., and Myers, C.,** Thiol-dependent DNA damage produced by anthracycline-iron complexes: the structure-activity relationships and molecular mechanisms, *Mol. Pharmacol.*, 27, 356, 1985.
122. **Mariam, Y. H. and Glover, G. P.,** Degradation of DNA by metalloanthracycline: requirement for metal ions, *Biochem. Biophys. Res. Commun.*, 136, 1, 1986.
123. **Demant, E. J. F. and Jensen, P. K.,** Destruction of phospholipids and respiratory-chain activity in pig-heart submitochondrial particles induced by an adriamycin-iron complex, *Eur. J. Biochem.*, 132, 551, 1983.
124. **Demant, E. J. F.,** NADH oxidation in submitochondrial particles protects respiratory chain activity against damage by adriamycin-Fe^{3+}, *Eur. J. Biochem.*, 137, 113, 1983.
125. **Adachi, T., Nagae, T., Ito, Y., Hirano, K., and Sugioka, M.,** Relation between cardiotoxic effect of adriamycin and superoxide anion radical, *J. Pharm. Dyn.*, 6, 114, 1983.
126. **Yasumi, M., Minaga, T., Nakamura, K., Kizu, A., and Ijichi, H.,** Inhibition of cardiac NADP-linked isocitrate dehydrogenase by adriamycin, *Biochem. Biophys. Res. Commun.*, 93, 631, 1980.
127. **Nicolay, K., Aue, W. P., Seelig, J., van Echteld, C. J. A., Ruigrok, T. J. C., and de Kruiff, B.,** Effects of the anti-cancer drug adriamycin on the energy metabolism of rat heart as measured by in vivo ^{31}P-NMR and implications for adriamycin-induced cardiotoxicity, *Biochim. Biophys. Acta*, 929, 5, 1987.
128. **Muhammed, H., Ramasarma, T., and Kurup, C. K. R.,** Inhibition of mitochondrial oxidative phosphorylation by adriamycin, *Biochim. Biophys. Acta*, 722, 43, 1983.
129. **Muhammed, H. and Kurup, C. K. R.,** Influence of ubiquinone on the inhibitory effect of adriamycin on mitochondrial oxidative phosphorylation, *Biochem. J.*, 217, 493, 1984.
130. **Pollakis, G., Goormaghtigh, E., and Ruysschaert, J.-M.,** Role of the quinone structure in the mitochondrial damage induced by antitumor anthracyclines. Comparison of adriamycin and 5-iminodaunorubicin, *FEBS Lett.*, 155, 267, 1983.
131. **Harris, R. N. and Doroshow, J. H.,** Effect of doxorubicin-enhanced hydrogen peroxide and hydroxyl radical formation on calcium sequestration by cardiac sarcoplasmic reticulum, *Biochem. Biophys. Res. Commun.*, 130, 739, 1983.
132. **Doroshow, J. H.,** Prevention of doxorubicin-induced killing of MCF-7 human breast cancer cells by oxygen-radical scavengers and iron-chelating agents, *Biochem. Biophys. Res. Commun.*, 135, 330, 1986.
133. **Sinha, B. K., Katki, A. G., Batist, G., Cowan, K. H., and Myer, C. E.,** Adriamycin-stimulated hydroxyl radical formation in human breast tumor cells, *Biochem. Pharmacol.*, 36, 793, 1987.
134. **Mavelli, I., Ciriolo, M. R., Dini, L., and Rotilio, G.,** Selective damage of Se-glutathione peroxidase by oxy radicals generated by daunomycin, in *Superoxide and Superoxide Dismutase in Chemistry, Biology, and Medicine*, Rotilio, G., Ed., Elsevier Science, Amsterdam, 1986, 425.
135. **Babson, J. R., Abell, M. S., and Reed, D. J.,** Protective role of the glutathione redox cycle against adriamycin-mediated toxicity in isolated hepatocytes, *Biochem. Pharmacol.*, 30, 2299, 1981.
136. **Reed, D. J.,** Regulation of reduced processes by glutathione, *Biochem. Pharmacol.*, 35, 7, 1986.
137. **Bredehorst, R., Panneerselvan, M., and Vogel, C. W.,** Doxorubicin enhances complement susceptibility of human melanoma cells by extracellular oxygen radical formation, *J. Biol. Chem.*, 262, 2034, 1987.
138. **Speyer, J. L., Green, M. D., Ward, C., Sanger, J., Kramer, E., Rey, M., Wernz, J. C., Blum, R. H., Meyers, M., Muggia, F. M., Ferrans, V., Stecy, P., Feit, F., Dubin, N., Jacquotte, A., Taubes, S., and London, C.,** A trial of ICRF-187 to selectively protect against chronic adriamycin cardiac toxicity: rationale and preliminary results of a clinical trial, in *Organ Directed Toxicities of Anticancer Drugs*, Hacker, M. P., Lazo, J. C., and Tritton, T. R., Eds., Martinus Nijhoff, Boston, 1988, 64.
139. **Ferrans, V. J., Herman, E. H., and Hamlin, R. L.,** Pretreatment with ICRF-187 protects against the chronic cardiac toxicity produced by very large cumulative doses of doxorubicin in beagle dogs, in *Organ Directed Toxicities of Anticancer Drugs*, Hacker, M. P., Lazo, J. C., and Tritton, T. R., Eds., Martinus Nijhoff, Boston, 1988, 56.
140. **Favalli, L., Lanza, E., Rozza, A., Poggi, P., Galimberti, M., and Villani, F.,** Significant reduction of delayed doxorubicin cardiotoxicity by ICRF-187 in rats, in *Organ Directed Toxicities of Anticancer Drugs*, Hacker, M. P., Lazo, J. C., and Tritton, T. R., Eds., Martinus Nijhoff, Boston, 1988, 224.

141. **Filppi, J. A., Imondi, A. R., and Wolgemuth, R. L.,** Characterization of the cardioprotective effect of (S)(+)-4,4′-propylene-2,6-piperazinedione (ICRF-187) on anthracycline cardiotoxicity, in *Organ Directed Toxicities of Anticancer Drugs,* Hacker, M. P., Lazo, J. C., and Tritton, T. R., Eds., Martinus Nijhoff, Boston, 1988, 225.
142. **Doroshow, J. H. and Schechter, J.,** Prevention of doxorubicin cardiac toxicity by dimethyl sulfoxide, in *Superoxide and Superoxide Dismutase in Chemistry, Biology and Medicine,* Rotilio, G., Ed., Elsevier Science, Amsterdam, 1986, 639.
143. **Willmott, N. and Cummings, J.,** Increased anti-tumor effect of adriamycin-loaded albumin microspheres is associated with anaerobic bioreduction of drug in tumor tissue, *Biochem. Pharmacol.,* 36, 521, 1987.
144. **Hacker, M. P., Lazo, J. S., Pritsos, C. A., and Tritton, T. R.,** Immobilized adriamycin: toxic potential in vivo and in vitro, in *Organ Directed Toxicities of Anticancer Drugs,* Hacker, M. P., Lazo, J. C., and Tritton, T. R., Eds., Martinus Nijhoff, Boston, 1988, 226.
145. **Wadler, S., Green, M. D., Basch, R., and Muggia, F. M.,** Lethal and sublethal effects of combination of doxorubicin and the bisdioxopiperazine (+)-1,2-bis(3,5-dioxopiperazinyl-1-yl)propane (ICRF 187), on murine sarcoma S 180 in vitro, *Biochem. Pharmacol.,* 36, 1495, 1987.
146. **Ripoll, E. A. P., Rama, B. N., and Webber, M. M.,** Vitamin E enhances the chemotherapeutic effects of adriamycin on human prostatic carcinoma cells in vitro, *J. Urol.,* 136, 524, 1986.
147. **Lown, J. W., Sondhi, S. M., Maudal, S. B., and Murphy, J.,** Synthesis and redox properties of chromophore modified glycosides related to anthracyclines, *J. Org. Chem.,* 47, 4304, 1982.
148. **Kharasch, E. D. and Novak, R. E.,** Bis(alkylamino)anthracenedione antineoplastic agent metabolic activation by NADPH-cytochrome P-450 reductase and NADH dehydrogenase: diminished activity relative to anthracyclines, *Arch. Biochem. Biophys.,* 224, 682, 1983.
149. **Kharasch, E. D. and Novak, R. F.,** Mitoxantrone and ametantrone inhibit hydroperoxide-dependent initiation and propagation reactions in fatty acid peroxidation, *J. Biol. Chem.,* 260, 10645, 1985.
150. **Frank, P. and Novak, R. E.,** Effects of anthrapyrazole antineoplastic agents on lipid peroxidation, *Biochem. Biophys. Res. Commun.,* 140, 797, 1986.
151. **Graham, M. A., Newell, D. R., Butler, J., Hoey, B., and Patterson, L.H.,** The effect of the anthrapyrazole antitumor agent CI 941 on rat liver microsome- and cytochrome P-450 reductase-mediated free-radical processes. Inhibition of doxorubicin activation in vitro, *Biochem. Pharmacol.,* 36, 3345, 1987.
152. **Powis, G. and Appel, P. L.,** Relationship of the single-electron reduction potential of quinones to their reduction by flavoproteins, *Biochem. Pharmacol.,* 29, 2567, 1980.
153. **Komiyama, T., Oki, T., and Inni, T.,** Activation of mitomycin C and quinone drug metabolism by NADPH-cytochrome P-450 reductase, *J. Pharmacol.-Dyn.,* 2, 407, 1979.
154. **Pritsos, C. A., Constantinides, P. P., Tritton, T. R., Heimbrook, D. C., and Sartorelli, A. C.,** Use of high-performance liquid chromatography to detect hydroxyl and superoxide radicals generated from mitomycin C, *Anal. Biochem.,* 150, 294, 1985.
155. **Doroshow, J. H.,** Mitomycin C-enhanced superoxide and hydrogen peroxide formation in rat heart, *J. Pharmacol. Exp. Ther.,* 218, 206, 1981.
156. **Pritsos, C. A. and Sartorelli, A. C.,** Generation of reactive oxygen radicals through bioactivation of mitomycin antibiotics, *Cancer Res.,* 46, 3528, 1986.
157. **Nakano, H., Sugioka, K., Nakano, M., Mizukami, M., Kimura, H., Tero-Kubota, S., and Ikegami, Y.,** Importance of Fe^{2+}-ADP and relative unimportance of HO· in the mechanism of mitomycin C-induced lipid peroxidation, *Biochim. Biophys. Acta,* 796, 285, 1984.
158. **Hoey, B. M., Butler, J., and Swallow, A. J.,** Reductive activation of mitomycin C, *Biochemistry,* 27, 2608, 1988.
159. **Andrews, P. A., Pan, S. S., and Bachur, N. R.,** Electrochemical reduction activation of mitomycin, *J. Am. Chem. Soc.,* 108, 4158, 1986.
160. **Pan, S.-S., Andrews, P. A., Glover, C. J., and Bachur, N. R.,** Reductive activation of mitomycin C and mitomycin C metabolites catalyzed by NADPH-cytochrome P-450 reductase and xanthine oxidase, *J. Biol. Chem.,* 259, 959, 1984.
161. **Peterson, D. M. and Fisher, J.,** Autocatalytic quinone methide formation from mitomycin C, *Biochemistry,* 25, 4077, 1986.
162. **Bachur, N. R., Gee, M. V., and Gordon, S. L.,** Enzymatic activation of actinomycin D (ACT-D) to free radical state, *Proc. Am. Assoc. Cancer Res.,* 19, 75, 1978.
163. **Nakazawa, H., Chou, F. E., Andrews, P. A., and Bachur, N. R.,** Chemical reduction of actinomycin D and phenoxazone analogues, *J. Org. Chem.,* 46, 1493, 1981.
164. **Sinha, B. K. and Cox, M. G.,** Stimulation of superoxide formation by actinomycin D and its N^2-substituted spin-labeled derivatives, *Mol. Pharmacol.,* 17, 432, 1980.
165. **Sehgal, R. K., Sengupta, S. K., Waxman, D. J., and Tauber, A. I.,** Enzymic and chemical reduction of 2-deaminoactinimycins to free radicals, *Anti-Cancer Drug Des.,* 1, 13, 1985.
166. **Hassett, D. J., Britigan, B. E., Svendsen, T., Rosen, G. M., and Cohen, M. S.,** Bacteria form intracellular free radicals in response to paraquat and streptonigrin. Demonstration of the potency of hydroxyl radical, *J. Biol. Chem.,* 262, 13404, 1987.

167. **Gutierrez, P. L., Friedman, R., and Bachur, N.**, Biochemical activation of AZQ([3,6-diaziridinyl]-2,5-bis[carboethoxyamino]-1,4-benzoquinone) to its free radical species, *Cancer Treat. Rep.*, 66, 339, 1982.
168. **Gutierrez, P. L. and Bachur, N. R.**, Free radicals in quinone containing antitumor agents. The nature of the diaziquone (3,6-diaziridinyl-2,5-bis(carboethoxyamino)-1,4-benzoquinone) free radical, *Biochim. Biophys. Acta*, 758, 37, 1983.
169. **Kikuchi, T., Sugiura, Y., and Komiyama, T.**, Generation of semiquinone radical from carbazilquinone by NADPH-cytochrome P-450 reductase, *Biochem. Pharmacol.*, 30, 1717, 1981.
170. **Gutierrez, P. L., Fox, B. M., Mossoba, M. M., Egorin, M. J., Nakazawa, H., and Bachur, R.**, Electron spin resonance of electrochemically generated free radicals from diaziquone and its derivatives, *Biophys. Chem.*, 22, 115, 1985.
171. **Gutierrez, P. L., Egorin, M. J., Fox, B. M., Friedman, R., and Bachur, N. R.**, Cellular activation of diaziquone [2,5-diaziridinyl-3,6-bis-(carboethoxyamino)-1,4-benzoquinone] to its free radical species, *Biochem. Pharmacol.*, 34, 1449, 1985.
172. **Gutierrez, P. L., Nayar, M. S. B., Nardino, R., and Callery, P. S.**, The chemical reduction of diaziquinone: products and free radical intermediates, *Chem.-Biol. Interact.*, 64, 23, 1987.
173. **Gutierrez, P. L., Biswal, S., Nardino, R., and Biswal, N.**, Reductive activation of diaziquone and possible involvement of free radicals and the hydroquinone dianion, *Cancer Res.*, 46, 5779, 1986.
174. **Butler, J., Hoey, B. M., and Lea, J. S.**, The reduction of anti-tumor diaziridinyl benzoquinones, *Biochim. Biophys. Acta*, 925, 144, 1987.
175. **Lown, J. W., Sim, S.-K., and Chen, H.-H.**, Hydroxyl radical production by free and DNA bound aminoquinone antibiotics and its role in DNA degradation. Electron spin resonance detection of hydroxyl radicals by spin trapping, *Can. J. Biochem.*, 56, 1042, 1978.
176. **Komiyama, T., Kikuchi, T., and Sugioka, Y.**, Generation of hydroxyl radical by anticancer quinone drugs, carbazilquinone, mitomycin C, aclacinomycin A and adriamycin in the presence of NADPH-cytochrome P-450 reductase, *Biochem. Pharmacol.*, 31, 3651, 1982.
177. **Gutteridge, J. H.**, Streptonigrin-induced deoxyribose degradation: inhibition by superoxide dismutase, hydroxyl radical scavengers and iron chelators, *Biochem. Pharmacol.*, 33, 3059, 1984.
178. **Gutteridge, J. M. C., Quinlan, G. J., and Wilkins, S.**, Mitomycin C-induced deoxyribose degradation inhibited by superoxide dismutase. A reaction involving iron, hydroxyl and semiquinone radicals, *FEBS Lett.*, 167, 37, 1984.
179. **Iyanagi, T. and Yamazaki, I.**, One-electron-transfer reactions in biochemical systems. III. One-electron reduction of quinones by microsomal flavin enzymes, *Biochim. Biophys. Acta*, 172, 370, 1969.
180. **Nakamura, S. and Yamazaki, I.**, One-electron-transfer reactions in biochemical systems. IV. A mixed mechanism in the reaction of milk xanthine oxidase with electron acceptors, *Biochim. Biophys. Acta*, 189, 29, 1969.
181. **Iyanagi, T. and Yamazaki, I.**, One-electron-transfer reactions in biochemical systems. V. Difference in the mechanism of quinone reduction by the NADH dehydrogenase and the NAD(P)H dehydrogenase (DT-diaphorase), *Biochim. Biophys. Acta*, 216, 282, 1970.
182. **Fischer, V., West, P. R., Nelson, S. D., Harrison, P. J., and Mason, R. P.**, Formation of 4-aminophenoxyl free radical from the acetominophen metabolite N-acetyl-p-benzoquinone imine, *J. Biol. Chem.*, 260, 11446, 1985.
183. **Hassan, H. M. and Fridovich, I.**, Intracellular production of superoxide radical and hydrogen peroxide by redox active compounds, *Arch. Biochem. Biophys.*, 196, 385, 1979.
184. **Lind, C., Hochstein, P., and Ernster, L.**, DT-Diaphorase as a quinone reductase: a cellular control devise against semiquinone and superoxide radical formation, *Arch. Biochem. Biophys.*, 216, 178, 1982.
185. **DiMonte, D., Ross, D., Bellomo, G., Eklöw, L., and Orrenius, S.**, Alterations in intracellular thiol homeostasis during the metabolism of menadione by isolated rat hepatocytes, *Arch. Biochem. Biophys.*, 235, 334, 1984.
186. **Orrenius, S., Rossi, L., Eklöw-Lastbom, L., and Thor, H.**, Oxidative stress in intact cells. A comparison of the effects of menadione and diquat in isolated hepatocytes, in *Free Radical Liver Injury*, Poli, G., Cheeseman, K. H., Diazani, M. U., and Slater, T. F., Eds., IRL Press, Oxford, 1985, 99.
187. **Talcott, R. E., Shu, H., and Wei, E. T.**, Dissociation of microsomal oxygen reduction and lipid peroxidation with the electron acceptors, paraquat and menadione, *Biochem. Pharmacol.*, 28, 665, 1979.
188. **Moore, G. A., O'Brien, P. J., and Orrenius, S.**, Menadione (2-methyl-1,4-naphthoquinone)-induced Ca^{2+} release from rat-liver mitochondria is caused by NAD(P)H oxidation, *Xenobiotica*, 16, 873, 1986.
189. **Frei, B., Winterhalter, K. H., and Richter, C.**, Menadione-(2-methyl-1,4-naphthoquinone)-dependent enzymatic redox cycling and calcium release by mitochondria, *Biochemistry*, 25, 4438, 1986.
190. **Thor, H., Smith, M. T., Hartzell, P., Bellomo, G., Jewell, S. A., and Orrenius, S.**, The metabolism of menadione (2-methyl-1,4-naphthoquinone) by isolated hepatocytes. A study of the implications of oxidative stress in intact cells, *J. Biol. Chem.*, 257, 12419, 1982.
191. **Wefers, H. and Sies, H.**, Hepatic low-level chemiluminescence during redox cycling of menadione and the menadione-glutathione conjugate: relation to glutathione and NAD(P)H:quinone reductase (DT-diaphorase) activity, *Arch. Biochem. Biophys.*, 224, 568, 1983.

192. **Ross, D., Thor, H., Threadgill, M. D., Sandy, M. S., Smith, M. T., Moldeus, P., and Orrenius, S.,** The role of oxidative processes in the cytotoxicity of substituted 1,4-naphthoquinones in isolated hepatocytes, *Arch. Biochem. Biophys.*, 248, 460, 1986.
193. **Smith, P. F., Alberts, D. W., and Rush, G. F.,** Role of glutathione reductase during menadione-induced NADPH oxidation in isolated rat hepatocytes, *Biochem. Pharmacol.*, 36, 3879, 1987.
194. **Gant, T. W., Rao, D. N. R., Mason, R. P., and Cohen, G. M.,** Redox cycling and sulfhydryl arylation: their relative importance in the mechanism of quinone cytotoxicity to isolated hepatocytes, *Chem.-Biol. Interact.*, 65, 157, 1988.
195. **Mansbach, C. M., II, Rosen, G. M., Rahn, C. A., and Strauss, K. E.,** Detection of free radicals as a consequence of rat intestinal cellular drug metabolism, *Biochim. Biophys. Acta*, 888, 1, 1986.
196. **Rosen, G. M., Freeman, B. A., and Mansbach, C. M., II,** Detection of superoxide as a consequence of cellular drug metabolism, in *Superoxide and Superoxide Dismutase in Chemistry, Biology and Medicine*, Rotilio, G., Ed., Elsevier Science, Amsterdam, 1986, 329.
197. **Rosen, G. M. and Freeman, B. A.,** Detection of superoxide generated by endothelial cells, *Proc. Natl. Acad. Sci. U.S.A.*, 81, 7274, 1984.
198. **Scarpa, M., Viglino, P., and Rigo, A.,** Stimulated superoxide ion fluxes in human erythrocytes, in *Superoxide and Superoxide Dismutase in Chemistry, Biology and Medicine*, Rotilio, G., Ed., Elsevier Science, Amsterdam, 1986, 388.
199. **Gerasimov, A. M. and Zakharov, A. S.,** Thiol-dependent superoxide radical generation induced by menadione and vikacol, *Biokhimia*, 50, 1872, 1985.
200. **Pritsos, C. A., Jensen, D. E., Pisani, D., and Pardini, R. S.,** Involvement of superoxide in the interaction of 2,3-dichloro-1,4-naphthoquinone with mitochondrial membranes, *Arch. Biochem. Biophys.*, 217, 98, 1982.
201. **Pritsos, C. A. and Pardini, R. S.,** A redox cycling mechanism of action for 2,3-dichloro-1,4-naphthoquinone with mitochondrial membranes and the role of sulfhydryl groups, *Biochem. Pharmacol.*, 33, 3771, 1984.
202. **Boveris, A., Docampo, R., Turrens, J. F., and Stoppani, A. O. M.,** Effect of β-lapachone on superoxide anion and hydrogen peroxide production in Trypanosoma cruzi, *Biochem. J.*, 175, 431, 1978.
203. **Docampo, R., Cruz, F. S., Boveris, A., Muniz, R. P. A., and Esquivel, D. M. S.,** Lipid peroxidation and the generation of free radicals, superoxide anion, and hydrogen peroxide in β-lapachone-treated *Trypanosoma cruzi* epimastigotes, *Arch. Biochem. Biophys.*, 186, 292, 1978.
204. **Docampo, R., Cruz, F. S., Boveris, A., Muniz, R. P. A., and Esquivel, D. M. S.,** β-Lapachone enhancement of lipid peroxidation and superoxide anion and hydrogen peroxide formation by sarcoma 180 ascites tumor cells, *Biochem. Pharmacol.*, 28, 723, 1979.
205. **Sugioka, K., Nakano, M., Naito, I., Tero-Kubota, S., and Ikegami, Y.,** Properties of a coenzyme, pyrroloquinoline quinone: generation of an active oxygen species during a reduction-oxidation cycle in the presence of NAD(P)H and O_2, *Biochim. Biophys. Acta*, 964, 175, 1988.
206. **Talcott, R. E., Smith, M. T., and Ciannini, D. O.,** Inhibition of microsomal lipid peroxidation by naphthoquinones: structure-activity relationships and possible mechanism of action, *Arch. Biochem. Biophys.*, 241, 88, 1985.
207. **Moore, G. A., Rossi, L., Nicotera, P., Orrenius, S., and O'Brien, P. J.,** Quinone toxicity in hepatocytes: studies on mitochondrial Ca^{2+} release induced by benzoquinone derivatives, *Arch. Biochem. Biophys.*, 259, 283, 1987.
208. **Blum, J. and Fridovich, I.,** Superoxide, hydrogen peroxide, and toxicity in two free-living nematode species, *Arch. Biochem. Biophys.*, 222, 35, 1983.
209. **Begleiter, A.,** Studies on the mechanism of action of quinone anticancer agents, *Biochem. Pharmacol.*, 34, 2629, 1985.
210. **Bellomo, G., Mirabelli, F., Dimonte, D., Richelmi, P., Thor, M., Orrenius, C., and Orrenius, S.,** Formation and reduction of glutathione protein mixed disulfphides during oxidative stress. A study with isolated hepatocytes and menadione, *Biochem. Pharmacol.*, 36, 1313, 1987.
211. **Smith, M. T. and Evans, C. G.,** Inhibitory effect of superoxide-generating quinones on superoxide dismutase, *Biochem. Pharmacol.*, 33, 3109, 1984.
212. **Biaglow, J. E.,** Cellular electron transfer and radical mechanisms for drug metabolism, *Radiat. Res.*, 86, 212, 1981.
213. **Crawford, D. R. and Schneider, D. L.,** Evidence that a quinone may be requested for the production of superoxide and hydrogen peroxide in neutrophils, *Biochem. Biophys. Res. Commun.*, 99, 1277, 1981.
214. **Gallin, J. I., Seligmann, B. E., Cramer, E. B., Schiffmann, E., and Fletcher, M. P.,** Effects of vitamin K on human neutrophil function, *J. Immunol.*, 128, 1399, 1982.
215. **Harbour, J. R. and Bolton, J. R.,** Superoxide formation in spinach chloroplasts: electron spin resonance detection by spin trapping, *Biochem. Biophys. Res. Commun.*, 64, 803, 1975.
216. **Baldwin, R. C., Past, A., MacGregor, J. T., and Hine, C. H.,** The rates of radical formation from the dipyridilium herbicides, paraquat, diquat and morfamquat in homogenates of rat lung, kidney and liver: an inhibitory effect of carbon monoxide, *Toxicol. Appl. Pharmacol.*, 32, 298, 1975.

217. **Sandy, M. S., Moldeus, P., Ross, D., and Smith, M. T.,** Role of redox cycling and lipid peroxidation in bipyridyl herbicide cytotoxicity. Studies with a compromised isolated hepatocyte model system, *Biochem. Pharmacol.*, 35, 3095, 1986.
218. **Hassan, H. M. and Fridovich, I.,** Superoxide radical and the oxygen enhancement of the toxicity of paraquat in *Escherichia coli*, *J. Biol. Chem.*, 253, 8143, 1978.
219. **Hassan, M. and Fridovich, I.,** Paraquat and *Escherichia coli*. Mechanism of production of extracellular superoxide radical, *J. Biol. Chem.*, 254, 10846, 1979.
220. **Hassan, H. M. and Moody, C. S.,** Superoxide dismutase protects against paraquat-mediated dioxygen toxicity and mutagenicity: studies in Salmonella typhimurium, *Can. J. Physiol. Pharmacol.*, 60, 1367, 1982.
221. **Yonei, S., Noda, A., Tachibana, A., and Akasaka, S.,** Mutagenic and cytotoxic effects of oxygen free radicals generated by methylviologen (paraquat) of Escherichia coli with different DNA-repair capacities, *Mutat. Res.*, 163, 15, 1986.
222. **Horton, J. K., Brigelius, R., Mason, R. P., and Bent, J. R.,** Paraquat uptake into freshly isolated rabbit lung epithelial cells and its reduction to the paraquat radical under anaerobic conditions, *Mol. Pharmacol.*, 29, 484, 1986.
223. **Bagley, A. C., Krall, J., and Lynch, R. E.,** Superoxide mediates the toxicity of paraquat for Chinese hamster ovary cells, *Proc. Natl. Acad. Sci. U.S.A.*, 83, 9189, 1986.
224. **Rabinowitch, H. D., Privalle, C. T., and Fridovich, I.,** Effect of paraquat on the green alga Dunaliella salina: protection by the mimic of superoxide dismutase, desferal-Mn(IV), *Free Rad. Biol. Med.*, 3, 125, 1987.
225. **Sinha, B. K., Singh, Y., and Krishna, C.,** Formation of superoxide and hydroxyl radicals from 1-methyl-4-phenylpyridinium ion (MPP^+): reductive activation by NADPH cytochrome P-450 reductase, *Biochem. Biophys. Res. Commun.*, 135, 583, 1986.
226. **Cojocel, C., Hannemann, J., and Baumann, K.,** Cephaloridine-induced lipid peroxidation initiated by reactive oxygen species as a possible mechanism of cephaloridine nephrotoxicity, *Biochim. Biophys. Acta*, 834, 402, 1985.
227. **Youngman, R. J. and Elstner, E. F.,** Oxygen species in paraquat toxicity: the crypto-OH radical, *FEBS Lett.*, 129, 265, 1981.
228. **Sutton, H. C. and Winterbourn, C. C.,** Chelated iron-catalyzed OH· formation from paraquat radicals and H_2O_2: mechanism of formate oxidation, *Arch. Biochem. Biophys.*, 235, 106, 1984.
229. **Winterbourn, C. C. and Sutton, H. C.,** HO· Production from H_2O_2 and enzymatically generated paraquat radicals: catalytic requirements and oxygen dependence, *Arch. Biochem. Biophys.*, 235, 116, 1984.
230. **Vile, G. F., Winterbourn, C. C., and Sutton, H. C.,** Radical-driven Fenton reactions: studies with paraquat, adriamycin, and anthraquinone-6-sulfonate and citrate, ATP, ADP, and pyrophosphate iron chelates, *Arch. Biochem. Biophys.*, 259, 616, 1987.
231. **Sandy, M. S., Moldeus, P., Ross, D., and Smith, M. T.,** Cytotoxicity of the redox cycling compound diquat in isolated hepatocytes: involvement of hydrogen peroxide and transition metals, *Arch. Biochem. Biophys.*, 259, 29, 1987.
232. **Thomas, C. E. and Aust, S. D.,** Reductive release of iron from ferritin by cation free radicals of paraquat and other bipyridines, *J. Biol. Chem.*, 261, 13064, 1986.
233. **Mason, R. P. and Holtzman, J. L.,** The mechanism of microsomal and mitochondrial nitroreductase. Electron spin resonance evidence for nitroaromatic free radical intermediates, *Biochemistry*, 14, 1626, 1975.
234. **Mason, R. P. and Holtzman, J. L.,** The role of catalytic superoxide formation in the O_2 inhibition of nitroreductase, *Biochem. Biophys. Res. Commun.*, 67, 1267, 1975.
235. **Sasame, H. A. and Boyd, M. R.,** Superoxide and hydrogen peroxide production and NADPH oxidation stimulated by nitrofurantoin in lung microsomes: possible implications for toxicity, *Life Sci.*, 24, 1091, 1979.
236. **Sealy, R. C., Swartz, H. M., and Olive, P. L.,** Electron spin resonance-spin trapping. Detection of superoxide formation during aerobic microsomal reduction of nitro-compounds, *Biochem. Biophys. Res. Commun.*, 82, 680, 1978.
237. **Docampo, R., Mason, R. P., Mottley, C., and Muntz, R. P. A.,** Generation of free radicals induced by nifurtimox in mammalian tissues, *J. Biol. Chem.*, 256, 10930, 1981.
238. **Docampo, R., Moreno, S. N. J., and Stoppani, A. O. M.,** Nitrofuran enhancement of microsomal electron transport, superoxide anion production and lipid peroxidation, *Arch. Biochem. Biophys.*, 207, 316, 1981.
239. **Moreno, S. N. J., Docampo, R., Mason, R. P., Leon, W., and Stoppani, A. O. M.,** Different behaviors of benznidazole as a free radical generator with mammalian and *Trypanosoma cruzi* microsomal preparations, *Arch. Biochem. Biophys.*, 218, 585, 1982.
240. **Masana, M., de Toranto, E. G. D., and Castro, J. A.,** Reductive metabolism and activation of benznidazole, *Biochem. Pharmacol.*, 33, 1041, 1984.

241. **Moreno, S. N. J., Mason, R. P., and Docampo, R.,** Reduction of nifurtimox and nitrofurantoin to free radical metabolites by rat liver mitochondria. Evidence of an outer membrane-located nitroreductase, *J. Biol. Chem.*, 259, 6298, 1984.
242. **Josephy, P. D., Palcic, B., and Skarsgard, L. D.,** Reduction of misonidazole and its derivatives by xanthine oxidase, *Biochem. Pharmacol.*, 30, 849, 1981.
243. **Rosen, G. M., Rauckman, E. J., Wilson, R. L., and Tschanz, C.,** Production of superoxide during the metabolism of nitrazepam, *Xenobiotica*, 14, 767, 1984.
244. **Yarlett, N., Rowlands, C. C., Evans, J. C., Yarlett, N. C., and Lloyd, D.,** Nitroimidazole- and oxygen-derived radicals detected by electron spin resonance in hydrogenosomal and cytosolic fractions from *Tritrichomonas vaginalis, Mol. Biochem. Parasitol.*, 24, 255, 1987.
245. **Docampo, R. and Stoppani, A. O. M.,** Generation of superoxide anion and hydrogen peroxide induced by nifurtimox in *Trypanosoma cruzi, Arch. Biochem. Biophys.*, 197, 317, 1979.
246. **Martin, W. J., Powis, G. W., and Kachel, D. L.,** Nitrofurantoin-stimulated oxidant production in pulmonary endothelial cells, *J. Lab. Clin. Med.*, 105, 23, 1985.
247. **Moreno, S. N. J., Mason, R. P., Muniz, R. P. A., Cruz, F. S., and Docampo, R.,** Generation of free radicals from metronidazole and other nitroimidazoles by Tritrichomonas foetus, *J. Biol. Chem.*, 258, 4051, 1983.
248. **Lloyd, D. and Pederson, J. Z.,** Metronidazole radical anion generation in vivo in Tritrichomonas vaginalis: oxygen quenching is enhanced in a drug-resistant strain, *J. Gen. Microbiol.*, 131, 87, 1985.
249. **Samuni, A., Bump, E. A., Mitchell, J. B., and Brown, J. M.,** Enhancement of misonidazole cytotoxicity by iron, *Int. J. Rad. Biol.*, 49, 77, 1986.
250. **Rao, D. N. R. and Mason, R. P.,** Generation of nitro radical anions of some 5-nitrofurans, 2- and 5-nitroimidazoles by norepinephrine, dopamine, and serotonin. A possible mechanism for neurotoxicity caused by nitroheterocyclic drugs, *J. Biol. Chem.*, 262, 11731, 1987.
251. **Dershwitz, M. and Novak, R. F.,** Generation of superoxide via the interaction of nitrofurantoin with oxyhemoglobin, *J. Biol. Chem.*, 257, 75, 1982.
252. **Buzzard, J. A. and Kopko, F.,** The flavin requirement and some inhibition characteristics of rat tissue glutathione reductase, *J. Biol. Chem.*, 238, 464, 1963.
253. **Dubin, M., Moreno, S. N. J., Martino, E. E., Docampo, R., and Stoppani, A. O. M.,** Increased biliary secretion and loss of hepatic glutathione in rat liver after nifurtimox treatment, *Biochem. Pharmacol.*, 32, 483, 1983.
254. **Kubow, S., Dubose, C. M., Janzen, E. G., Carlson, J. R., and Bray, T. M.,** The spin-trapping of enzymatically and chemically catalyzed free radicals from indolic compounds, *Biochem. Biophys. Res. Commun.*, 114, 168, 1983.
255. **Kubow, S., Janzen, E. G., and Bray, T. M.,** Spin-trapping of free radicals formed during in vitro and in vivo metabolism of 3-methylindole, *J. Biol. Chem.*, 259, 4447, 1984.
256. **Kubow, S., Bray, T. M., and Janzen, E. G.,** Spin-trapping studies on the effects of vitamin E and glutathione on free radical production induced by 3-methylindole, *Biochem. Pharmacol.*, 34, 1117, 1985.
257. **Maridonneau-Parini, J., Mirabelli, F., Richelmi, P., and Bellomo, G.,** Cytotoxicity of phenazine methosulfate in isolated rat hepatocytes is associated with superoxide anion production, thiol oxidation and alterations in intracellular calcium ion homeostasis, *Toxicol. Lett.*, 31, 175, 1986.
258. **Augusto, O., Alves, M. J. M., Colli, W., Filardi, L. S., and Brener, Z.,** Primaquine can mediate hydroxyl radical generation by Trypanosoma cruzi extracts, *Biochem. Biophys. Res. Commun.*, 135, 1029, 1986.
259. **Ames, J. R., Ryan, M. D., Klayman, D. L., and Kovacic, P.,** Charge transfer and oxy radicals in antimalarial action. Quinones, dapsone metabolites, metal complexes, iminium ions, and peroxides, *J. Free Rad. Biol. Med.*, 1, 353, 1985.
260. **Docampo, R., Moreno, S. N. J., and Mason, R. P.,** Generation of free radical metabolites and superoixde anion by the calcium indicators arsenazo III, antipyrylazo III, and murexide in rat liver microsomes, *J. Biol. Chem.*, 258, 14920, 1983.
261. **Moreno, S. N. J., Mason, R. P., and Docampo, R.,** Ca^{2+} and Mg^{2+}-enhanced reduction of arsenazo III to its anion free radical metabolite and generation of superoxide anion by an outer mitochondrial membrane azoreductase, *J. Biol. Chem.*, 259, 14609, 1984.
262. **Moreno, S. N. J., Mason, R. P., and Docampo, R.,** Reduction of the metallochromic indicators arsenazo III and antipyrylazo III to their free radical metabolites by cytoplasmic enzymes, *FEBS Lett.*, 180, 229, 1985.
263. **Moreno, S. N. J. and Docampo, R.,** Reduction of the metallochromic indicators murexide and tetramethylmurexide to their free radical metabolites by cytoplasmic enzymes and reducing agents, *Chem.-Biol. Interact.*, 57, 17, 1986.
264. **Docampo, R., Moreno, S. N. J., Muniz, R. P. A., Cruz, F. S., and Mason, R. P.,** Light-enhanced free radical formation and trypanocidal action of gentian violet (crystal violet), *Science*, 220, 1292, 1983.
265. **Teo, S., Pohl, L., and Halpert, J.,** Production of superoxide anion radicals during the oxidative metabolism of amino-chloramphenicol, *Biochem. Pharmacol.*, 35, 4584, 1986.

266. **Heikkila, R. E. and Cohen, G.,** 6-Hydroxydopamine: evidence for superoxide radical as an oxidative intermediate, *Science,* 181, 456, 1973.
267. **Cohen, G., Heikkila, R. E., and Macnamee, D.,** The generation of hydrogen peroxide, superoxide radical, and hydroxyl radical by 6-hydroxydopamine, dialuric acid, and related cytotoxic agents, *J. Biol. Chem.,* 249, 2447, 1974.
268. **Houee-Levin, C., Gardes-Albert, M., Ferradini, C., and Pucheault, J.,** Radiolysis study of the alloxan-dialuric acid couple. II. The autoxidation of dialuric acid, *Radiat. Res.,* 88, 20, 1981.
269. **Houee, C., Gardes, M., Pucheault, J., and Ferradini, C.,** Radical chemistry of alloxan-dialuric acid. Role of the superoxide radical, *Clin. Respir. Physiol.,* 17 (Suppl.), 21, 1981.
270. **Eyer, P. and Lengfelder, E.,** Radical formation during autoxidation of 4-dimethylaminophenol and some properties of the reaction products, *Biochem. Pharmacol.,* 33, 1005, 1984.
271. **Fatur, D. J., Sau, R. H. C., Davison, A. J., and Stich, H. F.,** Chemiluminescence in the oxidation of 6-hydroxydopamine: effects of lucigenin, scavengers of active oxygen, metal chelators and the presence of oxygen, *Photochem. Photobiol.,* 45, 413, 1987.
272. **Aver'yanov, A. A.,** Generation of superoxide anion radicals and hydrogen peroxide in the autoxidation of caffeic acid, *Biokhimia,* 46, 256, 1981.
273. **Aver'yanov, A. A.,** Generation of superoxide anion radical by phenols, *Biol. Nauki, (Moscow),* 39, N 7, 1984.
274. **Aver'yanov, A. A. and Ismailov, A. I.,** Superoxide radical involvement in gossypol autoxidation, *Biol. Nauki (Moscow),* N5, 76, 1986.
275. **Misra, H. P.,** Generation of superoxide free radical during the autoxidation of thiols, *J. Biol. Chem.,* 249, 2151, 1974.
276. **Munday, R.,** Toxicity of aromatic disulphides. I. Generation of superoxide radical and hydrogen peroxide by aromatic disulphides in vitro, *J. Appl. Toxicol.,* 5, 402, 1985.
277. **Saez, G., Thornalley, P. J., Hill, H. A. O., Hems, R., and Bannister, J. V.,** The production of free radicals during the autoxidation of cysteine and their effect on isolated rat hepatocytes, *Biochim. Biophys. Acta,* 719, 24, 1982.
278. **Munday, R.,** Studies on the mechanism of toxicity of the mycotoxin, sporidemin. I. Generation of superoxide radical by sporidemin, *Chem.-Biol. Interact.,* 41, 361, 1982.
279. **Munday, R.,** Generation of superoxide radical and hydrogen peroxide by 1,2,4-triaminobenzene, a mutagenic and myotoxic aromatic amine, *Chem.-Biol. Interact.,* 60, 171, 1986.
280. **Davis, G. and Thornalley, P. J.,** Free radical production from the aerobic oxidation of reduced pyridine nucleotides catalysed by phenazine derivatives, *Biochim. Biophys. Acta,* 724, 456, 1983.
281. **Michelson, A. M.,** Pro-oxidant effect of superoxide dismutase in the oxidation of rifamycin SV, *Free Rad. Res. Commun.,* 1, 185, 1986.
282. **Bandy, B. and Davison, A. J.,** Interactions between metals, ligands and oxygen in the autoxidation of 6-hydroxydopamine: mechanisms by which metal chelation enhances inhibition by superoxide dismutase, *Arch. Biochem. Biophys.,* 259, 305, 1987.
283. **Aver'yanov, A. A. and Lapikova, V. P.,** Involvement of active forms of oxygen in the mechanism of ferulic acid toxicity, *Izv. Akad. Nauk SSSR Ser. Biol.,* 521, 1985.
284. **Davison, A. J., Wilson, B. D., and Belton, P.,** Deterioration of axonal membranes induced by phenolic pro-oxidants. Roles of superoxide radicals and hydrogen peroxide, *Biochem. Pharmacol.,* 33, 3887, 1984.
285. **Robertson, P., Jr., Fridovich, S. E., Misra, H. P., and Fridovich, I.,** Cyanide catalyzes the oxidation of α-hydroxyaldehydes and related compounds: monitored as the reduction of dioxygen, cytochrome c, and nitroblue tetrazolium, *Arch. Biochem. Biophys.,* 207, 282, 1981.
286. **Shiga, T. and Imaizumi, K.,** Electron spin resonance study on peroxidase- and oxidase-reactions of horse radish peroxidase and methemoglobin, *Arch. Biochem. Biophys.,* 167, 469, 1975.
287. **Lasker, J. M., Sivarajah, K., Mason, R. P., Kalyanaraman, B., Abou-Donia, M. B., and Eling, T. E.,** A free radical mechanism of prostaglandin synthase-dependent aminopyrine demethylation, *J. Biol. Chem.,* 256, 7764, 1981.
288. **Eling, T. E., Mason, R. P., and Sivarajah, K.,** The formation of aminopyrine cation radical by the peroxidase activity of prostaglandin H synthase and subsequent reactions of the radical, *J. Biol. Chem.,* 260, 1601, 1985.
289. **Wilson, I., Wardman, P., Gohen, G. M., and Doherty, M. D. A.,** Reductive role of glutathione in the redox cycling of oxidizable drugs, *Biochem. Pharmacol.,* 35, 21, 1986.
290. **Kalyanaraman, B. and Sealy, R. C.,** Electron spin resonance-spin stabilization in enzymatic systems: detection of semiquinones produced during peroxidative oxidation of catechols and catecholamines, *Biochem. Biophys. Res. Commun.,* 106, 1119, 1982.
291. **West, P. R., Harman, L. S., Josephy, P. D., and Mason, R. P.,** Acetaminophen: enzymatic formation of a transient phenoxyl free radical, *Biochem. Pharmacol.,* 33, 2933, 1984.
292. **Ross, D. A., Albano, E., Nilsson, U., and Moldeus, P.,** Thiyl radicals—formation during peroxidase-catalyzed metabolism of acetaminophen in the presence of thiols, *Biochem. Biophys. Res. Commun.,* 125, 109, 1984.

293. **Potter, D. W. and Hinson, J. A.,** The 1- and 2-electron oxidation of acetaminophen catalyzed by prostaglandin H synthase, *J. Biol. Chem.*, 262, 974, 1987.
294. **Ross, D., Mehlhorn, R. J., Moldeus, P., and Smith, M. T.,** Metabolism of diethylstilbestrol by horseradish peroxidase and prostaglandin-H synthase. Generation of a free radical intermediate and its interaction with glutathione, *J. Biol. Chem.*, 260, 16210, 1985.
295. **Epe, B., Schiffmann, D., and Metzler, M.,** Possible role of oxygen radicals in cell transformation by diethylstilbestrol and related compounds, *Carcinogenesis*, 7, 1329, 1986.
296. **Baarnhielm, C. and Hansson, G.,** Oxidation of 1,4-dihydropyridines by prostaglandin synthase and the peroxidic function of cytochrome P-450. Demonstration of a free radical intermediate, *Biochem. Pharmacol.*, 35, 1419, 1986.
297. **Haim, N., Roman, J., Nemec, J., and Sinha, B. K.,** Peroxidative free-radical formation and O-demethylation of etoposide (VP-16) and teniposide (VM-26), *Biochem. Biophys. Res. Commun.*, 135, 215, 1986.
298. **Haim, N., Nemec, J., Roman, J., and Sinha, B. K.,** In vitro metabolism of etoposide (VP-16-213) by liver microsomes and irreversible binding of reactive intermediates to microsomal proteins, *Biochem. Pharmacol.*, 36, 527, 1987.
299. **Mottley, C. and Mason, R. P.,** An electron spin resonance study of free radical intermediates in the oxidation of indole acetic acid by horse radish peroxidase, *J. Biol. Chem.*, 261, 16860, 1986.
300. **Riley, P. A., Ed.,** *Hydroxyanisole. Recent Advances in Anti-Melanoma Therapy*, IRL Press, Oxford, 1984.
301. **Van De Straat, R., Bromans, R. M., Bosman, P., De Vries, J., and Vermeulen, N. P. E.,** Cytochrome P-450-mediated oxidation of substrates by electron-transfer; role of oxygen radicals and 1- and 2-electron oxidation of paracetamol, *Chem.-Biol. Interact.*, 64, 267, 1988.
302. **Peyster, A. D., Quintanilna, A., Packer, L., and Smith, M. T.,** Oxygen radical formation induced by gossypol in rat liver microsomes and human sperm, *Biochem. Biophys. Res. Commun.*, 118, 573, 1984.
303. **Manno, M., Ioannides, C., and Gibson, G. G.,** The modulation by arylamines of the in vitro formation of superoxide anion radicals and hydrogen peroxide by rat liver microsomes, *Toxicol. Lett.*, 25, 121, 1985.
304. **Nagata, C., Kodama, M., Kimura, T., and Nakayama, T.,** Mechanism of metabolic activation of carcinogenic aromatic amines, *Gann Monogr. Cancer Res.*, 30, 93, 1985.
305. **Noda, A., Noda, H., Ohno, K., Sendo, T., Misaka, A., Kanazawa, Y., Isobe, R., and Hirata, M.,** Spin-trapping of a free radical intermediate formed during microsomal metabolism of hydrazine, *Biochem. Biophys. Res. Commun.*, 133, 1086, 1985.
306. **Roy, D. and Liehr, J. G.,** Temporary decrease in renal quinone reductase activity induced by chronic administration of estradiol to male Syrian hamsters. Increased superoxide formation by redox cycling of estrogen, *J. Biol. Chem.*, 203, 3646, 1988.
307. **Beckman, J. S. and Siedow, J. N.,** Bactericidal agents generated by the peroxidase-catalysed oxidation of para-hydroquinones, *J. Biol. Chem.*, 260, 14604, 1985.
308. **Kyle, M. E., Miccadei, S., Nakae, D., and Farber, J. L.,** Superoxide dismutase and catalase protect cultured hepatocytes from the cytotoxicity of acetaminophen, *Biochem. Biophys. Res. Commun.*, 149, 889, 1987.
309. **Issels, R. D., Biaglow, J. E., and Gerweck, L. E.,** Role of activated oxygen species in the mechanism of cytotoxicity of cysteamine, in *Oxygen Radicals in Chemistry and Biology*, Bors, W., Saran, M., and Tait, D., Eds., Walter de Gruyter, Berlin, 1984, 665.
310. **Shroot, B. and Brown, C.,** Free radicals in skin exposed to dithranol and its derivatives, *Arzneim. Forsch.*, 36-2, 1253, 1986.
311. **Oberley, L. W. and Buettner, G. R.,** The production of hydroxyl radical by bleomycin and Iron (II), *FEBS Lett.*, 97, 47, 1979.
312. **Sugiura, Y.,** Production of free radicals from phenol and tocopherol by bleomycin-iron(II) complex, *Biochem. Biophys. Res. Commun.*, 87, 649, 1979.
313. **Buettner, G. R. and Oberley, L. W.,** The production of hydroxyl radical by tallysomycin and copper(II), *FEBS Lett.*, 101, 333, 1979.
314. **Sugiura, Y.,** The production of hydroxyl radical from copper(I) complex systems of bleomycin and tallysomycin: comparison with copper(II) and iron(II) systems, *Biochem. Biophys. Res. Commun.*, 90, 375, 1979.
315. **Melnyk, D. L., Horwitz, S. B., and Peisach, J.,** Redox potential of iron-bleomycin, *Biochemistry*, 20, 5327, 1981.
316. **Burger, R. M., Horwitz, S. B., Peisach, J., and Wittenberg, J. B.,** Oxygenated iron bleomycin. A short-lived intermediate in the reaction of ferrous bleomycin with O_2, *J. Biol. Chem.*, 254, 12299, 1979.
317. **Burger, R. M., Blanchard, J. S., Horwitz, B., and Peisach, J.,** The redox state of activated bleomycin, *J. Biol. Chem.*, 260, 15406, 1985.
318. **Antholine, W. E. and Petering, D. H.,** On the reaction of iron bleomycin with thiols and oxygen, *Biochem. Biophys. Res. Commun.*, 90, 384, 1979.
319. **Sugiura, Y.,** Bleomycin-iron complexes. Electron spin resonance study, ligand effect, and implication for active mechanism, *J. Am. Chem. Soc.*, 102, 5208, 1980.

320. **Albertini, J.-P., Garnier-Suillerot, A., and Tosi, L.,** Iron-bleomycin-DNA-system. Evidence of a long-lived bleomycin-iron-oxygen intermediate, *Biochem. Biophys. Res. Commun.*, 104, 557, 1982.
321. **Goldstein, S. and Czapski, G.,** Oxidation-reduction reactions of iron bleomycin in the absence and presence of DNA, *Free Rad. Res. Commun.*, 2, 259, 1987.
322. **Giloni, L., Takeshita, M., Johnson, F., Iden, C., and Grollman, A. P.,** Bleomycin-induced strand scission of DNA. Mechanism of deoxyribose cleavage, *J. Biol. Chem.*, 256, 8608, 1981.
323. **Wu, J. C., Kozarich, J. W., and Stubbe, J.,** Mechanism of bleomycin: evidence for a rate-determining 4'-hydrogen abstraction from poly(dA-dU) associated with the formation of both free base propenals, *Biochemistry*, 24, 7562, 1985.
324. **McGall, G. H., Rabow, L. E., Stubbe, J., and Kozarich, J. W.,** Incorporation of ^{18}O into glycolic acid obtained from the bleomycin-mediated degradation of DNA. Evidence for 4'-radical trapping by $^{18}O_2$, *J. Am. Chem. Soc.*, 109, 2836, 1987.
325. **Lown, J. W. and Sim, S.,** The mechanism of the bleomycin-induced cleavage of DNA, *Biochem. Biophys. Res. Commun.*, 77, 1150, 1977.
326. **Sugiura, Y., Suzuki, T., Kuwahara, J., and Tanaka, H.,** On the mechanism of hydrogen peroxide-, superoxide- and ultraviolet light-induced DNA cleavages of inactive bleomycin-iron(III) complex, *Biochem. Biophys. Res. Commun.*, 105, 1511, 1982.
327. **Scheulen, M. E. and Kappus, H.,** The activation of oxygen by bleomycin is catalyzed by NADPH-cytochrome P-450 reductase in the presence of iron ions and NADPN, in *Oxygen Radicals in Chemistry and Biology*, Bors, W., Saran, M., and Tait, D., Eds., Walter de Gruyter, Berlin, 1984, 425.
328. **Mahmutoglu, I. and Kappus, H.,** Oxy radical formation during redox cycling of the bleomycin-iron(III) complex by NADPH-cytochrome P-450 reductase, *Biochem. Pharmacol.*, 34, 3091, 1985.
329. **Kappus, H., Mahmutoglu, I., Kostrucha, J., and Scheulen, M. E.,** Liver nuclear NADPH-cytochrome P-450 reductase may be involved in redox cycling of bleomycin-iron(III), oxy radical formation and DNA damage, *Free Rad. Res. Commun.*, 2, 271, 1987.
330. **Mahmutoglu, I. and Kappus, H.,** Redox cycling of bleomycin-Fe(III) by an NADH-dependent enzyme, and DNA damage in isolated rat liver nuclei, *Biochem. Pharmacol.*, 36, 3677, 1987.
331. **Gutteridge, J. M. C. and Fu, X.-C.,** Enhancement of bleomycin-iron free radical damage to DNA by antioxidants and their inhibition of lipid peroxidation, *FEBS Lett.*, 123, 71, 1981.
332. **Ekimoto, H., Takahashi, K., Matsuda, A., Takita, T., and Umezawa, H.,** Lipid peroxidation by bleomycin-iron complexes in vitro, *J. Antibiot.*, 38, 1077, 1985.
333. **Gutteridge, J. M. C. and Fu, X.-C.,** Protection of iron-catalyzed free radical damage to DNA and lipids by copper(II)-bleomycin, *Biochem. Biophys. Res. Commun.*, 99, 1354, 1981.
334. **Conley, N. S., Yarbro, J. W., Ferrari, H. A., and Zeidler, R. B.,** Bleomycin increases superoxide anion generation by pig peripheral alveolar macrophages, *Mol. Pharmacol.*, 30, 48, 1986.
335. **Reynolds, E. S. and Mosleu, M. T.,** Free radical damage in liver, in *Free Radicals in Biology*, Vol. 4, Pryor, W. A., Ed., Academic Press, New York, 1980, 49.
336. **Albano, E., Tomasi, A., Cheeseman, K. H., Vannini, V., and Dianzani, M. U.,** Use of isolated hepatocytes for the detection of free radical intermediates of halogenated hydrocarbons, in *Free Radicals in Liver Injury*, Poli, G., Cheeseman, K. H., Dianzani, M. U., and Slater, T. F., Eds., IRL Press, Oxford, 1985, 7.
337. **Tomasi, A., Albano, E., Banni, S., Botti, B., Corongiu, F., Dessi, M. A., Iannone, A., Vanini, V., and Dianzini, M. U.,** Free radical metabolism of carbon tetrachloride in rat liver mitochondria. A study of the mechanism of activation, *Biochem. J.*, 246, 313, 1987.
338. **Junqueira, V. B. C., Simizu, K., Videla, L. A., and Barros, S. B. D.,** Dose-dependent study of the effects of acute lindane administration on rat liver superoxide anion production, antioxidant enzyme activities and lipid peroxidation, *Toxicology*, 41, 193, 1986.
339. **Urquhart, A. J. and Elder, G. H.,** Hexachlorobenzene-induced oxygen activation by mouse liver microsomes: comparison with phenobarbitone and 20-methylcholanthrene, *Biochem. Pharmacol.*, 36, 3795, 1987.
340. **Montellano, P. R. O. and Watanabe, M. D.,** Free radical pathways in the in vitro hepatic metabolism of phenelzine, *Mol. Pharmacol.*, 31, 213, 1987.
341. **Goldberg, I. H.,** Free radical mechanisms in neocarzinostatin-induced DNA damage, *Free Rad. Biol. Med.*, 3, 41, 1987.
342. **Walker, P. D. and Shah, S. V.,** Evidence suggesting a role for hydroxyl radical in gentamicin-induced acute renal failure in rats, *J. Clin. Invest.*, 81, 334, 1988.
343. **Balny, C. and Douzou, P.,** Production of superoxide ions by photosensitization of dyes, *Biochem. Biophys. Res. Commun.*, 56, 386, 1974.
344. **Martin, J. P. and Logsdon, N.,** Oxygen radicals mediate cell inactivation by acridine dyes, fluorescein, and lucifer yellow CH, *Photochem. Photobiol.*, 46, 45, 1987.
345. **Jahnke, L. S. and Frenkel, A. W.,** Evidence for the photochemical production of superoxide mediated by saponified chlorophyll, *Biochem. Biophys. Res. Commun.*, 66, 144, 1975.

346. **Nishikimi, M., Yamada, H., and Yagi, K.,** Generation of superoxide anion with a photoreduced phenoxazine, *Photochem. Photobiol.*, 27, 269, 1978.
347. **Felix, C. C., Hyde, J. S., Sarna, T., and Sealy, R. C.,** Melanin photoreactions in aerated media: electron spin resonance evidence for production of superoxide and hydrogen peroxide, *Biochem. Biophys. Res. Commun.*, 84, 335, 1978.
348. **Buettner, G. R. and Oberley, L. W.,** The apparent production of superoxide and hydroxyl radicals by hematoporphyrin and light as seen by spin-trapping, *FEBS Lett.*, 121, 161, 1980.
349. **Buettner, G. R. and Hall, R. D.,** Superoxide, hydrogen peroxide and singlet oxygen in hematoporphyrin derivative-cysteine, -NADH and light systems, *Biochim. Biophys. Acta*, 923, 501, 1987.
350. **Carmichael, A. J., Mossoba, M. M., and Riesz, P.,** Photogeneration of superoxide by adriamycin and daunomycin, *FEBS Lett.*, 164, 401, 1983.
351. **Carmichael, A. J., Samuni, A., and Riesz, P.,** Photogeneration of superoxide and decarboxylated peptide radicals by carboquone, mitomycin-C and streptonigrin—an electron spin resonance and spin trapping study, *Photochem. Photobiol.*, 41, 635, 1985.
352. **Martin, J. P., Jr., Colina, K., and Logsdon, N.,** Role of oxygen radicals in the phototoxicity of tetracyclines towards Escherichia coli B, *J. Bacteriol.*, 169, 2516, 1987.
353. **Joshi, P. C. and Pathak, M. A.,** Production of singlet oxygen and superoxide radicals by psoralens and their biological significance, *Biochem. Biophys. Res. Commun.*, 112, 638, 1983.
354. **Pathak, M. A. and Joshi, P. C.,** Production of active oxygen species ($^1O_2$ and O_2^-) by psoralens and ultraviolet radiation (320—400 nm), *Biochim. Biophys. Acta*, 798, 115, 1984.
355. **Afanas'ev, I. B., Korkina, L. G., Suslova, T. B., and Soodaeva, S. K.,** Are quinones producers or scavengers of superoxide ion in cell?, *Arch. Biochem. Biophys.*, 281, 245, 1990.
356. **Przybyszewski, W. M. and Malec, J.,** Hydroxyurea, methotrexate, and adriblastine can mediate nonenzymatic reduction of nitroblue tetrazolium with NADH which is inhibited by superoxide dismutase, *Biochem. Pharmacol.*, 36, 3312, 1987.
357. **Nohl, H.,** Identification of the site of adriamycin-activation in the heart cell, *Biochem. Pharmacol.*, 37, 2633, 1988.
358. **Rumyantseva, G. V. and Weiner, L. M.,** Redox transformations of quinone antitumor drug in liver microsomes, *FEBS Lett.*, 234, 459, 1988.
359. **Rajagopalan, S., Politi, P. M., Sinha, B. K., and Myers, C. E.,** Adriamycin-induced free radical formation in the perfused rat heart: implications for cardiotoxicity, *Cancer Res.*, 48, 4766, 1988.
360. **Pedersen, J. Z., Marcocci, L., Rossi, L., Mavelli, I., and Rotilio, G.,** First electron spin resonance evidence for the generation of the duanomycin free radical and superoxide by red blood cell membranes, *Ann. N.Y. Acad. Sci.*, 551, 121, 1988.
361. **Minotti, G.,** Adriamycin-dependent release of iron from microsomal membranes, *Arch. Biochem. Biophys.*, 268, 398, 1989.
362. **Monteiro, H. P., Vile, G. F., and Winterbourn, C. C.,** Release of iron from ferritin by semiquinone, anthracycline, bipyridyl, and nitroaromatic radicals, *Free Rad. Biol. Med.*, 6, 587, 1989.
363. **Griffin-Green, E. A., Zaleska, M. M., and Erecinska, M.,** Adriamycin-induced lipid peroxidation in mitochondria and microsomes, *Biochem. Pharmacol.*, 37, 3071, 1988.
364. **Vile, G. F. and Winterbourn, C. C.,** Adriamycin-dependent peroxidation of rat liver and heart microsomes catalyzed by iron chelates and ferritin. Maximum peroxidation at low oxygen partial pressures, *Biochem. Pharmacol.*, 37, 2893, 1988.
365. **Vile, G. F. and Winterbourn, C. C.,** Microsomal reduction of low-molecular-weight Fe^{3+} chelates and ferritin enhancement by adriamycin, paraquat, menadione, and anthraquinone 2-sulfonate and inhibition by oxygen, *Arch. Biochem. Biophys.*, 267, 606, 1988.
366. **Geetha, A., Sankar, R., and Devi, C. S. S.,** Effect of α-tocopherol on peroxidative membrane damage caused by doxorubicin: an in vitro study in human erythrocytes, *Ind. J. Exp. Biol.*, 27, 274, 1989.
367. **Thayer, W. S.,** Evaluation of tissue indicators of oxidative stress in rats treated chronically with adriamycin, *Biochem. Pharmacol.*, 37, 2189, 1988.
368. **Praet, M., Calderon, P. B., Pollakis, G., Roberfroid, M., and Ruysschaert, J. M.,** A new class of free radical scavengers reducing adriamycin mitochondrial toxicity, *Biochem. Pharmacol.*, 37, 4617, 1988.
369. **Hasinoff, B. B. and Davey, J. P.,** Adriamycin and its iron(III) and copper-(II) complexes. Glutathione-induced dissociation; cytochrome c oxidase inactivation and protection; binding to cardiolipin, *Biochem. Pharmacol.*, 37, 3663, 1988.
370. **Ganey, P. E., Kauffman, F. C., and Thurman, R. G.,** Oxygen-dependent hepatotoxicity due to doxorubicin: role of reducing equivalent supply in perfused rat liver, *Mol. Pharmacol.*, 34, 695, 1989.
371. **Keizer, H. G., Vanrijn, J., Pinedo, H. M., and Joenje, H.,** Effect of endogenous glutathione, superoxide dismutase, catalase, and glutathione peroxidase on adriamycin tolerance of Chinese hamster ovary cells, *Cancer Res.*, 48, 4495, 1988.
372. **Cervantes, A., Pinedo, H. M., Lankeima, J., and Schuurhuis, G. J.,** The role of oxygen-derived free radicals in the cytotoxicity of doxorubicin in multidrug resistant and sensitive human ovarian cancer cells, *Cancer Lett.*, 41, 169, 1988.

373. **Alegria, A. E., Samuni, A., Mitchell, J. B., Riesz, P., and Russo, A.,** Free radicals induced by adriamycin-sensitive and adriamycin-resistant cells: a spin-trapping study, *Biochemistry,* 28, 8653, 1989.
374. **Sinha, B. K., Dusre, L., Collins, C., and Myers, C. E.,** Resistance to paraquat and adriamycin in human breast tumor cells: role of free radical formation, *Biochim. Biophys. Acta,* 1010, 304, 1989.
375. **Mimnaugh, E. G., Dusre, L., Atwell, J., and Myers, C. E.,** Differential oxygen radical susceptibility of adriamycin-sensitive and adriamycin-resistant MCF-7 human breast tumor cells, *Cancer Res.,* 49, 8, 1989.
376. **Sinha, B. K., Mimnaugh, E. G., Rajagopalan, S., and Meyers, C. E.,** Adriamycin activation and oxygen free radical formation in human breast tumor cells: protective role of glutathione peroxidase in adriamycin resistance, *Cancer Res.,* 49, 3844, 1989.
377. **Grankvist, K., Steudahl, U., and Henriksson, R.,** Comparative study of demethoxydaunorubicin with other anthracyclines on generation of oxygen radicals and clonogenic survival of fibroblasts, *Pharmacol. Toxicol.,* 65, 40, 1989.
378. **Kolodziejczyk, P., Reszka, K., and Lown, J. W.,** Enzymatic oxidative activation of 5-iminodaunorubicin. Streptophotometric and electron paramagnetic resonance studies, *Biochem. Pharmacol.,* 38, 803, 1989.
379. **Bravo-Cuellar, A., Balercia, G., Levesque, J. P., Liu, X. H., Osculati, F., and Orbach-Arbouys, S.,** Enhanced activity of peritoneal cells after aclacinomycin injection: effect of pretreatment with superoxide dismutase on aclacinomycin-induced cytological alterations and antitumoral activity, *Cancer Res.,* 49, 1578, 1989.
380. **Napetschnig, S. and Sies, H.,** Generation of photoemissive species by mitomycin C redox cycling in rat liver microsomes, *Biochem. Pharmacol.,* 36, 1617, 1987.
381. **Politi, P. M., Rajagopalan, S., and Sinha, B. K.,** Free-radical formation by mitomycin C and its novel analogs in cardiac microsomes and the perfused rat heart, *Biochim. Biophys. Acta,* 992, 341, 1989.
382. **Pritsos, C. A., Keyes, S. R., and Sartorelli, A. C.,** Effect of the superoxide dismutase inhibitor, diethyldithiocarbamate, on the cytotoxicity of mitomycin antibiotics, *Cancer Biochem. Biophys.,* 10, 289, 1989.
383. **Mahmutoglu, I. and Kappus, H.,** Redox cycling of bleomycin-Fe(III) and DNA degradation by isolated NADH-cytochrome b_5 reductase: involvement of cytochrome b_5, *Mol. Pharmacol.,* 34, 578, 1988.
384. **Turner, M. J., Bazarth, C. H., and Strauss, K. E.,** Evidence for intracellular superoxide formation following the exposure of guinea pig enterocytes to bleomycin, *Biochem. Pharmacol.,* 38, 85, 1989.
385. **Moseley, P. L.,** Augmentation of bleomycin-induced DNA damage in intact cells, *Am. J. Physiol.,* 257, C882, 1989.
386. **Flitter, W. D. and Mason, R. P.,** The enzymatic reduction of actinomycin D to a free radical species, *Arch. Biochem. Biophys.,* 267, 632, 1988.
387. **Nomoto, K., Okabe, T., Suzuki, H., and Tanaka, N.,** Mechanism of action of lactoquinomycin A with special reference to the radical formation, *J. Antibiot.,* 41, 1124, 1988.
388. **Dalal, N. S. and Shi, X.,** On the formation of oxygen radicals by fredericamycin A and implications to its anticancer activity; an ESR investigation, *Biochemistry,* 28, 748, 1989.
389. **Sinha, B. K., Eliot, H. M., and Kalayanaraman, B.,** Iron-dependent hydroxyl radical formation and DNA damage from a novel metabolite to the clinically active antitumor drug VP-16, *FEBS Lett.,* 227, 240, 1988.
390. **Kalyanaraman, B., Nemec, J., and Sinha, B. K.,** Characterization of free radicals produced during oxidation of etoposide (VP-16) and its catechol and quinone derivatives. An ESR study, *Biochemistry,* 28, 4839, 1989.
391. **Rao, D. N., Takahashi, N., and Mason, R. P.,** Characterization of a glutathione conjugate of the 1,4-benzosemiquinone-free radical formed in rat hepatocytes, *J. Biol. Chem.,* 263, 17981, 1988.
392. **Moore, G. A., Rossi, L., Nicotera, P., Orrenius, S., and O'Brien, P. J.,** Quinone toxicity in hepatocytes: studies on mitochondrial Ca^{2+} release induced by benzoquinone derivatives, *Arch. Biochem. Biophys.,* 259, 283, 1987.
393. **Thor, H., Mirabelli, F., Salis, A., Cohen, G. M., Bellomo, G., and Orrenius, S.,** Alterations in hepatocyte cytoskeleton caused by redox cycling and alkylating quinones, *Arch. Biochem. Biophys.,* 262, 397, 1988.
394. **van Ommen, B., Voncken, J. W., Müller, F., and van Bladeren, P. J.,** The oxidation of tetrachloro-1,4-benzoquinone by microsomes and purified cytochrome P 450b. Implications for covalent binding to protein and involvement of reactive oxygen species, *Chem.-Biol. Interact.,* 65, 247, 1988.
395. **Livingstone, D. R., Martinez, P. G., and Winston, G. W.,** Menadione-stimulated oxyradical formation in digestive gland microsomes of the common mussel, Mytilus edulis L, *Aquat. Toxicol.,* 15, 213, 1989.
396. **Buffinton, G. D., Öllinger, K., Brunmark, A., and Cadenas, E.,** DT-Diaphorase catalyzed reduction of 1,4-naphthoquinone derivatives and glutathionyl-quinone conjugates. Effect of substituents on autoxidation rates, *Biochem. J.,* 257, 561, 1989.
397. **Öllinger, K., Llopis, J., and Cadenas, E.,** Study of the redox properties of naphthazarin (5,8-dihydroxy-1,4-naphthoquinone) and its glutathionyl conjugate in biological reactions: one- and two-electron enzymatic reduction, *Arch. Biochem. Biophys.,* 275, 514, 1989.

398. **Lea, J. S., Garner, H. J., Butler, J., Hoey, B. M., and Ward, T. H.,** The lack of correlation between toxicity and free radical formation of two diaziridinyl benzoquinones, *Biochem. Pharmacol.*, 37, 2023, 1988.
399. **Gutierrez, P. L.,** Mechanism(s) of bioreductive activation. The example of diaziquone (AZQ), *Free Rad. Biol. Med.*, 6, 405, 1989.
400. **Castro, G. D., Lopez, A., and Castro, J. A.,** Evidence for hydroxyl free radical formation during paraquat but not for nifurtimox liver microsomal biotransformation. A dimethylsulfoxide scavenging study, *Arch. Toxicol.*, 62, 355, 1988.
401. **Clejan, L. and Cederbaum, A. I.,** Synergistic interactions between NADPH-cytochrome P-450 reductase, paraquat and iron in the generation of active oxygen radicals, *Biochem. Pharmacol.*, 38, 1779, 1989.
402. **Sugimoto, H., Matsuzaki, S., Hamana, K., Nagamine, T., Yamada, S., Suzuki, M., and Kobayashi, S.,** Superoxide dismutase and α-tocopherol suppress the paraquat-induced elevation of N^1-acetylspermidine and putrescine in primary culture of adult rat hepatocytes, *Life Sci.*, 45, 2365, 1989.
403. **Tsokos-Kuhn, J. O.,** Lethal injury by diquat redox cycling in an isolated hepatocyte model, *Arch. Biochem. Biophys.*, 265, 415, 1988.
404. **Korytowski, W., Feix, C. C., and Kalyanaraman, B.,** Oxygen activation during the interaction between MPTP metabolites and synthetic neuromelamin—an ESR-spin trapping, optical, and oxidase electrode study, *Biochem. Biophys. Res. Commun.*, 154, 781, 1988.
405. **Rossetti, Z. L., Sotgiu, A., Sharp, D. E., Hadjiconstantinou, M., and Neff, N. H.,** 1-Methyl-4-phenyl-1,2,3,6-tetrahydropyridine (MPTP) and free radicals in vitro, *Biochem. Pharmacol.*, 37, 4573, 1988.
406. **Rao, D. N., Jordan, S., and Mason, R. P.,** Generation of nitro radical anions of some 5-nitrofurans, and 2- and 5-nitroimidazoles by rat hepatocytes, *Biochem. Pharmacol.*, 37, 2907, 1988.
407. **Washburn, P. C. and DiGiulio, R. T.,** Nitrofurantoin-stimulated superoxide production by channel catfish (Ictalurus punctatus) hepatic microsomal and soluble fractions, *Toxicol. Appl. Pharmacol.*, 95, 363, 1988.
408. **Washburn, P. C. and DiGiulio, R. T.,** Stimulation of superoxide production by nitrofurantoin, p-nitrobenzoic acid and m-dinitrobenzene in hepatic microsomes of three species of freshwater fish, *Environ. Toxicol. Chem.*, 8, 171, 1989.
409. **Rossi, L., Silva, J. M., McGirr, L. G., and O'Brien, P. J.,** Nitrofurantoin-mediated oxidative stress cytotoxicity in isolated rat hepatocytes, *Biochem. Pharmacol.*, 37, 3109, 1988.
410. **Farber, J. L., Leonard, T. B., Kyle, M. E., Nakae, D., Serroni, A., and Rogers, S. A.,** Peroxidation-dependent and peroxidation-independent mechanisms by which acetaminophen kills cultured rat hepatocytes, *Arch. Biochem. Biophys.*, 267, 640, 1988.
411. **Birge, R. B., Bartoloue, J. B., Nishanian, E. V., Bruno, M. K., Mangold, J. B., Cohen, S. D., and Khairallah, E. A.,** Dissociation of covalent binding from the oxidative effects of acetaminophen. Studies using dimethylated acetaminophen derivatives, *Biochem. Pharmacol.*, 37, 3383, 1988.
412. **Augusto, O., Schreiber, J., and Mason, R. P.,** Direct ESR detection of a free radical intermediate during the peroxidase-catalyzed oxidation of the antimalarial drug primaquine, *Biochem. Pharmacol.*, 37, 2791, 1988.
413. **Fletcher, K. A., Barton, P. F., and Kelly, J. A.,** Studies on the mechanisms of oxidation in the erythrocyte by metabolites of primaquine, *Biochem. Pharmacol.*, 37, 2683, 1988.
414. **Wataya, Y., Yamana, K., Hiramoto, K., Ohtsuka, Y., Okuba, Y., Negishi, K., and Hayatsu, H.,** Generation of intracellular active oxygens in mouse FM3A cells by 3-hydroxyamino-1-methyl-5H-pyrido 4,3-b indole, the activated Trp-P-Z, *Gann*, 79, 576, 1988.
415. **Albano, E. and Tomasi, A.,** Spin trapping of free radical intermediates produced during the metabolism of isoniazid and prioniazid in isolated hepatocytes, *Biochem. Pharmacol.*, 36, 2913, 1987.
416. **Albano, E., Tomasi, A., Coria-Gatti, L., and Iannoni, A.,** Free radical activation of monomethyl and dimethyl hydrazines in isolated hepatocytes and liver microsomes, *Free Rad. Biol. Med.*, 6, 3, 1989.
417. **Stroo, W. E. and Schaffer, S. W.,** Furazolidone-enhanced production of free radicals by avian cardiac and hepatic microsomal membranes, *Toxicol. Appl. Pharmacol.*, 98, 81, 1989.
418. **Dutton, D. R., Reed, G. A., and Parkinson, A.,** Redox cycling of resorufin catalyzed by rat liver microsomal NADPH-cytochrome P-450 reductase, *Arch. Biochem. Biophys.*, 268, 605, 1989.
419. **Docampo, R., Moreno, S. N. J., and Mason, R. P.,** Generation of superoxide anion and hydrogen peroxide during redox cycling of 5-(4-nitrophenyl)-penta-2,4-dienal by mammalian microsomes and enzymes, *Chem.-Biol. Interact.*, 65, 123, 1988.
420. **Munday, R.,** Dialuric acid autoxidation. Effects of transition metal on the reaction rate and on the generation of "active oxygen" species, *Biochem. Pharmacol.*, 37, 409, 1988.
421. **Winterbourn, C. C. and Munday, R.,** Glutathione-mediated redox cycling of alloxan. Mechanisms of superoxide dismutase inhibition and of metal-catalyzed hydroxyl radical formation, *Biochem. Pharmacol.*, 38, 271, 1989.
422. **Eichler, W. and Schertel, B.,** Dihydroorotate induces calcium release from rat liver mitochondria: a contribution to the mechanism of alloxan-induced calcium release, *Biol. Chem. Hoppe-Seyler*, 369, 1287, 1988.

423. **Nukatsuka, M., Sakurai, H., and Kawada, J.,** Microsomal reduction of alloxan produces alloxan anion radicals, *Naturwissenschaften,* 76, 574, 1989.
424. **Sakurai, K. and Miura, T.,** Generation of free radicals by alloxan in the presence of bovine serum albumin: a role of protein sulfhydryl groups in alloxan cytotoxicity, *Biochem. Int.,* 19, 402, 1989.
425. **Miura, T. and Sakurai, K.,** Iron release from ferritin by alloxan radical, *Life Sci.,* 43, 2145, 1988.
426. **Sakurai, K. and Miura, T.,** Iron release from ferritin and generation of hydroxyl radical in the reaction system of alloxan with reduced glutathione; a role of ferritin in alloxan toxicity, *Chem. Pharm. Bull.,* 36, 4534, 1988.
427. **Pileblad, E., Slivka, A., Bratvold, D., and Cohen, G.,** Studies on the autoxidation of dopamine: interaction with ascorbate, *Arch. Biochem. Biophys.,* 263, 447, 1988.
428. **Gee, P. and Davison, A. J.,** Intermediates in the aerobic autoxidation of 6-hydroxydopamine: relative importance under different reaction conditions, *Free Rad. Biol. Med.,* 6, 271, 1989.
429. **Mueller, K., Wiegrebe, W., and Younes, M.,** Formation of active oxygen species by dithranol. III. Dithranol, active oxygen species, and lipid peroxidation in vivo, *Arch. Pharmacol.,* 320, 59, 1987.
430. **Mueller, K. and Kappus, H.,** Hydroxyl radical formation by dithranol, *Biochem. Pharmacol.,* 37, 4277, 1988.
431. **Monteiro, H. P., Abdalla, D. S. P., Augusto, O., and Bechara, E. J. H.,** Free radical generation during δ-aminolevulinic acid autoxidation: induction by hemoglobin and connections with porphyrinpathies, *Arch. Biochem. Biophys.,* 271, 206, 1989.
432. **Munday, R.,** Generation of superoxide radical, hydrogen peroxide and hydroxyl radical during the autoxidation of N,N,N′,N′-tetramethyl-p-phenylenediamine, *Chem.-Biol. Interact.,* 65, 155, 1988.
433. **Valoti, M., Sipe, H. J., Jr., Sgaragli, G., and Mason, R. P.,** Free radical intermediates during peroxidase oxidation of 2-t-butyl-4-methoxyphenol, 2,6-di-t-butyl-4-methylphenol, and related phenol compounds, *Arch. Biochem. Biophys.,* 269, 423, 1989.

Chapter 4

PRODUCTION OF OXYGEN RADICALS BY CELLS

I. INTRODUCTION

In preceding chapters, the production of oxygen radicals in cells during the oxidation of various natural compounds and xenobiotics as well as in the enzymatic processes catalyzing the superoxide production in the presence or the absence of xenobiotics was discussed. It has been shown that the generation of oxygen radicals occurs in mitochondria, microsomes, nuclei, and the cytosol. All these processes are responsible for the intracellular production of oxygen radicals. Obviously, the extracellular release of oxygen radicals will depend on their ability to traverse the cytosol and the plasmalemma. Certainly, only the superoxide ion, due to its relatively long lifetime (see Volume I) may, in principle, exit from the cell, since the hydroxyl radical (HO·) and the perhydroxyl radical (HOO·) are species too active to make such a trip. (At present, there is no indication of the release of crypto-hydroxyl radicals from the cells. If they are hydroperoxy iron complexes, they must have a sufficiently long lifetime and be able to penetrate lipid membranes as was proposed for the explanation of toxic effects of anthracycline antibiotics [Chapter 3].)

The permeability of lipid membranes by the superoxide ion has been studied in several works.[1-4] It was concluded that the superoxide ion is able to traverse the membranes from erythrocyte ghosts, soybean and egg yolk phospholipids, and chloroplast thylakoids by way of the anion channel. However, the fraction of $O_2^{\overline{\cdot}}$ which was able to cross the lipid membrane was small. For example, in the case of soybean phospholipid membranes, the permeability coefficient for the superoxide ion was equal to $2.1 \cdot 10^{-6}$ cm · s^{-1}, and the amount of $O_2^{\overline{\cdot}}$ crossing the membrane was probably less than 8%.[3] In the case of egg phosphatidylcholine membranes, the permeability coefficient for the superoxide ion was even smaller ($7.6 \cdot 10^{-8}$ cm · s^{-1}).[4] Therefore, the extracellular release of oxygen radicals originating from the inside cellular components must, as a rule, be negligible.

Indeed, it is now well known that the cells able to generate oxygen radicals have corresponding enzymes placed in the plasmalemma, first of all, NADPH oxidase.[5-8] There are two other potential generators of oxygen radicals: lipoxygenase and cyclooxygenase. The contributions of these enzymes in the production of oxygen radicals depend on the cell type and will be further discussed in detail.

The production of oxygen radicals was first shown for the most important professional phagocytes, neutrophils.[9] It was found that "respiratory burst" in phagocytes (a sudden increase in oxygen consumption occurring after the interaction of phagocytes with stimuli) is a consequence of one-electron reduction of dioxygen to a superoxide ion. The respiratory burst in neutrophils and other phagocytes during phagocytosis is induced by such natural stimuli as opsonized microorganisms, immunoglobulins, leukotriene B_4 (LTB_4, produced by stimulated macrophages), *N*-formylated oligopeptides, etc.[10] At present, other numerous compounds and particles are shown also to be able to activate phagocytes. It was proposed that the superoxide release by neutrophils during phagocytosis is a key phenomenon in unspecific host defense and in the evolution of the inflammatory process.

It was shown later that other phagocytes (eosinophils, basophils, blood monocytes, Kupffer cells, etc.), as well as some other cells (endothelial cells, mesangial cells, leukemic cells, lymphoid tumor cells, fibroblasts, epithelial cells, etc.), are also able to produce oxygen radicals. Another group of oxygen radical producers is plant cells, which developed specific enzymatic systems to generate oxygen radicals.

Now we will consider the cell types able to produce oxygen radicals.

II. NEUTROPHILS

A. DETECTION OF THE SUPEROXIDE ION PRODUCED BY STIMULATED NEUTROPHILS

Neutrophils, eosinophils, and basophils are polymorphonuclear leukocytes (PMNs), neutrophils being the most important group of leukocytes (~97% of the total number of PMNs). Owing to that, neutrophils were most thoroughly studied as the producers of oxygen radicals. In 1973, Babior et al.[9] showed that human neutrophils reduced cytochrome c. The reduction was doubled in the presence of latex particles and was inhibited by superoxide dismutase (SOD). On these grounds, it was concluded that stimulated neutrophils produce the superoxide ion. This conclusion was confirmed in a great many experimental works. The SOD-inhibitable cytochrome c reduction was observed in the presence of human neutrophils[11-13] and mouse, guinea pig, or rabbit leukocytes.[14,15] For the stimulation of respiratory burst and superoxide production, such stimuli as opsonized zymosan,[14] live and killed bacteria,[11,13] and latex particles[12,15] were applied. It was found that neither 2,4-dinitrophenol, an uncoupler of oxidative phosphorylation, nor cyanide and azide, heme enzyme inhibitors, affect superoxide production by neutrophils,[13] which excludes the participation of mitochondria in the generation of the superoxide ion and indicates the existence in neutrophils of special enzyme(s) responsible for this process.

Since the reduction of cytochrome c by the superoxide ion (Reaction 1) leads to the regeneration of dioxygen, the addition of cytochrome c to neutrophils must decrease the dioxygen uptake by neutrophils.

$$O_2^{\cdot-} + \text{cytochrome c(III)} \rightarrow O_2 + \text{cytochrome c(II)} \qquad (1)$$

Indeed, Babior[16] observed a decrease in net dioxygen uptake by zymosan-stimulated neutrophils in the presence of cytochrome c. A similar effect was found later for FMNL-stimulated neutrophils.[17] However, Segal and Meshulam[18] did not find any effect of cytochrome c on dioxygen consumption by stimulated neutrophils. Green et al.[19] supposed that a true value of dioxygen consumption may be distorted under certain conditions due to the establishment of an O_2 diffusion layer at the outer surface of the PMN membrane.

There are various modifications of the cytochrome c assay in its application to the detection of the superoxide ion generated by phagocytes. For example, acetylated cytochrome c was used for the detection of $O_2^{\cdot-}$ produced by myristate-stimulated guinea pig PMNs.[20] Markert and Allaz[21] developed a technique for continuous and simultaneous detection of superoxide ion production and dioxygen consumption by zymosan-stimulated leukocytes using SOD-inhibitable cytochrome c reduction and polarography with a Clark-type electrode. It should be noted that one of the major shortcomings of the cytochrome c assay is an uncertainty in the optimal cytochrome c concentration. Thus, it was shown[12] that an increase in the cytochrome concentration from 18 to 180 μM led to increasing the amount of cytochrome c reduced by phagocytizing granulocytes to 2 to 2.5 times.

In addition to the SOD-inhibitable cytochrome c reduction, other traditional assays for the detection of the superoxide ion were applied by studying the production of oxygen radicals by neutrophils. As always, the NBT reduction can be used, although this method seems to be less specific than cytochrome c reduction, since not only SOD but also inactivated SOD and albumins inhibited the nitroblue tetrazolium (NBT) reduction by PMNs.[22] Briggs et al.[23] developed a new technique for detecting the superoxide ion produced by neutrophils based on manganese-dependent, SOD-inhibitable diaminobenzidine oxidation.

Very important methods of the detection of oxygen radical production by leukocytes are based on the chemiluminescence (CL) technique and spin trapping. Actually, even before the work of Babior et al.,[9] who introduced the cytochrome c assay for the detection of superoxide production by PMNs, Allen et al.[24] have shown that human leukocytes stimulated by opsonized, heat-killed *Proteus shermanii* or opsonized live *Staphylococcus aureus* bacteria induced spontaneous CL. In subsequent work,[25] it was shown that both spontaneous CL and NBT reduction by phagocytically challenged PMNs were partially inhibited by SOD. This fact was interpreted as evidence of participation of the superoxide ion in CL stimulation, although a key role in this process was thought to be played by singlet oxygen.

McPhail et al.[26] have shown that the particulate fraction isolated from human neutrophils generated spontaneous CL inhibited by SOD and catalase. On these grounds, it was concluded that neutrophils produce a superoxide ion and hydroxyl radicals. SOD also inhibited the CL induced by normal and myeloperoxidase-deficient PMNs.[27] There are other works in which spontaneous CL was used for superoxide detection, but the most important results were later obtained by applying two sensitizers: luminol and lucigenin.

In 1976, Allen and Loose[28] studied the oxygen burst in human and rabbit PMNs and rabbit alveolar and peritoneal macrophages with the aid of the luminol-dependent CL technique. It was found that phagocytes stimulated with *Escherichia coli* bacteria produced the luminol-dependent CL, which had the greatest intensity in the case of human PMNs. The CL response was inhibited by SOD and benzoate. Later on, lucigenin, a more specific amplifier of the superoxide-induced CL, was applied.[29] It was found that lucigenin greatly enhanced the CL induced by the particulate fraction isolated from PMA-(12-*O*-myristate 13-acetate)-stimulated human neutrophils. CL was suppressed by SOD (50%) and was insensitive to cyanide. In contrast to luminol-dependent CL, lucigenin-dependent CL was independent of myeloperoxidase.

Recently, an ultrasensitive video intensifier microscopy was applied for studying the dynamics of oxygen radical production from opsonized, zymosan-stimulated human neutrophils measured by the luminol-dependent CL.[30] These experiments demonstrated the heterogeneity of neutrophils with respect to oxidative metabolism.

It is of great importance that both sensitizers, luminol and lucigenin, apparently characterize two different modes of the generation of oxygen radicals. It has been proposed[31] that the luminol-dependent CL measures the production of oxygen radicals by the peroxide-peroxidase-dependent reaction in neutrophils, since the cells without myeloperoxidase (isolated from donors with myeloperoxidase deficiency) and cytoplasts (granule-depleted cells) produced only a small amount of chemiluminescence, and the xanthine-xanthine oxidase system, a classic generator of the superoxide ion, did not stimulate the luminol-dependent CL. On the other hand, neutrophils stimulated by soluble stimuli (*N*-formyl-methionyl-leucyl-phenylalanine [FMLP], platelet activating factor [PAF], LTB_4, A 23187, and PMA) produce strong SOD-inhibitable, lucigenin-dependent CL, and lucigenin does not induce CL in the myeloperoxidase-H_2O_2-Cl^- system.[32] Selectivity of lucigenin-dependent CL as the superoxide assay is supported by the resemblance of the kinetics of cytochrome c reduction and the lucigenin CL produced by neutrophils.[32] Contrary to that, the kinetics of luminol- and lucigenin-dependent Cl produced by zymosan- and latex-stimulated PMNs were not identical.[33]

One may therefore suggest that the lucigenin- and luminol-dependent CL actually characterize the production of different types of oxygen radicals by different enzymes: the superoxide ion by NADPH oxidase and hydroxyl or crypto-hydroxyl radicals by myeloperoxidase. In practice, however, the luminol-dependent CL is usually considered as an assay for all oxygen radicals generated by cells, including the superoxide ion, since the CL response is inhibited by SOD. Probably, these data may be interpreted as follows: the luminol-dependent CL is induced by all oxygen radicals and the lucigenin-dependent CL is induced by the superoxide ion alone.

It was also proposed[31,34] that the luminol-dependent CL permits distinguishing the extra- and intracellular production of oxygen radicals. It was concluded that extracellular response reached a maximum after 1 to 2 min and intracellular response reached a maximum after 5 to 7 min in the case of FMLP-stimulated human neutrophils. The ratio between the extra- and intracellular production of oxygen radicals by leukocytes dependent on the presence of cytochalasin B, Ca^{2+}, and Na^{+}.

In recent work, Gyllenhammer[35] confirmed that the early luminol-dependent CL produced by LTB_4- and FMLP-stimulated human neutrophils is due to an extracellular generation of oxygen radicals, presumably the superoxide ion. It was shown that the luminol-dependent CL was inhibited by SOD, catalase, NaN_3, and mannitol. Contrary to the data obtained in Reference 31, the SOD-inhibitable, luminol-dependent CL was also observed in the cell-free xanthine-xanthine oxidase system. The peak of luminol-dependent CL coincided with the linear phase of the cytochrome c reduction by neutrophils.

Recently, a new sensitizer, 2-methyl-6[*p*-methoxyphenyl]-3,7-dihydroimidazo-[1,2-α]pyrazin-3-one (MCLA), was applied for the enhancement of CL produced by opsonized, zymosan-stimulated granulocytes and monocytes.[36] It is believed that MCLA is a more effective sensitizer than lucigenin.

Green et al.[37] were apparently the first workers who registered the ESR spectra of DMPO-OH and DMPO-OOH spin adducts produced by stimulated neutrophils in the presence of DMPO. Depending on stimuli, DMPO-OH or a mixture of DMPO-OH and DMPO-OOH were obtained. SOD inhibited the formation of all electron spin resonance (ESR) spectra; therefore, one may suppose that DMPO-OH was formed as a result of the DMPO-OOH decomposition. Later on, the DMPO-OOH spin adduct was obtained during the oxidation of NADPH by NADPH oxidase isolated from PMA-stimulated guinea pig neutrophils.[38] SOD inhibited the DMPO-OOH formation. The SOD-inhibitable DMPO-OH formation was also observed upon the stimulation of human PMNs by antibody-coated *Trypanosoma cruzi* cells.[39]

Another type of the spin-trapping method was proposed by Rosen et al.,[40] who used the oxidation of 2-ethyl-1-hydroxy-2,5,5-trimethyl-3-oxazolidine (OXANOH) into the stable OXANO· free radical as a method for the detection of the superoxide ion produced by stimulated neutrophils. This method must of course be less specific than the traditional spin-trapping technique as, besides $O_2^{\overline{\cdot}}$, many other active species can oxidize OXANOH. In addition, this method apparently underestimates the superoxide concentration in comparison with cytochrome c reduction, although the last method also may give a diminished value of the rate of superoxide production.[40]

Makino et al.[41] developed a new method of simultaneous measurement of superoxide production, dioxygen consumption, and H_2O_2 production by neutrophils with the aid of diacetyldeuteroheme-substituted horseradish peroxidase (HRP). Here, the superoxide ion was detected via the formation of HRP compound III.

B. THE MECHANISM OF SUPEROXIDE PRODUCTION BY NEUTROPHILS

1. NADPH Oxidase Is a Main Generator of Superoxide Production by Neutrophils

It was already pointed out that mitochondria cannot be an origin of superoxide release by neutrophils, as heme inhibitors and uncouplers of oxidative phosphorylation do not affect the production of oxygen radicals, which confirms the conclusions made at the consideration of the permeability of lipid membranes by the superoxide ion (see Section I). The same is apparently true for other intracellular components (microsomes, the cytosol, etc.). Recently, Jones et al.[42] confirmed that the level of xanthine oxidase in human neutrophils is small in comparison with the flux of the superoxide ion produced during neutrophil activation. In addition, allopurinol, an inhibitor of xanthine oxidase, did not inhibit superoxide production by neutrophils.

It has now been ascertained that a main generator of the superoxide ion by human neutrophils is NADPH oxidase, which catalyzes the one-electron reduction of dioxygen via Reaction 2.

$$NADPH + H^+ + O_2 \rightarrow NADP^+ + 2\,O_2^{\bar{}} + 2\,H^+ \quad (2)$$

In resting cells NADPH oxidase is "dormant", and neutrophils produce insignificant amounts of the superoxide ion. (Some indications of superoxide production by resting cells obtained in earlier works are apparently erroneous.) Upon activation with various stimuli (see below), neutrophils begin to produce oxygen radicals (the superoxide ion and, under certain conditions, hydroxyl radicals). NADPH oxidase may be isolated from stimulated neutrophils in active form.[43,44] As early as 1975, Babior et al.[43] showed that lysates from human granulocytes produced a superoxide ion in the presence of NADPH and NADH.

The process of activating NADPH oxidase in neutrophils is very complicated and depends on the nature of the stimuli. This process is of great importance for phagocyte metabolism and therefore was studied in many works which have been recently reviewed.[6-8] We will not repeat these reviews and will consider only the principal steps of the activation process. It is now generally accepted that NADPH oxidase consists of two major components: a flavoprotein and the b-type cytochrome. For example, the superoxide-generating fraction from normal human neutrophils contains 0.25 nmol flavin adenine dinucleotide (FAD) per milligram protein and 0.28 nmol cytochrome b per milligram protein.[45] These components are assembled into an electron-transfer chain, transferring electrons from NADPH to dioxygen to form the superoxide ion. It was proposed that the electron-transfer chain may also contain quinones (as a third component), but there seems to be no reliable support for this proposal.

Thus the simplest model for NADPH oxidase is Reaction 3:

$$NADPH \rightarrow \text{flavoprotein} \rightarrow \text{cytochrome b} \rightarrow O_2 \quad (3)$$

The one-electron reduction potential of cytochrome b in neutrophils (cytochrome b_{558}) is (−0.225) — (−0.245) V;[8] therefore, the one-electron reduction of dioxygen ($E^\circ = -0.16$ V) by cytochrome b_{558} is a thermodynamically favorable process. It should be noted that there are data which contradict this simple model of NADPH oxidase, and there are, as well, its other more complicated models. These data and models are considered in detail in Reference 8.

It has been shown[46] that the majority of superoxide-forming activity in stimulated neutrophils is in the membrane fraction. However, two basic components of NADPH oxidase, flavoprotein and cytochrome b_{558} are apparently distributed asymmetrically in plasmalemma and granules. It was concluded[47] that about 75% of the total amount of cytochrome b_{558} is present in granule membranes, and therefore the activation of dormant enzyme requires the translocation of cytochrome b_{558} to plasmalemma by membrane fusion. However, the cytochrome b_{558} translocation does not seem to be a single (or necessary) pathway for superoxide generation.[7]

The activity of NADPH oxidase at the cell surface was shown in the experiments with *p*-diazobenzene sulfonic acid.[48] This compound, which is known to react predominantly with proteases of the external cell membrane, inhibited superoxide production by cytochalasin B- and cytochalasin A-stimulated human PMNs. It has been proposed[49] that NADPH oxidase is positioned in plasmalemma in the *trans* configuration with the superoxide-generating site located on the extracellular side of the outer membrane and the NADPH electron-accepting site located on the inner, cytoplasmic side of the membrane. Such a configuration of NADPH oxidase was ascertained in the experiments with human PMNs in which the $O_2^{\bar{}}/O_2$ ratio decreased from 0.95 in the case of stimulation by Con A to 0.20 upon the stimulation by phagocytizable, serum-treated zymosan. A decrease in the $O_2^{\bar{}}/O_2$ ratio was due to the impossibility of $O_2^{\bar{}}$ release after the superoxide-generating site disposed on the outer side of the plasmalemma was carried out inside the vesicle during the phagocytosis of zymosan particles. It was also found that exogenous NADPH did not stimulate intact cells, obviously due to the dislocation of the NADPH site on the inner side of the membrane.

The first step of the activation of NADPH oxidase is the binding of stimuli to neutrophils. In accord with up-to-date data,[6-8] the binding of external stimuli to their receptors on the surface of neutrophils induces important modifications in the membrane and the cytosol. The stimulus-receptor interaction triggers a series of biochemical events usually described as transduction, which includes calcium entry and calcium mobilization from intracellular stores, phospholipid hydrolysis, transmembrane ion fluxes, protein phosphorylation, intracellular translocation of enzymes, etc. Different stimuli may stimulate the NADPH oxidase activation in a different way. Below, the pathways of superoxide production by neutrophils activated by various stimuli will be considered.

2. Stimulation of NADPH Oxidase by Phagocytosable Particles

The production of oxygen radicals by neutrophils was first studied during the phagocytosis of bacteria, viruses, and various aggregate materials. It was found that respiratory burst and superoxide production are associated with the act of engulfment of particles. Superoxide production measured as the SOD-inhibitable cytochrome c reduction or the SOD-inhibitable CL was detected upon stimulation of human neutrophils with heat-killed, opsonized *Proteus shermanii* and opsonized live *Staphylococcus aureus* bacteria,[24] opsonized latex particles,[12,24,50] immunoglobulin G (IgG)-coated latex particles,[37] *E. coli* cells,[13] antibody-coated *T. cruzi* cells,[39] opsonized zymosan,[15,44] etc. Similar results were obtained with zymosan-stimulated mouse and guinea pig PMNs.[14] Superoxide production by zymosan-stimulated PMNs may be enhanced (primed) by phosphatidylethanolamine[51] or by the substances secreted by human adenocarcinoma cells.[52] Recently it was also shown[53] that the production of oxygen radicals by human neutrophils as well as by monocytes and alveolar macrophages may be stimulated by unopsonized surface-adherent bacteria (*S. aureus*).

Although the production of oxygen radicals is an important and obligatory event in phagocytosis development, the phagocytosis itself is not an obligatory condition of oxygen radical generation. Thus it was found[54] that the inhibition of the phagocytosis of human PMNs with cytochalasin B actually did not affect the PMN capacity to produce a superoxide ion after activation with aggregated IgG (agg IgG) or serum-treated zymosan. (In fact, superoxide production was greater than that by normal PMNs.) It was concluded that cytochalasin B increased superoxide production by preventing the phagocytic vacuole formation in which a part of O_2^- is dismuted by endogenous SOD.

Extracellular release of the superoxide ion depends on the stimulus/cell ratio. It was found[55] that at the very low stimulus/cell ratio, neutrophils activated by opsonized polystyrene beads did not release the superoxide ion into the bulk medium, apparently due to the neutrophils having time to complete vacuole formation before releasing superoxide ion. When a stimulus/cell ratio exceeded the critical value depending on the bead size, neutrophils became "overstressed" and released the superoxide ion before the vacuole formation was completed.

Hallett and Campbell[50] concluded that, in contrast to the stimulation of rat PMNs by the soluble stimulus FMLP (see below), which resulted in an increase in intracellular calcium concentration, the stimulation of PMNs with unopsonized latex particles was independent of an enhancement of the calcium level. However, it was shown later[56,57] that extracellular calcium increased superoxide production by the neutrophils activated by opsonized zymosan or heat-aggregated IgG. An increase in superoxide production was inhibited by calcium chelators (EDTA or EGTA) and calcium antagonists (verapamil, nifedrine, etc.).[57]

3. Stimulation of NADPH Oxidase by Calcium-Dependent Soluble Stimuli

There is a group of stimuli whose action is closely bound to an increase in the Ca^{2+} concentration in the cytosol. It is believed that the rise in calcium concentration due to increasing the calcium influx or the rapid mobilization of calcium from intracellular stores is an important intermediate step in the activation process of NADPH oxidase. Therefore, it was proposed that this group of stimuli consisting of chemotactic peptides, calcium ionophores, agg IgG, etc., activates NADPH oxidase via a Ca-dependent pathway. The mechanism of calcium participating in the activation process is uncertain. Suzuki et al.[58] proposed that Ca^{2+} ions enhance the catalytic activity of NADPH oxidase by binding to the enzyme.

The most studied stimulus of this group is *N*-formyl-methionyl-leucyl-phenylalanine (FMLP), a chemotactic peptide. In neutrophils, FMLP induces the hydrolysis of phosphatidylinositol 4,5-biphosphate (PIP_2) leading to the generation of 1,2-diacylglycerol (DG) and inositol 1,4,5-triphosphate (IP_3). Hydrolysis of PIP_2 is thought to be an important step in the activation process as DG is an activator of protein kinase (see below) and IP_3 is a trigger of the release of calcium from intracellular stores.[59] However, it was found[60] that low concentrations of FMLP stimulate superoxide production by neutrophils only after cytoskeletal rearrangement induced by cytochalasin B. These findings are believed to be of physiological significance as they may be relevant to the neutrophil response to chemotactic agents such as DGs, whose action is probably mimicked by FMLP.

Calcium ionophores, ionomycin, and A 23187 stimulate superoxide production by mobilization of Ca^{2+} ions from intracellular stores. A similar stimulation mechanism is also proposed for lectins, leukotriene B_4, fatty acids, aggregated IgG, and platelet activating factor (1-*O*-hexadecyl-2-acetyl-sn-glycero-3-phosphorylcholine, PAF). The superoxide production stimulated by FMLP is enhanced (primed) by other compounds or particles which may or may not be the independent stimuli. Thus the FMLP-stimulated neutrophils were primed by ionomycin,[61] the tumor necrosis factor (TNF, a 17,000-Da protein formed upon exposure of macrophages to endotoxin),[62-65] aliphatic amines (putrescine, spermidine, and spermine),[66] bile salts (lithocholate, deoxycholate, and chenodeoxycholate),[67] and the granulocyte-macrophage colony-stimulating factor (GM-CSF).[68] It was proposed that these priming agents may mobilize the intracellular calcium (TNF),[62] enhance the availability of external calcium (the bile salts),[67] or induce the phosphorylation of specific neutrophil proteins (TNF).[65]

LTB_4 (12[R]-dihydroxyeicosatetraenic acid, a product of arachidonic acid lipoxygenation), may stimulate superoxide production by neutrophils in the presence of calcium and cytochalasin B or only in the presence of calcium.[69] It was proposed that LTB_4 acts as an ionophore, increasing Ca^{2+} concentration. LTB_4 is also able to prime superoxide production by FMLP-stimulated neutrophils.[70] It is important that LTB_4 is produced *in vivo* by stimulated neutrophils and mononuclear phagocytes. Another product of lipoxygenase reaction, 5-HETE (5-hydroxy-6,8,11,14-eicosatetraenonate), also stimulates superoxide production by neutrophils supposedly via the enhancement of calcium level.[71]

K^+, Rb^+, and Cs^+ induced superoxide production by guinea pig neutrophils.[72] Since the intracellular calcium antagonist TMB-8 (8-[*N*,*N*-diethylamino]-octyl-3,4,5-trimethoxybenzene) inhibited superoxide production, it was concluded that the alkaline metal-stimulated $O_2^{\bar{}}$ formation is independent of extracellular calcium, but depends on calcium mobilization from intracellular stores. Superoxide production by human neutrophils stimulated by fluoride ion apparently also depends on calcium.[73]

4. Stimulation of NADPH Oxidase by Activators of Protein Kinase C

An important event in the stimulation of superoxide production by neutrophils is the activation of the phospholipid-dependent enzyme protein kinase C. This enzyme phosphorylates NADPH oxidase, which is an obligatory step of its activation process. The physiological inductor of protein kinase C is DG, formed during hydrolysis of PIP_2. DG is thought to enhance the affinity of protein kinase C for Ca^{2+} and phosphatidylserine, resulting in the enzyme activation.[74] A role of DG as the protein kinase inductor is mimicked by such important stimuli as phorbol esters (tumor promoters), the most important inductor among them being 12-*O*-myristate 13-acetate (PMA). Phorbol esters are direct activators of protein kinase C.[5,59,75]

Superoxide production by PMA-stimulated neutrophils depends on Ca^{2+} and Mg^{2+} and is inhibited by EDTA.[76] At the same time, PMA triggers O_2^- generation without an increase in cytosolic calcium.[59] It is believed[5] that similarly to DG, phorbol esters activate protein kinase C by increasing the affinity of the enzyme for calcium. Recently, it was also found[77] that PMA is able to stimulate the translocation of protein kinase C to the particulate fraction. An increase in particulate-associated protein kinase C was in parallel with increasing superoxide production. It should be noted that the translocation of protein kinase C to the particulate fraction was also observed when PMNs were stimulated with opsonized zymosan, A 23187, and OAG (1-oleoyl-2-acetylglycerol).

Among other activators of protein kinase C which stimulate superoxide production are mezerein, a tumor promoter,[71] and synthetic DGs (sn-1,2-dihexanoylglycerol, sn-1,2-dioctanoylglycerol, and OAG).[78,79] Synthetic diglycerols stimulated superoxide production by neutrophils after pretreatment with cytochalasin B. It is interesting that PMA and OAG equally activated protein kinase C, but superoxide production upon PMA-stimulation was nearly eight times greater than the O_2^- formation in the case of OAG stimulation.[80] In accord with the mechanism proposed, the inhibitor of protein kinase C H-7 (1-[5-isoquinolinesulfonyl]-2-methylpiperazine) inhibited superoxide production by PMA-stimulated human neutrophils.[75,81]

Indomethacin enhanced superoxide production by OAG- and A 23187-stimulated human neutrophils, apparently via the inhibition of DG kinase.[82] There are contradictory data concerning the ability of protein kinase inhibitors H-7 and C-I to suppress superoxide production by neutrophils stimulated by FMLP and calcium ionophore A 23187. Although such effects were observed in some works,[75,83] they are apparently not always observed[81] and depend on temperature.[83] As in the case of FMLP stimulation, TNF primed superoxide production by PMA-stimulated neutrophils.[63] It should be noted that PMA is able to restore the production of oxygen radicals by FMLP- and wheat germ agglutinin-stimulated PMNs after its inhibition by immune complexes.[84] Thymol possibly also belongs to the activators of protein kinase C, since superoxide production by thymol-stimulated guinea pig neutrophils was inhibited by trifluoroperazine, an inhibitor of protein kinase C.[85]

It has been shown[86,87] that the inhalation of anesthetic halothane (2-bromo-2-chloro-1,1,1-trifluoroethane) inhibits superoxide production by neutrophils. However, Tsuchiya et al.[88] recently found that halothane stimulated superoxide production by 12-*O*-tetradecanoylphorbol-13-acetate (TPA)- and FMLP-activated guinea pig intraperitoneal neutrophils, apparently via the activation of protein kinase C because its inhibitor, H-7, inhibited halothane-stimulated superoxide production. The origin of this discrepancy is uncertain.

5. Stimulation of NADPH Oxidase by Other Soluble Stimuli

There are a great many other soluble compounds which are able to stimulate the production of oxygen radicals by neutrophils. It has already been pointed out that the products of lipoxygenation of arachidonic acid LTB_4 and 5-HETE stimulate superoxide production by neutrophils, possibly increasing the calcium level. However, arachidonic acid itself directly stimulates NADPH oxidase. It was proposed[89,90] that arachidonic acid and other *cis*-unsaturated acids disorder the gel-like regions of neutrophil membrane, thereby affecting the proteins (phosphatidylinositol-specific phospholipase and protein kinase C) involved in the stimulation of superoxide production in these regions. Arachidonic acid apparently also enhances calcium transport across lipid membranes[91] and mobilizes NADPH oxidase from granule membranes.[47] Since a well-known detergent, sodium dodecyl sulfate (SDS) is able to substitute arachidonate, the last must manifest a detergent like effect on membranes. It was also proposed[92] that the stimulation of superoxide production by arachidonic acid requires the preliminary protein phosphorylation.

It was previously assumed[90] that only *cis*-polyunsaturated fatty acids (for example, arachidonate and linoleate) are able to stimulate release of the superoxide ion from neutrophils. However, as long ago as 1978, Kakinuma and Minakami[20] showed that the saturated acid myristate stimulated superoxide production by guinea pig PMNs measured via the reduction of acetylated cytochrome c. Later on, it was shown[93,94] that saturated, *cis*-unsaturated, and *trans*-unsaturated fatty acids are equally active in the stimulation of superoxide production by intact neutrophils and cell-free preparations under Ca^{2+}-depleted conditions. It was proposed that the earlier data concerning the inability of saturated and *trans*-unsaturated acids to activate neutrophils are due to the fact that the experiments were carried out in the calcium-containing media. Indeed, the inclusion of calcium in the media strongly inhibited the activation of neutrophils by saturated and *trans*-unsaturated acids, but did not affect it in the case of *cis*-unsaturated acids.

This conclusion agrees with the findings[95] that the stimulation of superoxide production by guinea pig neutrophils with arachidonate and linoleate took place simultaneously with an increase in the intracellular Ca^{2+} level, whereas myristate stimulated the generation of $O_2^{\cdot-}$ without affecting the content of intracellular calcium. It was therefore proposed that unsaturated fatty acids increased the level of intracellular calcium and enhanced the protein kinase C activity. Badwey et al.[96] proposed that the mechanism of stimulation of superoxide release from neutrophils by all-*trans*-retinal and other retinoids is the same (disordering of the membrane gel-like regions) as in the case of *cis*-unsaturated acids.

It has been shown[97] that superoxide production by guinea pig granulocytes can be stimulated by anionic detergents (digitonin and deoxycholate). A similar activating effect was observed upon the addition of SDS to guinea pig peritoneal PMNs[98] and a cell-free system of pig neutrophils.[99] Ginsburg et al.[100,101] have studied the stimulation of superoxide production in human blood leukocytes by polycations. It was found that the stimulating effect of poly-L-histidine (PHSTD) depended on the presence of extracellular calcium, since calcium channel blockers depressed superoxide production. PHSTD turned out to be one of the most potent stimulators of superoxide production (about 16 nmol/10^6 cells/min[101]). The authors proposed that PHSTD activates superoxide production at the cell surface as a soluble agent. Effective stimulation of superoxide release by leukocytes was also observed when bacteria (streptococci) opsonized the cationic polyelectrolytes (PHSTD and poly-L-arginine) were used.[100] It was suggested that there is a special "recognition" by the leukocyte membrane of certain amino acids which are responsible for the activation of the oxygen burst. Stimulation of superoxide production by human neutrophils with polyarginine and polyanetholesulfonate was markedly enhanced by cytochalasins.[102]

In 1975, Nakagawara and Minakami[103] showed that cytochalasin E, an inhibitor of cytokinesis and membrane movement, stimulated superoxide production by guinea pig leukocytes. In subsequent work[104] these authors found that the $O_2^{\cdot-}$ release by cytochalasin E-stimulated human neutrophils is greatly enhanced by Con A or D_2O and is inhibited by microtubule-disrupting agents (colchicine or vinblastine). Therefore, it was concluded that superoxide production by neutrophils is controlled by the cytoskeletal components and the microfilament-microtubule system. In contrast to cytochalasin E, cytochalasins A and B suppressed superoxide release by neutrophils. At the same time, a mixture of FMLP and cytochalasin B is a powerful stimulator of superoxide production by rat neutrophils.[105]

Phosphatidylserine is an effective stimulator of superoxide production by solubilized NADPH oxidase.[106] Its effect was not influenced by Ca^{2+}, EGTA, or dioctanoylglycerol. Therefore, phosphatidylserine is apparently a direct activator of NADPH oxidase. Dieldrin increased superoxide production by glycogen-elicited rat peritoneal neutrophils.[107] Preincubation of neutrophils with suprathreshold concentrations of PMA, FMLP, and A 23187 greatly enhanced dieldrin-stimulated superoxide production.

Sharp et al.[108] have shown that β-endorphin stimulated superoxide production by human PMNs within the range of 10^{-14} to 10^{-8} *M* with a maximum response at 10^{-12} *M*. Its stimulating effect may be compared with that of FMLP, for which the maximum response corresponds to 10^{-8} M. β-Endorphin supposedly activates PMNs through stereoselective opiate receptors. It has also been shown that neutrophil superoxide production is stimulated by the antigen-antibody complex,[98] Con A,[98] lipopolysaccharides (endotoxins) and the components of the complement system,[109] tumor promoters palytoxin and thapsigargin,[110] interleukins,[111] recombinant human interferon,[112] hexachlorocyclohexane,[113] PAF (a mixture of acetylglyceryl ether phosphoglycerides),[114] fluoride anion,[115] calf skin-soluble collagen,[116] phosphatidic acid,[117] the antileprosy agent clofazimine,[118] nicotine,[119] and chlorhexidine.[120] It is interesting that chlorhexidine inhibited FMLP stimulation and promoted the PMA stimulation of neutrophils.

An important factor which apparently greatly affects the oxygen burst in neutrophils is adherence. Adherence is a physiological factor, since the circulating neutrophils must adhere to vessel walls before they can migrate into tissues. It has been shown[121] that adherence of human PMNs on glass sharply changes the release of the superoxide ion. Human neutrophils adherent on protein-coated nylon fibers and stimulated by FMLP, ionophore A 23187, or TNF produced a considerably greater amount of $O_2^{\bar{}}$ in comparison with the suspended cells.[122] On these grounds, it was suggested that the attachment of neutrophils to endothelial cells or connective tissue substances must strongly increase their ability to produce oxygen radicals. However, the efficiency of superoxide production by neutrophils can be affected by the presence in endothelial cells of antioxidative enzymes. Thus, it has been shown[123] that superoxide production by PMNs attached to the endothelium is less than superoxide production by PMNs attached to plastic due to the presence of SOD in endothelial cells.

The combinations of various stimuli may manifest various summary effects on the production of oxygen radicals by neutrophils. It has been shown[124] that a mixture of FMLP or $C5_a$ with Con A synergistically stimulated superoxide production by human neutrophils, while arachidonic acid + FMLP, $C5_a$, or Con A stimulated neutrophils additively and PMA + FMLP, $C5_a$, or Con A acted nonadditively. Similarly, suboptimal concentrations of TPA and calcium ionophore A 23187 stimulated synergistically superoxide production by guinea pig neutrophils,[125] and the combination of low concentrations of A 23187 and a synthetic DG, OAG, significantly increased the production of oxygen radicals by neutrophils (whereas each of these stimuli at these concentrations did not).[126]

6. Kinetics, Duration, and Other Features of Superoxide Production by Stimulated Neutrophils

Different mechanisms of neutrophil activation lead to the difference in kinetics of oxygen radical production in the presence of various stimuli. Studying the lucigenin-dependent CL produced by human neutrophils, Gyllenhammer[32] has shown that there is a difference between the kinetics of superoxide production by neutrophils activated by stimuli acting through surface receptors (LTB_4, FMLP, and PAF) and by stimuli depending on protein kinase C (PMA). The first group of stimuli activated superoxide production immediately after their addition, while there was a lag period of about 90 s upon PMA stimulation. Stimulation by ionophore A 23187 gave an intermediate result. Similar results were obtained in subsequent works.[127,128] It was proposed that the biggest lag period observed upon PMA stimulation is due to the membrane association of protein kinase C, which in this case precedes superoxide production.

It has long been known that the respiratory burst in stimulated neutrophils is a self-limiting phenomenon, because both oxygen uptake and superoxide production begin to decline 20 to 30 min after activation. Jandle et al.[129] proposed that the termination of oxygen radical production is explained by the destruction of NADPH oxidase through the action of myeloperoxidase. It was found that myeloperoxidase inhibitors plus cytochalasin B prevented the decline in superoxide production by zymosan-activated neutrophils. It has also been shown[130] that myeloperoxidase decreased the duration, but not the magnitude, of $O_2^{\overline{\cdot}}$ production, presumably via the inactivation of NADPH oxidase by HOCl. It was supposed that myeloperoxidase may regulate superoxide production *in vivo,* since this enzyme is released from neutrophils simultaneously with the activation of respiratory burst. Dahlgren and Lock[131] have shown that the respiratory burst in human neutrophils is regulated by extracellular peroxidase in the case of FMLP and PMA stimulation, but not during the ionomycin-dependent activation of neutrophils. These authors believe that this phenomenon is explained by the difference between extra- and intracellular production of oxygen radicals.

Another mechanism of the limitation of superoxide production by neutrophils was proposed by Henderson et al.[132] It is known that the generation of the superoxide ion by NADPH oxidase depolarizes the membrane potential[133] and enhances pH (in the case of FMLP-stimulated human neutrophils, a pH value increased from 7.22 to 7.80).[134] It was proposed[132] that NADPH oxidase activity is limited by the efflux of protons which prevents a massive depolarization of the membrane potential and is a charge compensator for superoxide production. It was also found[135] that superoxide production by guinea pig neutrophils depended on the Na^+,K^+ concentrations in the medium; the absence of Na^+ triggered the respiratory burst, whereas K^+ affected the duration of superoxide production.

The total yield of superoxide production by neutrophils depends on the stimuli applied and other experimental conditions. For example, a mean superoxide production was equal to 1.3 to 1.7 n*M*/min per 10^6 cells for latex-stimulated human neutrophils,[12] 217 $\pm$ 99 n*M*/min per 10^6 cells for PMA-stimulated human neutrophils,[136] and 13 $\pm$ 4 n*M*/min per 10^6 cells for FMLP-stimulated rat neutrophils in the presence of cytochalasin B.[105] The yield of the superoxide ion depends on pH, diminishing with a decrease in pH values. Thus the yield of $O_2^{\overline{\cdot}}$ produced by zymosan-stimulated neutrophils at pH 6.0 was equal to 11% of yield at pH 7.5.[137] Recently, using the luminol- and lucigenin-dependent CL as well as the cytochrome c reduction assay, Gyllenhammer[138] confirmed that the superoxide yield produced by FMLP- or PMA-stimulated neutrophils increased with pH elevation from 6.5 to 9.2. The same dependence was observed for the xanthine-xanthine oxidase system. Therefore, one may suppose that an increase in the $O_2^{\overline{\cdot}}$ yield at elevated pH values is explained by an increase in the lifetime of the superoxide ion. The rate of superoxide production depends on the O_2 consumption, but only at low tension of dioxygen (about 1% O_2[137] and less than 60 μM O_2).[105]

It is usually accepted that the superoxide ion is the only primary product of the reduction of dioxygen by granulocytes and is a precursor of hydrogen peroxide formed during its dismutation. Thus Bannister et al.[139] have shown that DMPO doubled the rate of dioxygen consumption, indicating that all H_2O_2 was formed as a result of superoxide dismutation. Later on, Loschen[140] also concluded that practically all dioxygen consumption by PMA-stimulated leukocytes is due to superoxide production. This conclusion was based on the findings that epinephrine roughly doubled dioxygen consumption due to the reduction of all $O_2^{\overline{\cdot}}$ to H_2O_2, whereas the radical cation of 2,2′-azino-di(3-ethylbenzothiazoline-6-sulfonic acid) practically abolished dioxygen uptake due to the reoxidation of all $O_2^{\overline{\cdot}}$ into O_2. The same conclusion was reached by Makino et al.,[41] who found that peroxidase Compound III was formed stoichiometrically to O_2 consumption at the addition of diacetyldeuteroheme-substituted HRP to PMA-stimulated neutrophils, i.e., all dioxygen consumed was converted to a superoxide ion.

Contrary to that, based on the relationship between superoxide production, NADPH oxidation, and the O_2 consumption by NADPH oxidase from PMA-stimulated human neutrophils, Green and Wu[141] concluded that similar to xanthine oxidase, NADPH oxidase has at least two separate redox sites of dioxygen reduction, namely, a site of univalent reduction of dioxygen producing O_2^- and a site of divalent dioxygen reduction producing H_2O_2. Earlier, Hoffstein et al.[142] came to the same conclusion, based on a dependence of the O_2^-/H_2O_2 ratio on the nature of the stimuli. (Soluble IgG aggregates, FMLP, and A 23187 mostly stimulated superoxide production, whereas immune complexes activated the release of both the superoxide ion and hydrogen peroxide.) Thus at present, this question seems to be unresolved, although one may suppose that the stoichiometric conversion of dioxygen consumed to a superoxide ion[41] proves the existence only of the one-electron reduction site in NADPH oxidase and that a dependence of the O_2^-/H_2O_2 ratio on experimental conditions may be explained by partial dismutation of the superoxide ion.

7. Inhibition of Neutrophil Superoxide Production

There are two possible modes of the inhibition of neutrophil superoxide production: the interaction of inhibitors with the superoxide ion and the inhibition of the one from various steps of the NADPH oxidase activation process or the enzyme activity. Thus the inhibitors of the first type must be scavengers of the superoxide ion. The inhibitory action of SOD on superoxide production by neutrophils is well known. It was also shown[143] that the SOD-mimicked compound Cu(II)(3,5-diisopropylsalicylic acid)$_2$ strongly inhibited the production of oxygen radicals by TPA-stimulated human neutrophils. Among other free radical scavengers, ascorbic acid,[144-146] ascorbyl palmitate,[144] α-tocopherol,[144] α-tocopheryl acetate,[147] human serum albumin,[148] menadione,[149] and glucocorticosteroids[150] are shown to inhibit superoxide production by FMLP-, zymosan-, and latex bead-stimulated neutrophils. It was found[144] that the efficiency of free radical scavengers increased with their lipophilicity; therefore, ascorbyl palmitate was a more effective inhibitor than ascorbic acid.

A substantially larger group of compounds includes various inhibitors of NADPH oxidase and the enzymes participating in the activation of NADPH oxidase or affecting its activity. An important role in the activation process belongs to adenine compounds (ATP, ADP, AMP, and adenosine). The effects of these compounds depend on the applied stimuli and experimental conditions. At physiological concentrations, ATP augments superoxide production by FMLP- and PMA-stimulated neutrophils.[151-155] ATP apparently is unable to stimulate superoxide production by itself, but it significantly increases the rate of O_2^- formation when neutrophils were pretreated with cytochalasin B. It may be explained by the cytochalasin-mediated Ca^{2+} mobilization, as the stimulation of superoxide production by ATP is evidently due to an increase in cytosolic calcium mobilization.[152,154] It is important that the stimulatory ATP concentrations are close to those attained extracellularly at the sites of platelet thrombus formation.[153] Ward et al.[156] concluded that stimulation by platelets of neutrophil superoxide production is due to the release of adenine nucleotides (ATP or ADP).

On the other hand, ATP may manifest an inhibitory action. For example, the inhibition of superoxide production by PMA-stimulated human neutrophils with fructose 1,6-diphosphate (FDP) was explained by the FDP-stimulation of ATP synthesis.[157] In contrast to ATP, other adenine nucleotides and adenosine are mainly inhibitors of superoxide production by neutrophils. Thus adenosine inhibited superoxide production by FMLP-stimulated human neutrophils,[154,158] supposedly via the interaction with a specific cell surface receptor.[154,159] Ward et al.[151] reported the stimulatory action of ATP, ADP, AMP, and adenosine on the formation of the superoxide ion by FMLP-activated rat neutrophils; however, all these compounds did not affect superoxide production by phorbol ester- and immune complex-stimulated neutrophils. Cyclic AMP also inhibited superoxide production by FMLP-stimulated human neutrophils.[160] In contrast to the conclusion made in Reference 156, McGarrity et al.[161] explained the inhibitory action of platelet lysate and release products on superoxide production by FMLP-stimulated neutrophils by the presence in them of adenine nucleotides.

Fantone and Kinnes[162] have shown that prostaglandins 15-M-PtE$_1$ (15-[S]-11-methyl-prostaglandin E$_1$) and PGI$_2$ reversibly inhibited superoxide production by FMLP-stimulated neutrophils, but did not affect the rate of O_2^- formation in the case of PMA- or arachidonate-stimulation. It was proposed that the inhibitory effect of prostaglandins is due to an increase in the intracellular AMP level. Superoxide production by FMLP-stimulated neutrophils was also inhibited by prostaglandins PGD_2, leucine enkephalin, and methionine enkephalin.[163] Recently, Gryglewski et al.[164] confirmed the inhibitory effect of prostaglandins E_1, E_2, 6-keto-E_1, and D_2 on superoxide production by FMLP-stimulated human PMNs. PGD_2 and PGE_1 also inhibited the zymosan-stimulated production of oxygen radicals. However, prostacyclin and iloprost (a carbacyclin analog of prostacyclin) did not affect the production of oxygen radicals by both FMLP- and zymosan-stimulated PMNs.

There are many other compounds which are able to inhibit the release of oxygen radicals by neutrophils. Among them are unsaturated aldehydes (acrolein and crotonaldehyde),[165] *N*-ethyl maleimide,[166] histamine,[167] antibiotics doxorubicin,[168,169] rhein (4,5-dihydroxy-anthraquinone-2-carboxylic acid),[170] and coumermycin,[171] hydrocortisone and methylprednisolone,[172] dehydroepiandrosterone and the synthetic steroid 16α-Br-epiandrosterone,[173] retinol and retinol acetate (in contrast to retinoic acid, which stimulates superoxide production),[136] and a new immunomodulatory agent, fanetizole mesylate.[174] These compounds apparently affect calcium-dependent steps of the NADPH oxidase activation or (in the case of unsaturated aldehydes) react with sulfhydryl groups. Many nonsteroidal anti-inflammatory drugs and β-adrenergic agonists including naloxone,[175] salbutanol, fenoterol, isoprenaline,[176] auranofin, phenylbutazone, sulfasalazine, piroxicam, primaquine, indomethacin,[177] nimesulide,[178] antiarthritic gold-containing drugs,[180] diazepam,[179] the local anesthetic agent tetracaine-HCl,[181] and inhalation anesthetics (halothane, enflurane, and isoflurane)[86,87] are also inhibitors of leukocyte superoxide production.

The inhibitory effect of sulfasalazine is apparently due to the occurrence in this compound of the azo group, since a similar inhibitory effect was found in the case of the other azocompound, olsalazine.[182] Nonsteroidal anti-inflammatory drugs efficiently inhibited superoxide production in the case of FMLP-stimulation, but were relatively weak inhibitors (or did not inhibit at all) the IgG aggregates-stimulated PMNs.[177] Taniguchi and Takanaka[183] have shown that many antihistaminic agents, adrenergic β-antagonists, and antiarrythmic drugs inhibited PMA- and FMLP-stimulated superoxide production by guinea pig peritoneal PMNs. The inhibitor of cyclooxygenase indomethacin slightly affected the PMA-stimulated O_2^- production, but efficiently inhibited it in the case of FMLP stimulation.

Calcium channel antagonists, verapamil, diltiazem, and nisoldipine inhibited superoxide production by PMA-stimulated neutrophils.[184] It was proposed that the inhibitory action of these compounds was not explained by blocking the calcium flux because they affect both the catalytic activity and the activation of NADPH oxidase. Flament et al.[185] have also proposed that the inhibitory effect of the 5-lipoxygenase inhibitor piriprost on FMLP-stimulated superoxide production by human neutrophils did not involve lipoxygenase inhibition, but was due to the interaction with the FMLP receptor.

Ebselen (2-phenyl-1,2-benisoselenazol-3[2H]-one) inhibited superoxide production by TPA-stimulated guinea pig PMNs.[186] Its inhibitory effect may be a result of the direct interaction with NADPH oxidase or of a decrease in reduced glutathione since ebselen possesses glutathione peroxidase-like activity. Antibiotics erythromycin and roxithromycin inhibited FMLP, A 23187, and benoxaprofen stimulation of superoxide production by human neutrophils (but not stimulation by PMA and opsonized zymosan).[187] It was proposed that the antioxidative effect of these antibiotics was an origin of increasing neutrophil migration.

Me_2SO (DMSO) and Me_2SO_2 inhibited the production of oxygen radicals by opsonized zymosan- and PMA-stimulated human neutrophils.[188] Although DMSO is a well-known scavenger of hydroxyl radicals, the inhibitory effects of it and Me_2SO_2 were not due to their antiradical activity, as both compounds did not react with O_2^- in the cell-free system. Hurst et al.[189,190] proposed that the inhibition by chloroquine and hydroxychloroquine (antirheumatic drugs) of superoxide release by FMLP-stimulated human PMNs is due to suppressing the FMLP-induced hydrolysis of phosphoinositides. Bilirubin was a stronger inhibitor of superoxide production by neonatal PMNs stimulated with Con A + cytochalasin D than by PMA, which probably indicates its interaction with the membrane.[191]

Gennaro and Romeo[192] have shown that the inhibitors of the exchange of anions across the plasma membranes SIIS (4-acetamido-4′-isothio-cyanostilbene-2,2-disulfonate) and ANS (1-anilino-8-naphthalene sulfonate) inhibited release of the superoxide ion by phagocyting granulocytes exposed to *Bacillus mycoides*. On these grounds it was concluded that the superoxide ion exits from the granulocyte through an anion channel. Younes and Rolbke[193] have shown that the depletion of intracellular glutathione with phorone results in the inhibition of superoxide production by 50% in PMA-stimulated human granulocytes. It was therefore suggested that many xenobiotics which are able to interact with glutathione may affect the phagocytic activity of granulocytes.

The inhibitors of various enzymes may also influence superoxide production by neutrophils. Thus, the selective thromboxane synthetase inhibitor U 63557A significantly reduced superoxide production by FMLP-, zymosan-activated serum-, and A 23187-stimulated human PMNs in the presence of platelets.[194] The inhibitory effect of U 63557A is apparently explained by its stimulation of the prostaglandin PGI_2 release in the presence of platelets. The phospholipase A_2 inhibitor, 4-*p*-bromophenacyl bromide, inhibited $O_2^{\bar{\cdot}}$ production by human neutrophils stimulated by clofazimine (an antileprosy agent), while the protein kinase C inhibitor, H-7, did not.[118] Muid et al.[195] have studied the effect of a "cocktail" consisting of an inhibitor of DG kinase, R 59022, an inhibitor of DG lipase, RHC 80267, and an inhibitor of phospholipase A_2 (indomethacin or sodium meclofenamate) on superoxide production by stimulated human neutrophils. There was an insignificant inhibitory effect in the case of opsonized zymosan and even the stimulation in the case of a fluoride ion or gamma-hexachlorocyclohexane. Since the inhibitors applied inhibit the main pathways of arachidonate formation, these data apparently indicate that arachidonate can participate only in receptor-mediated transduction with the C3b/Fc receptor stimulus (opsonized zymosan).

The inhibitor of thiol protease NCO-700 (a compound related to L-*trans*-epoxysuccinylpeptides) inhibited FMLP-stimulated superoxide production by rabbit leukocytes, which indicates a possible role of thiol protease in the activation of NADPH oxidase.[196] However, NCO-700 also inhibited the SOD-inhibitable, luminol-dependent CL produced by xanthine oxidase. Therefore, this compound is also a scavenger of the superoxide ion.

C. PRODUCTION OF HYDROXYL RADICALS BY NEUTROPHILS

The production of a superoxide ion by xanthine oxidase and other enzymes usually triggers the formation of hydroxyl or crypto-hydroxyl radicals when iron ions or complexes are present (Chapter 1). Therefore, one may expect that the release of a superoxide ion by neutrophils must result in the generation of other active oxygen radicals if iron ions or complexes and hydrogen peroxide are present at the site of $O_2^{\bar{\cdot}}$ formation. On the other hand, a well-known oxygen-dependent cytotoxic activity of neutrophils and other phagocytes is easily explained by the formation of high reactive hydroxyl (or crypto-hydroxyl) radicals. Indeed, as early as 1975, Salin and McCord[197] showed that the viability of human PMNs phagocyting *E. coli* cells increased in the presence of SOD, mannitol, and catalase. As mannitol is a classic HO· scavenger, it was concluded that hydroxyl radicals are formed during phagocytosis.

This proposal was supported in subsequent works. Auclair et al.[198] have shown that the scavengers of hydroxyl radicals (Tris®, mannitol, and benzoate) inhibited NADPH oxidation and dioxygen uptake by the particulate (granule-rich) fraction isolated from zymosan-stimulated human leukocytes. The same fraction released ethylene from methional and 2-keto-methylthiobutyric acid (another test for detecting hydroxyl radicals); the reaction was inhibited by benzoate, ethanol, and mannitol.[199] The formation of ethylene from 2-keto-4-methylthiobutyric acid was also observed after the stimulation of guinea pig leukocytes with digitonin, linoleic acid, or bacteria.[200] Human leukocytes activated by opsonized zymosan oxidized ^{14}C benzoate to form $^{14}CO_2$.[201] The reaction was inhibited by SOD, catalase, azide, and mannitol. On these grounds the participation of hydroxyl radicals in the decarboxylation of benzoate was proposed. The presence of iron ions or complexes is apparently obligatory for the generation of hydroxyl radicals by neutrophils. For example, Thomas et al.[202] have shown that phagocyting PMNs are able to oxidize linoleyl alcohol only in the presence of iron salts.

All the above studies used indirect methods for detecting hydroxyl radicals. However, the spin-trapping technique was also successfully used for the detection of hydroxyl radicals produced by neutrophils. Thus, an ESR spectrum of a DMPO-OH spin adduct was recorded after adding DMPO to human neutrophils stimulated by IgG-coated latex particles or zymosan.[37] In the case of stimulation by PMA, a mixture of DMPO-OH and DMPO-OOH spin adducts was obtained. SOD inhibited the formation of spin adducts in all cases. A mixture of DMPO-OH and DMPO-OOH spin adducts was also formed by neutrophil NADPH oxidase in the presence of transferrin and hydrogen peroxide.[203]

Notwithstanding much evidence of the production of hydroxyl radicals by leukocytes obtained in earlier works, these results were later questioned. It is known that the DMPO-OH spin adduct may be formed not only in the reaction of HO· with DMPO, but also as a result of DMPO-OOH decomposition. Britigan and co-workers[204-206] made an attempt to estimate the contribution of the last process in the total yield of DMPO-OH. At the beginning, it was concluded[204] that PMA-stimulated human neutrophils generate hydroxyl radicals in the presence of iron ions, but zymosan-stimulated neutrophils do not, since in the last case SOD inhibited DMPO-OH formation and catalase did not. However, in both cases an attempt to convert HO· radicals to methyl radicals by the addition of DMSO gave only insignificant amounts of DMPO-Me. This fact indicates that most DMPO-OH adducts are formed from DMPO-OOH. It was proposed[205] that the extensive release of lactoferrin into the extracellular environment during lactoferrin degranulation (which accompanies neutrophil activation) may be a reason of inhibition of hydroxyl radical production, since lactoferrin chelates iron ions to form an inactive complex (Chapter 2). It has also been shown[207,208] that carefully prepared neutrophils stimulated by opsonized zymosan, PMA, and fatty acids produced at 20°C only a DMPO-OOH spin adduct. An increase in temperature up to 37°C induced the conversion of DMPO-OOH to DMPO-OH.

There are other data supporting doubts in the possibility of HO· generation by leukocytes. It was found[209] that DMPO-OH is formed in the reaction of DMPO with sodium hypochlorite. Since HOCl is formed by myeloperoxidase of neutrophils, which is active during the respiratory burst, it is possible that hypochlorite may at least in part be responsible for the DMPO-OH formation at the neutrophil activation. Winterbourn[210] has also shown that myeloperoxidase inhibits the production of hydroxyl radicals in the model system (xanthine plus xanthine oxidase) by destroying hydrogen peroxide. It was therefore concluded that hydroxyl radical production by neutrophils seems to be unlikely due to the release of myeloperoxidase during neutrophil activation.

Although the above data seem to make doubtful the possibility of the production of hydroxyl radicals by leukocytes, many important findings cannot be explained if we agree that leukocytes are incapable of generating these radicals. For example, Umeda et al.[211] have shown that cytotoxic activity of human neutrophils against cultural human tumor cells was partially inhibited by catalase and hydroxyl radical scavengers (DMSO, ethanol, etc.). This fact was interpreted as evidence of the participation of hydroxyl radicals in cell killing. It was also concluded that hydroxyl radicals may be produced by lipoxygenase and not NADPH oxidase as NDGA, a lipoxygenase inhibitor, efficiently suppressed the PMN-mediated cytolysis. It may be true, but one should remember that NDGA, a polyphenolic compound, is also an effective scavenger of hydroxyl radicals.

In contrast to the results obtained in References 204 and 205, Chiba et al.[212] have shown that the DMPO-Me spin adduct was formed in the presence of DMSO when guinea pig leukocytes were activated by cerebroside sulfur ester (a component of myelin membranes). Thus in this case, the DMPO-OH spin adduct must at least in part arise from hydroxyl radicals. Kleinhaus and Barefoot[213] estimated the concentration of hydroxyl radicals (detected as the DMPO-OH spin adducts) produced by opsonized zymosan- and PMA-stimulated human neutrophils which was equal to (0.35 to 0.55) nmol/10^6 cells. Recently, it was proposed[214] that a failure to detect hydroxyl radicals on the basis of DMPO-OH formation may be due to the destruction of the DMPO-OH formed by the superoxide ion. These authors observed a direct reaction between the superoxide ion produced by neutrophils and DMPO-OH or DMPO-Me spin adducts.

Additional evidence in favor of the generation of hydroxyl radicals or other active oxygen radicals different from the superoxide ion by leukocytes or macrophages was obtained with the aid of the CL technique. It was already pointed out that the difference in the lucigenin- and luminol-dependent CL produced by leukocytes may be a consequence of the formation of different active oxygen species. It has been shown[215] that human granulocytes stimulated by Con A produced both lucigenin- and luminol-dependent CL, whereas phytohemagglutin (PHA)-stimulated granulocytes produced only luminol-dependent CL. On these grounds, one may propose that active species responsible for CL in PHA-stimulated granulocytes are not superoxide ions. However, there is no reliable evidence that they are hydroxyl radicals. It has been shown[142] that stimuli can affect the $O_2^{\overline{\cdot}}/H_2O_2$ ratio during the activation of neutrophils. Probably due to this, stimuli may also change a ratio between the lucigenin- and luminol-dependent CL responses without producing new oxygen species.

III. EOSINOPHILS AND BASOPHILS

As was mentioned earlier, eosinophils and basophils form a small fraction (about 2 to 3%) of the total amount of granulocytes, which makes studying their oxidative metabolism more difficult. Eosinophils have important regulatory functions, and numerous diseases are associated with eosinophilia in blood and tissue. The most simple method of studying eosinophils is the isolation of eosinophils from the blood of patients with eosinophilia. Using this method, DeChatelet et al.[216] have shown that in spite of a lower rate of phagocytosis, the oxidative metabolism of phagocyting eosinophils is substantially greater than that of phagocyting neutrophils. It is believed that as in neutrophils, NADPH oxidase is a principal generator of oxygen radicals in eosinophils.

For activation of NADPH oxidase in eosinophils, particular stimuli (opsonized zymosan)[216-218] and soluble stimuli (PMA,[217,218] TPA, gamma-interferon (IFN), TNF,[219] and PAF[220] were applied. Zymosan-stimulated eosinophils and neutrophils generated approximately equal amounts of superoxide ion measured by CL or the cytochrome c reduction. Yoshie et al.[219] have shown that gamma-IFN, alpha-IFN, and TNF greatly increased the luminol-dependent CL induced by TPA-stimulated eosinophilic cells from leukemia line Eol-1. CL response was inhibited by SOD.

Since eosinophils from patients with eosinophilia may possess abnormalities in morphology and biochemistry, Shult et al.[217] and Petreccia et al.[218] have studied superoxide production by eosinophils isolated from the blood of normal donors. Again, superoxide production by opsonized, zymosan-stimulated eosinophils and neutrophils was nearly identical, but in the case of PMA-stimulation, a rate of superoxide formation by eosinophils was nearly twice that for neutrophils.[217] It was proposed that elevated production of the superoxide ion by PMA-stimulated eosinophils may be a consequence of an increase in the number or affinity of PMA receptors. Petreccia et al.[218] also showed that eosinophils displayed greater initial rates and duration of superoxide production than neutrophils. It was proposed that, unlike neutrophils, eosinophils probably have the preassembled respiratory burst oxidase which is responsible for superoxide production by resting cells. However, such an activity can be an artifact due to surface disturbance during cell preparation.

We found only one work concerning the production of oxygen radicals by basophils. Henderson and Kaliner[221] have shown that human leukemia basophils produced a superoxide ion upon immunologic (anti-human IgE) and nonimmunologic (nontoxic histamine-released agent) stimulation.

IV. MONOCYTES

Similarly to PMNs (neutrophils, eosinophils, and basophils), blood mononuclear cells (monocytes) are phagocytes and also are able to produce oxygen radicals. It has been shown that human blood monocytes produce oxygen radicals upon stimulation with opsonized zymosan, latex particles, aggregated human IgG, PMA, FMLP, A 23187, and PAF.[222-226] PMA and FMLP stimulation of superoxide production by monocytes was primed with thalidomide.[227] Similarly, gamma-IFN primed superoxide production by dexamethasone-treated human monocytes stimulated with opsonized zymosan, PMA, or live bacteria.[237]

For detection of oxygen radicals, spontaneous,[228] luminol-dependent,[229] and lucigenin-dependent CL,[230,231] as well as cytochrome c reduction,[223,227,230,232] were applied. Weiss et al.[223] used the NBT reduction that, in his opinion, permits separation of the intra- and extracellular production of the superoxide ion. It was found[222,224] that monocytes are significantly less effective generators of oxygen radicals than neutrophils. It is probably explained by the elevated SOD activity in monocytes which is about three times greater than that in neutrophils.[231] In spite of the similarity of the mechanisms of superoxide production by monocytes and PMN leukocytes, there seem to be some differences. Thus, Leonard et al.[233] have shown that the duration of superoxide production by human monocytes stimulated by FMLP is about 3 min, while it is equal to 15 to 20 min in the case of PMA stimulation. It was proposed that the FMLP activation of monocytes is mediated by at least two intermediates, one of them being inositol triphosphate or DG. Adenosine inhibited superoxide production stimulated by FMLP, but not by PMA.

Stimulation of human monocytes by Con A and A 23187 depended on the cytoplasmic ionized calcium.[234] Thus, the stimulation of human monocytes by Con A induced an increase in cytoplasmic calcium at 31 $\pm$ 6 s and the onset of superoxide generation at 61 $\pm$ 9 s. At the same time, Con A did not stimulate superoxide production in calcium-depleted cells until calcium was restored to the incubation medium. Costa-Casnellie et al.[235] proposed that the stimulation of monocytes by phorbol esters and Con A occurs by different pathways. TPA (13-tetradecanoate phorbol acetate) induced the translocation of protein kinase C from the cytosol to the membrane. It is believed that a tight association of protein kinase C with the membrane is required for superoxide production by TPA-stimulated monocytes. Contrary to this, the Con A stimulation of monocytes resulted in the shift of this enzyme from the membrane to the cytosol. Therefore, the stimulation of superoxide production by Con A (which was greatly enhanced by cytochalasin B) was independent of the tightly bound protein kinase C level.

The luminol-dependent CL induced by human monocytes was inhibited by the myeloperoxidase inhibitor azide by 80% and partly by SOD and catalase.[236] On these grounds, the authors proposed that the luminol-dependent CL produced by monocytes depends on myeloperoxidase and is in part of intracellular origin.

Various inhibitors of superoxide production by blood monocytes were described. 4-Aminoquinolines (chloroquine [CQ] and hydroxychloroquine [HCQ]) and 9-aminoacridine (mepacrine [MP]), which are applied for the treatment of rheumatoid arthritis and systemic lupus erythematosus, selectively inhibited superoxide production by human blood monocytes.[232] CQ and HCQ did not affect superoxide production by TPA-, A 23187-, and fluoride-stimulated monocytes, while MP inhibited it in the case of TPA and A 23187 stimulation. CQ and MP inhibited superoxide production by opsonized zymosan- and FMLP-stimulated monocytes, and HCQ inhibited only zymosan-stimulated superoxide production. It was proposed that the inhibitory effect of 4-aminoquinolines is due to their interaction with the surface receptors.

The anti-inflammatory drug piroxicam inhibited the superoxide release by FMLP-stimulated monocytes, but had no effect on superoxide production by opsonized zymosan-, PMA-, or A 23187-stimulated monocytes.[225] It was concluded that piroxicam inhibits neither protein kinase C nor NADPH oxidase, but interferes with FMLP receptor binding. Albumin inhibited superoxide production by TPA-, A 23187-, and ionomycin-stimulated monocytes supposedly due to the binding of these stimuli.[239] Superoxide production by monocytes was suppressed in rabbits fed with cholesterol, which may be a consequence of the effect of cholesterol on monocyte plasma membranes.[240]

The production of hydroxyl radicals by monocytes was studied only casually. Geffner et al.[241] have shown that nonspecific monocyte cytotoxicity induced by precipitating immune complexes or soluble heat-aggregated IgG was partially inhibited by SOD, catalase, mannitol, benzoate, and ethanol. On these grounds, it was proposed that active oxygen radicals (probably hydroxyl radicals) may participate in the lysis of nonsensitized target cells by activated monocytes. Heales et al.[242] proposed that the inhibition of luminol-dependent CL in zymosan-stimulated human monocytes by the reduced pterins (tetrahydrobiopterin, dihydrobiopterin, and dihydroncopterin) is due to their activity as hydroxyl radical scavengers. Recently, Britigan et al.[243] have shown that monocytes stimulated by PMA and opsonized zymosan formed the DMPO-OH spin adducts in the presence of ferric ions. There was no HO· formation when monocytes were activated without iron ions. In contrast, PMNs did not produce hydroxyl radicals with or without iron ions. These authors proposed that the formation of hydroxyl radicals by monocytes, but not by PMNs, is explained by the absence of lactoferrin in mononuclear phagocytes.

V. MACROPHAGES

Macrophages are monocyte-derived tissue phagocytes which, similar to granulocytes and monocytes, are able to produce oxygen radicals after stimulation with particulate and soluble stimuli. In 1975, DeChatelet et al.[15] did not detect superoxide production by opsonized zymosan-stimulated rabbit peritoneal and alveolar macrophages, using the SOD-inhibitable cytochrome c reduction as an assay for the superoxide ion. However, in the same year, Drath and Karnovsky[14] detected superoxide production by mouse and guinea pig alveolar and peritoneal macrophages. These discrepancies are probably explained by the fact that macrophages, as monocytes, produce in general a smaller amount of oxygen radicals than neutrophils (superoxide production by PMA-stimulated human alveolar macrophages and blood monocytes is approximately identical).[244] Due to this, the cytochrome c assay cannot always be used for the detection of a superoxide ion produced by macrophages. Indeed, Oyanagui[245] was obliged to use NADH oxidation in the presence of lactate dehydrogenase for detecting the superoxide ion produced by guinea pig macrophages, since the cytochrome c reduction was not a sufficiently sensitive method.

Nonetheless, in other works the SOD-inhibitable reduction of cytochrome c was widely used for the detection of the superoxide ion produced by macrophages.[226,246-253] Ando et al.[254] applied the SOD-inhibitable NBT reduction for measuring superoxide production by rabbit alveolar macrophages activated with bronchial lavage fluid. The spin-trapping technique with DMPO as a spin trap was applied for the detection of oxygen radicals produced by Bacillus-Calmette-Guérin-activated macrophages.[255] However, as in the case of neutrophils, a very promising method of superoxide detection is the CL technique with the use of luminol,[256-263] lucigenin,[264] or other sensitizers (such as a *Cypridina* luciferin analog, 2-methyl-6-phenyl-3,7-dihydroimidazo [1,2a]pyrazin-3-one).[265]

As in the case of blood phagocytes, NADPH oxidase is apparently a main superoxide generator in macrophages (the possibility of the production of oxygen radicals in phagocytes by lipoxygenase and cyclooxygenase is discussed below). This enzyme was isolated from mixed membrane particulate fraction of rat pulmonary macrophages.[266] However, the activation mechanism of NADPH oxidase in macrophages probably differs in some extent from that in neutrophils. Thus, Sakata et al.[267] have shown that the release of arachidonic acid rather than the activation of protein kinase C is the most important factor of the activation process of NADPH oxidase in macrophages. These authors also proposed that arachidonate may activate NADPH oxidase without activating protein kinase C, since the stimulation of superoxide production by arachidonate in a cell-free system was not inhibited by the protein kinase inhibitor H-7. The difference in the activation mechanisms of NADPH oxidase in macrophages and leukocytes is possibly explained by the difference in the intracellular location of cytochrome b, which is located in specific granules in leukocytes and in plasma membrane in macrophages.

Contrary to this, Tsunawaki and Nathan[268] concluded that there is no evidence of the fact that the release of arachidonic acid is an obligatory step in the activation of NADPH oxidase in mouse peritoneal macrophages, as nontoxic concentrations of exogenous arachidonate induced no superoxide production from intact macrophages. However, Ida et al.[269] recently confirmed the importance of the arachidonate-mediated pathway of NADPH oxidase activation. These authors found that the stimulation of macrophages by IgG_2 immune complexes resulted in both superoxide production and the release of arachidonic acid and its metabolites (prostaglandin E_2 and thromboxane B_2). Although the activation of protein kinase C by arachidonate probably contributed to the activation of NADPH oxidase, the arachidonate release seemed to play a major role in FcγR(receptor)-mediated superoxide production. In contrast, the activation of protein kinase C was apparently more important in the case of superoxide production by TPA-activated macrophages.

There seems to be a relationship between membrane potential changes and superoxide production in macrophages. Both phenomena increase during macrophage activation.[270] Hishinuma et al.[271] proposed that macrophages possess two pathways of stimulation of O_2^- production: a glucose-independent, NADPH-supplying pathway and a glucose-dependent pathway, since the superoxide-specific CL sensitized by the *Cypridina* luciferin analog was inhibited by the glycolic inhibitor 2-deoxyglucose only by 60%. Unexpectedly, Esterline and Trush[272] recently concluded that resting rat alveolar macrophages are able to produce a superoxide ion (measured by the lucigenin-dependent CL). It was concluded that the superoxide ion is generated by mitochondria and that lucigenin is able (in contrast to previous conclusions) to enter the cells and to interact with mitochondria.

As stimuli of superoxide production by alveolar and peritoneal macrophages, opsonized zymosan,[246,248,252,265,273] PMA,[246,273] and TPA[269] were used. Lectins (Con A, PHA, and wheat germ agglutinin) also stimulated superoxide production by guinea pig peritoneal macrophages.[250] The highest rate of superoxide release was obtained in the case of pretreatment with cytochalasin E. PAF induced superoxide production by monocyte-derived macrophages[226] and bone marrow-derived macrophages.[264] Its synthetic analog 1-*O*-octadecyl-2-*O*-methyl-*rac*-glycero-3-phosphocholine was incapable of stimulating the production of oxygen radicals by itself, but potentiated the PMA-induced release of the superoxide ion.[264] Unlike PMA and opsonized zymosan, PAF was also able to stimulate superoxide production at late stages of macrophage maturation.[226]

Human milk macrophages released oxygen radicals upon stimulation with opsonized zymosan or PMA.[273] The production of oxygen radicals could be primed by bacterial lipopolysaccharides. Superoxide production by mouse peritoneal macrophages was stimulated by *cis*-platin[274] and a monosaccharide precursor of *E. coli* lipid A.[275] Similarly, human C-reactive protein stimulated superoxide production by guinea pig peritoneal macrophages.[276] Bacterial lipopolysaccharides enhanced PMA-stimulated superoxide production by macrophages from normal rats, but have no effect on superoxide production by macrophages from tumor-bearing rats.[277]

Korkina et al.[257-259] have studied the effect of polyelectrolytes on the activation of peritoneal macrophages. It was found that polyanions (polyacrylates and polyphosphates) did not activate rat peritoneal macrophages, whereas synthetic (polymeric ammonium salts) and natural (protamine and polymyxin B) polycations stimulated the production of oxygen radicals by macrophages measured by luminol-dependent CL. It was proposed that the macrophage activation by polycations is a consequence of multiple electrostatic binding of a stimulus molecule to negatively charged receptors of cytoplasmic membrane.

Important stimuli of macrophage activation are also fibrogenic dusts. These stimuli were studied in detail[260-263] in connection with the investigation of the role of oxygen radicals in developing the dust-associated diseases. These authors found that the exposure of peritoneal macrophages to fibrogenic dust particles and fibers (crystalline quartz, amorphous silica condensate, chrysotile- and crocidolite-asbestos) induced the luminol-dependent CL inhibited by SOD, catalase, and hydroxyl radical scavengers. It was concluded that the structure of the oxygen radicals formed depended on the stimuli applied: quartz mainly stimulated the formation of the superoxide ion, whereas asbestos fibers seemed to stimulate the formation of both the superoxide ion and hydroxyl radicals. These results were confirmed in subsequent works. Case et al.[278] have found that the SOD-inhibitable cytochrome c reduction by hamster alveolar macrophages was stimulated by chrysotile and crocidolite fibers. Similarly, the stimulation of superoxide production by rat, hamster, and rabbit alveolar macrophages in the presence of mineral dusts (crocidolite, erionite, sepiolite, etc.) has been shown by Hansen and Mossman[279] and by Wilhelm et al.[280]

As in the case of PMNs, the activating effect of stimuli on macrophages can be primed by various compounds. Thus, priming by immunomodulators such as lymphokine supernatants, alpha- and gamma-IFN, polysaccharides, 6-*O*-stearoyl muramyl dipeptide, and somatotropin of PMA- and zymosan-stimulated macrophages has been shown.[281-284] Priming by gamma-IFN (the primary T cell product), somatotropin, and some other immunomodulators can be an important physiological process of macrophage activation. PMA stimulation of superoxide production by mouse peritoneal macrophages was enhanced by low concentrations of mercury chloride.[285]

Anticancer antibiotic bleomycin increased the production of oxygen radicals (measured as the cytochrome c reduction or the lucigenin- and luminol-dependent CL) generated by PMA- and latex-stimulated alveolar and peritoneal macrophages.[286,287] It was proposed that the priming effect of bleomycin may be explained by its interaction with macrophage membrane, the generation of the superoxide ion by the iron-bleomycin complex,[286] or by the acceleration of electron transfer via the NADPH oxidase system due to the binding of bleomycin to cytochrome b.[287]

DiGregorio et al.[288] applied an electrooptical method based on measuring the NBT reduction for the detection of the superoxide ion generated by single rat pulmonary alveolar macrophages. It was found that the respiratory burst was enhanced after the attachment of macrophages to the bottom of the culture dishes. Thus, as in the case of neutrophils, adherence stimulates superoxide production by macrophages.

Different types of macrophages may respond in different ways to the stimuli applied. Phillips and Hamilton[448] have shown that whereas murine resident peritoneal macrophages generated a superoxide ion in response to PMA and zymosan, murine bone marrow-derived macrophages (BMM) produced a superoxide ion only in the case of activation by zymosan, but not by PMA. However, BMM produced a superoxide ion in the presence of PMA when they were primed by cytokines (TNF, gamma-INF, etc.). It was therefore concluded that unprimed BMM were lacking in the PMA-dependent signaling pathway that was corrected by cytokines.

The macrophages of the central nervous system (the microglia) produced a superoxide ion upon stimulation with opsonized zymosan and PMA, but not by Con A or FMLP.[289] The amount of the superoxide ion produced by the microglia was equivalent to the O_2^- concentrations generated by neutrophils and monocytes. Strokes et al.[247] have shown that alveolar macrophages from Bacillus-Calmette-Guérin-vaccinated rabbits (which possess an increased efficiency of phagocytosis) produced significantly higher levels of the superoxide ion than alveolar macrophages from normal rabbits during the phagocytosis of zymosan particles. Both alveolar and interstitial rat lung macrophages produced oxygen radicals upon stimulation with PMA, IgA-immune complex, or IgG-immune complex.[290]

In 1976, Oyanagui[245] for the first time studied the inhibitory effect of anti-inflammatory drugs and some other compounds on superoxide production by peritoneal macrophages. The inhibitory effect of these compounds on O_2^- production by macrophages was compared with the inhibition of the generation of O_2^- by xanthine oxidase. This permitted separation of the inhibition due to the interaction of drugs with oxygen radicals (i.e., the antiradical activity of compounds) from the inhibition of activation of NADPH oxidase. It was found that for most anti-inflammatory drugs (diclofenac sodium, oxyphenbutazone, phenylbutazone, mefenamic acid, aspirin, etc.), IC_{50} values obtained in the experiments with macrophages were significantly lower than those obtained in the xanthine-xanthine oxidase system. The same was true for cytochalasin B, pyrogallol, and L-epinephrine. At the same time, the inhibition by ascorbate, chlorpromazine, and luminol was partly or completely due to the interaction of these compounds with oxygen radicals. The inhibitory effect of ascorbic acid on superoxide production by alveolar and peritoneal macrophages was also shown in subsequent works.[291,292]

In addition to SOD, whose inhibitory effect on the production of oxygen radicals by macrophages was shown in many works, other scavengers of free radicals are also effective inhibitors. Thus, it has been shown[292] that the flavonoid rutin and NDGA were very effective inhibitors of the production of oxygen radicals by asbestos-stimulated peritoneal macrophages. Both compounds directly reacted with oxygen radicals since they manifested the similar inhibitory effect in cell-free systems.

Ebselen (PZ 51, 2-phenyl-1,2-benzisoselenzol-3-[2H]-one) and its sulfur analog PZ 25 inhibited superoxide production by zymosan-stimulated mouse peritoneal macrophages[293] and by FMLP- and PMA-stimulated guinea pig alveolar macrophages.[253] Superoxide production was measured by the luminol-dependent CL and the cytochrome c reduction, respectively. It was proposed that the inhibitory effect of ebselen on superoxide production by neutrophils and macrophages may be due to its glutathione peroxidase-like activity[186] or peroxide scavenging action.[293] However, PZ 25, a sulfur analog of ebselen, was devoid of glutathione peroxidase-like activity, but nonetheless inhibited the generation of a superoxide ion. Therefore, the inhibitory effects of both compounds should be explained by other reasons, possibly by the direct inhibition of protein kinase C.[253]

Anticancer drugs (doxorubicin, vincristine, and dexamethasone) inhibited superoxide production and suppressed the antibody-dependent cellular cytotoxicity induced by PMA-stimulated rat peritoneal macrophages.[294]

The Ca^{2+} channel blocker verapamil inhibited the Con A-stimulated respiratory burst in rat alveolar macrophages in the presence of extracellular calcium.[253] It was also shown that both the calcium-independent PMA-stimulation of superoxide production and the calcium-dependent Con A and A 23187 stimulation of alveolar macrophages were inhibited by hyperoxia (O_2 exposure). *p*-Nitrophenyl-*p'*-guanidinobenzoate (NPGB), a protease inhibitor, inhibited the formation of a superoxide ion by PMA-stimulated peritoneal macrophages supposedly by blocking the activation of the membrane-bound NADPH oxidase or by acting on the active enzyme.[295] NPGB did not react with the superoxide ion as it did not inhibit superoxide production by xanthine oxidase. In contrast to previous findings,[296] Zeidler and Conley[252] found that adenosine inhibited the production of oxygen radicals by zymosan-stimulated pig alveolar macrophages supposedly by suppression of the methylation of membrane-bound phospholipids. As in the case of PMNs, acrolein and crotonaldehyde inhibited superoxide production by PMA-stimulated rat alveolar macrophages.[165] Diphenylene-iodonium[297] and cepharanthine[298] were inhibitors of the generation of oxygen radicals by PMA-stimulated peritoneal macrophages. Superoxide production by macrophages was also inhibited by activated mast cells or mast cell granules supposedly due to scavenging of the superoxide ion by mast cell granule-bound SOD.[299]

VI. GENERATORS OF OXYGEN RADICALS IN PHAGOCYTES, OTHER THAN NADPH OXIDASE

NADPH oxidase is undoubtedly a major generator of oxygen radicals in PMNs, monocytes, and macrophages. However, there are other enzymes which may be responsible for the production and release of oxygen radicals by phagocytes, namely, lipoxygenase and cyclooxygenase. We already discussed the ability of these enzymes to generate a superoxide ion (Chapter 2), which is apparently formed as a result of the one-electron oxidation of NADH or NADPH. The use of specific inhibitors led some authors to the conclusion that these enzymes may be alternative sources of oxygen radicals in phagocytes. Thus, Smith and Weidemann[300] have shown that eicosatetraenoic acid, an inhibitor of lipoxygenase and cyclooxygenase, inhibited the calcium-dependent CL induced by activated peritoneal macrophages, while indomethacin, an inhibitor of only cyclooxygenase, did not. On these grounds, it was proposed that the calcium-dependent component of CL is generated by lipoxygenase. Contrary to this, the inhibitors of both cyclooxygenase (aspirin and indomethacin) and lipoxygenase (NDGA) inhibited the luminol-dependent CL of zymosan-stimulated human PMNs and monocytes.[301]

Ginsburg et al.[302] have shown that the lipoxygenase inhibitors strongly suppressed superoxide production by human neutrophils stimulated with lipoteichoic-antilipoteichoic acid complexes. Hume et al.[255] concluded that the lipoxygenase pathway of oxygen radical production is of importance for the PMA-activated Bacillus-Calmette-Guérin-stimulated mouse peritoneal macrophages. This conclusion was based on the fact that in the presence of the spin trap DMPO, spin adducts of carbon-centered free radicals were obtained in addition to the DMPO-OOH and DMPO-OH spin adducts. Lewis et al.[303] also proposed that the release of oxygen radicals by macrophages may occur via the lipoxygenase pathway.

It was found[256] that the formation of hydroxyl radicals (measured as the ethylene formation from 2-keto-4-thiomethylbutiric acid) and the luminol-dependent CL induced by PMA-stimulated murine peritoneal macrophages in the presence of murine alpha-interferon was inhibited by the lipoxygenase inhibitor NDGA. At the same time, the luminol-dependent CL was not affected by SOD and catalase. On these grounds, it was assumed that hydroxyl radicals were formed by the lipoxygenase pathway of arachidonate metabolism. Cheung et al.[304] also concluded that lipoxygenase is an origin of hydroxyl radicals detected as the DMPO-OH spin adducts upon the arachidonate stimulation of human neutrophils. The formation of DMPO-OH was inhibited by SOD, mannitol, and the lipoxygenase inhibitors NDGA and *N*-ethylmaleimide. The production of hydroxyl radicals was also demonstrated in the model cell-free system soybean lipoxygenase-arachidonate.

The inhibitory analysis has, of course, some shortcomings. For example, as was pointed out above, the inhibitory action of the lipoxygenase inhibitor piriprost on superoxide generation by neutrophils apparently is not due to the participation of lipoxygenase in the production of oxygen radicals.[185] The inhibitory effect of NDGA (a polyphenolic compound) may also be due to the direct interaction of $O_2^{\cdot -}$ and HO· with this compound.[292] Nonetheless, lipoxygenase and possibly cyclooxygenase can apparently participate in the production of oxygen radicals by phagocytes, although their contributions may depend on the type of phagocyte (see also below).

From the membrane fraction of PMA-stimulated porcine neutrophils, Sakane et al.[305] isolated a new enzyme, NADPH-cytochrome c reductase, which is believed to oxidize NADPH to form a superoxide ion when it is combined with cytochrome b_{559}. It was proposed that it is another enzyme responsible for the respiratory burst in neutrophils which, in contrast to NADPH oxidase containing only FAD, has both FAD and FMN.

VII. THE PRODUCTION OF OXYGEN RADICALS BY OTHER CELLS

It has been long and rightly believed that the generation of oxygen radicals by whole cells is an essential property of migrated phagocytes, granulocytes, monocytes, and macrophages in which the respiratory burst may be stimulated by various physiological and nonphysiological stimuli. However, subsequent investigations showed that many other cells, including fixed tissue phagocytes, and nonphagocytes, can produce oxygen radicals. It seems that there are three different pathways for the generation of oxygen radicals by these cells. (1) The cells may contain the same enzymes (NADPH oxidase, lipoxygenase, and cyclooxygenase) which generate oxygen radicals in phagocytes. (2) The cells may develop the NADPH oxidase activity as a result of differentiation induced by some stimuli. (3) The cells may produce oxygen radicals by the enzymatic or nonenzymatic pathways different from those in phagocytes. Unfortunately, in many cases, the origin of the production of oxygen radicals in these cells are uncertain.

Glomerular mesangial cells possess some properties of phagocytes, including the capacity to produce oxygen radicals after stimulation. It has been shown that cultured mesangial cells from rat glomeruli produced a superoxide ion (measured by the SOD-inhibitable cytochrome c reduction) upon stimulation with zymosan,[306] immune complexes,[307] and complement membrane attack complexes.[308] Baud et al.[306] proposed that stimulated mesangial cells produced oxygen radicals via the lipoxygenase pathway. It has also been shown[309] that isolated rat glomeruli (containing epithelial and mesangial cells) produced the luminol-dependent CL in response to serine proteases, trypsin and chymotrypsin. CL was inhibited by the scavengers of the superoxide ion and hydroxyl radicals (SOD, benzoate, and tryptophan, but not catalase). Production of the superoxide ion by mesangial cells stimulated by complement membrane attack complexes (28 to 68 nmol $O_2^{\cdot -}$ per milligram cell protein per 30 min) was comparable to that produced by stimulated macrophages and neutrophils.[308] It was proposed that the generation of oxygen radicals may be of importance in developing the immune-mediated glomerulonephritis in which mesangial cells play the role of effectors.

Another type of fixed tissue phagocytosing cells able to produce oxygen radicals are endothelial cells, which are integral constituents of blood vessels. It was proposed that release of the superoxide ion by these cells may induce the changes in microvascular environment and stimulate chronic inflammation. Matsubara and Ziff[310] have shown that PMA and the calcium ionophore A 23187 synergistically increased the SOD-inhibitable reduction of cytochrome c by human endothelial cells. The stimulation effect of PMA on superoxide production suggests the involvement of protein kinase C. Cytokines (interleukin 1 and recombinant gamma-INF) also stimulated superoxide production by human endothelial cells.[311] Release of the superoxide ion by endothelial cells sharply increase during phagocytosis of polystyrene microspheres or chylomicron-size lipid particles as well as upon stimulation with PMA.[312]

Endothelial cells also released the superoxide ion when the cells were subjected to anoxia and reoxygenation.[313,314] Zweier et al.[314] proposed that $O_2^{\cdot-}$ and HO· produced by bovine endothelial cells subjected to anoxia and reoxygenation and detected as DMPO-OOH and DMPO-OH spin adducts are generated by xanthine oxidase, since the production of oxygen radicals was inhibited by the xanthine oxidase inhibitor oxypurinol. Unfortunately, the authors did not discuss the question of permeability of plasmalemma by oxygen radicals generated in the cytosol.

Oxygen radicals are also generated by the cells which became capable of most phagocyte functions as a result of differentiation along the granulocyte pathway. The most studied cells of this type are human leukemic cells (HL-60 cells) induced to differentiate by DMSO. These cells produce a superoxide ion upon stimulation with PMA.[205,315,316] Superoxide production by PMA-stimulated HL-60 cells was equal to 0.4 to 7 nmol/min per 10^6 cells and was comparable to that by PMNs (11 nmol/min per 10^6 cells). It has been shown[315] that the HL-60 cells contain NADPH oxidase.

Peterson et al.[317] have shown that well-differentiated murine lymphoid tumor cells produced a superoxide ion which was detected by the NBT reduction. These authors believe that it is spontaneous unstimulated superoxide production by nonphagocytic cells. The NBT reduction was stimulated by gamma-linoleic acid and decreased when the cells were loaded with SOD. Human leukemic blast cells produced a superoxide ion upon stimulation with PMA.[318] Human B-lymphoblast cells developed by transformation with Epstein-Barr virus (EBV) from human peripheral blood lymphocytes are similar to monocytes. Thus, they produced the lucigenin-dependent CL upon stimulation with PMA.[319] CL was a consequence of superoxide formation since it was inhibited by SOD, but was not affected by catalase and mannitol. It was supposed that the EBV transformation uncovers genetic information which leads to developing the NADPH oxidase activity. It is of interest that in contrast to PMA stimulation, FMLP and A 23187 were incapable of stimulating production of oxygen radicals by EBV-transformed B cells.[320]

It has been shown[321-323] that PMA induces differentiation of human monocyte-macrophage-like histiocytic leukemia U 937 cells which was accompanied by development in these cells of the superoxide-generating capacity upon the appropriate stimulation. Indeed, the PMA-induced U 937 cells generated a superoxide ion in the presence of 4β-phorbol-12,13-didecanoate (PDD), OAG, serum,[321,323] or Con A.[322] In the case of Con A stimulation, superoxide production by the U 937 cells exceeded by 10 to 20 times superoxide production by Con A-stimulated monocytes and neutrophils. The lucigenin-dependent CL induced by serum-, PDD-, and OAG-stimulated human U 937 cells was inhibited by the protein kinase inhibitor H-7 and the calmodulin antagonist W-7.[321] It has been proposed[323] that superoxide production by these and some other nonphagocytic cells upon stimulation with mitogens may perform other physiological functions than superoxide production by phagocytes, being, for example, a major signal for the activation of Na^+/H^+ antiport.

Walker carcinosarcoma cells stimulated by FMLP produced the luminol-dependent CL which was inhibited by SOD, catalase, mannitol, and the cyclooxygenase inhibitor flurbiprofen.[324] In contrast, the lipoxygenase inhibitor NDGA did not affect CL. The cells also reduced acetylated cytochrome c. It was proposed that carcinosarcoma cells may produce oxygen radicals by the cyclooxygenase pathway.

Das et al.[325] have found that polyunsaturated fatty acids (PUFA) (arachidonic, linoleic, etc.) increased the reduction of NBT by breast cancer cells, but not normal human fibroblasts. It was proposed that the enhancement of the production of oxygen radicals by PUFA in cancer cells may be due to the low SOD and glutathione peroxidase activities in these cells. Rieder et al.[326] have shown that guinea pig Kupffer cells produced a superoxide ion in response to the phagocytosis of zymosan. In comparison with resident peritoneal macrophages, Kupffer cells generated about 60% O_2^-. Twice as high a superoxide release was achieved upon the stimulation of these cells with lipopolysaccharide from *Salmonella minnesota*, muramyl dipeptide, or recombinant gamma-INF.

The possibility of generation of oxygen radicals by lymphocytes was discussed in several works. Attention was mostly attached to the large granular lymphocytes named natural killers (NK), which are able to lyse certain tumor cells without prior sensitization. Suthanthiran et al.[327] have shown that the activity of NK cells was inhibited by the scavengers of hydroxyl radicals (DMSO, thiourea, dimethylurea, benzoic acid, etc.), but not by SOD or catalase. In addition, it was substantially inhibited by NDGA, a lipoxygenase inhibitor. On these grounds, it was proposed that NK cells produced hydroxyl radicals by the lipoxygenase pathway. These findings and the importance of the lipoxygenase pathway for the generation of hydroxyl radicals were completely confirmed in subsequent works.[328,329] However, it was recently shown[330] that although antioxidants (*n*-propyl gallate and catechin) do inhibit the NK-mediated cell killing, there is no evidence of the formation of DMPO-OH spin adducts in the early postbinding phase of NK cytolysis. Therefore, these authors proposed that free radical scavengers may inhibit NK cytolysis by a nonradical mechanism. However, this conclusion does not seem very convincing since there are many reasons why DMPO-OH adducts may not be formed or rapidly destroyed in the cell systems. In the case of the nonradical mechanism, the effect of lipoxygenase inhibitors remains also unexplained.

Chaudhri et al.[331,332] have shown that antioxidants and chelators inhibited the activation and proliferation of T-lymphocytes induced by alloantigen or a combination of PMA and the calcium ionophore ionomycin. On these grounds, it was concluded that oxygen radicals must be produced during the early stages of T-lymphocyte activation. However, Melinn and McLaughlin[333] found that although both T-lymphocytes and non-T-lymphocytes reduced NBT upon Con A stimulation, they did not induce the lucigenin-dependent CL that strongly refuted the possibility of superoxide production.

Markus et al.[334] have shown that aggregated and unaggregated platelets produced a superoxide ion which was detected via the cytochrome c and NBT reduction. Similarly, Henderson and Kaliner[221] have found that upon immunologic (anti-human IgE) and nonimmunologic (nontoxic histamine-releasing agent) stimulation, rat peritoneal mast cells and human lung mast cells produced a superoxide ion detected by CL and cytochrome c reduction. A superoxide ion was also generated by prokaryotic cells (*Streptococcus faecalus*), which produced the SOD-inhibitable CL,[335] and by bacterial suspensions of *Brevibacterium flavum*, *Mycobacterium lacticola*, and *Micrococcus glutamicus*, which reduced cytochrome c, oxidized epinephrine, and stimulated the luminol-dependent CL.[336]

Fischer and Adams[337] have shown that TPA- and mezerein (a second-stage tumor promoter)-stimulated mouse epidermal cells produced the luminol-dependent CL inhibited by SOD and the SOD mimetic Cu(II)(3,4-diisopropylsalicylic acid)$_2$, but not catalase and mannitol, which indicated the production of a superoxide ion. Superoxide production was also inhibited by lipoxygenase inhibitors (NDGA and benoxaprofen), but not the cyclooxygenase inhibitors indomethacin and flurbiprofen. On these grounds, it was concluded that an enzyme responsible for superoxide production by epidermal cells is lipoxygenase.

Unstimulated monkey arterial smooth-muscle cells and human arterial smooth-muscle cells produced a superoxide ion with a rate of 0.3 to 3.0 nmol/min/mg protein, which is equal to 1 to 13% of superoxide production by PMA-stimulated human monocytes.[338] In contrast to stimulated neutrophils and monocytes, generation of the superoxide ion by smooth-muscle cells was continuous. The mechanism of superoxide production by these cells remains obscure.

In Chapter 2 we discussed production of the superoxide ion in cells induced by various xenobiotics. In contrast to the generation of oxygen radicals by NADPH oxidase or lipoxygenase which takes place in plasmalemma, xenobiotics stimulate formation of the superoxide ion inside the cells, for example, in mitochondria. The studies of permeability of lipid membranes by $O_2^{\overline{\cdot}}$ (see above) showed that extracellular release of the superoxide ion originating from the inside cellular components must be negligible. Recent findings confirm this conclusion. Kahl et al.[339] have studied release of the superoxide ion by rat hepatocyte suspensions after the uptake of quinones using the lucigenin-dependent CL. It was found that the observed CL is apparently produced by the damaged cells which generate the superoxide ion at a much greater rate than intact cells. Therefore, these experiments give no evidence of superoxide release by intact hepatocytes. A similar conclusion was achieved by studying intracellular superoxide production by rat cerebral cortical astrocytes.[340] It has been shown that in the presence of arachidonate and other PUFA, these cells sharply increased intracellular NBT reduction, evidently due to formation of the superoxide ion. However, the SOD-inhibitable NBT reduction by intact cells was insignificant, obviously due to the impermeability of SOD into the cells and the absence of superoxide release outside the cells.

Aitken and Clarkson[341] have shown that the calcium ionophore A 23187 greatly stimulated the production of oxygen radicals by human spermatozoa. Oxygen radical production depended on the presence of calcium and was not affected by mitochondrial inhibitors (antimycin and rotenone) or the scavengers of hydroxyl radicals. However, cytochrome c reduced the generation of oxygen radicals by 50% therefore, spermatozoa apparently produce a superoxide ion upon A 23187 stimulation.

Takahashi et al.[342] recently found that the MCLA-dependent CL emitted by sea urchin eggs during fertilization was inhibited by SOD. The authors concluded that the generator of the superoxide ion in this process was probably ovoperoxidase of cortical granules.

VIII. PLANT CELLS

In preceding sections we have considered the production of oxygen radicals by animal cells. Many plant cells are also able to produce oxygen radicals, but the mechanisms of oxygen radical production in plants differ considerably from those in animal tissues. The most important generators of oxygen radicals in plants are apparently chloroplasts.[343,344] The superoxide ion was detected in illuminated chloroplasts on the basis of the oxidation of sulfite,[345] epinephrine,[346] ascorbate,[347-349] and hydroxylamine[350] and the reduction of cytochrome c.[346,351,352] Some of these reactions were inhibited by SOD. In the pioneer work of 1975, Harbour and Bolton[353] described the detection of a superoxide ion in illuminated chloroplasts by the formation of a DMPO-OOH spin adduct. In subsequent works the DMPO-OOH spin adducts were obtained during the illumination of chloroplasts and photosystem I particles,[354] in pea chloroplasts in the presence and absence of artificial photosystem I electron acceptors such as paraquat,[355] and in lupine chloroplasts under normal and pathological conditions.[356,357] The superoxide ion is formed not only in chloroplasts from high plants, but also by the photoreaction center of photosynthetic bacteria.[358]

It was proposed[343] that the superoxide ion is formed at the reducing side of photosystem I of chloroplasts. The compound responsible for the one-electron reduction of dioxygen is probably the reductant with a reduction potential of about -0.530 V which is formed during the illumination of chloroplasts. A major candidate for the low-potential reductant is ferredoxin.[349,344,350,355] Artificial photosystem I electron acceptors such as paraquat are also very effective generators of the superoxide ion.[349,350] It has also been shown[348] that superoxide production by chloroplasts is stimulated by ascorbate which simultaneously scavenges oxygen radicals. The last reaction is inhibited by SOD. The rate of superoxide production by spinach chloroplasts is about 15 μM/mg chlorophyll per hour.[343] Asada et al.[343] concluded that the superoxide ion is not formed during the photooxidation of water by photosystem II. However, Allen and Hall[347] have shown that it may be generated by photosystem II in the presence of paraquat which is reduced by hydrogen peroxide. Upham and Jahnke[359] proposed that a superoxide ion formed in photooxidation of chloroplast thylakoids promotes the formation by the Fenton reaction of hydroxyl radicals responsible for subsequent photoinhibition.

In addition to superoxide production dependent on one-electron reduction of dioxygen by photosystems I and II, cells from higher plants have superoxide-producing enzymatic systems similar to those in animal cells. Such a catalytic system apparently presents in the cell walls. It was proposed[360,361] that the superoxide ion is formed in cell walls as a result of dioxygen reduction by NADH in the presence of peroxidase and coniferylalcohol. The participation of peroxidase in the generation of the superoxide ion was also assumed by studying the SOD-inhibitable oxidation of epinephrine by intact pea roots[362] and rice leaves.[363]

On the other hand, Doke[364,365] proposed that superoxide production by plant cells is catalyzed by NADPH oxidase located in the plasma membrane. Recently, it was also proposed[366] that radish plasma membranes contain NADPH cytochrome c oxidoreductase and NADPH oxidase, which reduce dioxygen to a superoxide ion. This proposal was based on the fact that dioxygen consumption and NADPH oxidation by radish plasmalemma-enriched fractions were inhibited by SOD and catalase. However, it was concluded that analogy of these enzymes with NADPH oxidase of leukocytes seems to be doubtful because the NADPH oxidation by radish plasma membranes was inhibited by EDTA. The participation of other enzymes in oxygen radical production by plant cells cannot be excluded. Thus, Lynch and Thompson[367] proposed that superoxide production by the microsomal membranes and the cytosol from bean cotyledons is catalyzed by lipoxygenase as it was inhibited by the lipoxygenase inhibitor U 28938. Both cytosolic lipoxygenase activity and superoxide production (measured as the formation of tyron semiquinone) increased when senescence progressed.

Up-to-date studies show that an important mode of the activation of superoxide-generating systems in plant cells is the response of a plant to infection.[361] In this case, it is tempting to assume that the enzymatic systems producing the superoxide ion in plants are similar to those of phagocytes. It has been shown[368] that aged potato tuber slices produced the superoxide ion after Tungus penetration of potato cells by incompatible, but not compatible, races of *Phytophthora infestans*. It was proposed that superoxide production may be an origin of hypersensitive cell death. Superoxide production also sharply increased after infection of potato leaves with *P. infestans*,[372] of tobacco leaves with tobacco mosaic virus,[369] and of rice leaves with the pathogenic fungus *Pyricularia oryzae*,[363] after infestation of the roots of resistant tomato cultivars with the nematode *Meloidogyne incognita*,[370] etc. It is of interest that an increase in the capacity of iron-deficient bean roots to reduce ferric complexes (Turbo reductase) is also mediated by the superoxide ion.[371]

In conclusion, it should be noted that the activation of superoxide-generating enzymatic systems of plants is a very complicated process which is now still poorly understood. There are at least two stages of superoxide production by plant tissues during the host-parasite interaction:[372] at the first stage, the superoxide ion is generated before parasite penetration, possibly due to the contact of plant cells with some substances released by a parasite; at the second stage, a much greater stimulation of superoxide production takes place during the incompatible interaction of plant cells with a parasite. Other peculiarities of the production of oxygen radicals by plant cells are to be studied in future works.

IX. ADDITIONS

It was recently proposed[373] that lysosomal fusion is a prerequisite for the induction of luminol-dependent CL during activation of granulocytes. Vilim and Wilhelm[374] concluded that luminol-dependent CL does not permit discrimination among various oxygen species produced by activated phagocytes. Based on studies of solubilized activated neutrophil

NADPH oxidase, Ellis et al.[375] confirmed that a main pathway for electron transport in this enzyme is NADPH → FAD → cyt b_{-245} → O_2. Pilloud et al.[376] have studied the parameters of activation of superoxide-generating NADPH oxidase from bovine neutrophils in a cell-free system. Pou et al.[377] have shown that the superoxide ion produced inside the zymosan-stimulated neutrophils can be trapped by DMPO. For removing extracellular $O_2^{\bar{\cdot}}$, polyethylene glycol-modified SOD was used.

Ohsaka et al.[378] have studied the relationship between superoxide production, membrane depolarization, and an increase in cytoplasmic calcium in activated human granulocytes. Dahlgren[379] compared the effects of SOD and catalase on luminol-dependent CL and cytochrome c reduction by FMLP- and PMA-stimulated human PMNs. Greenwald et al.[380] applied a test based on the degradation of deoxyribose for detection of hydroxyl radicals produced by PMNs.

Ligeti et al.[381] have studied the activation of neutrophil NADPH oxidase in the cell-free system, consisting of a particulate fraction, the cytosol, arachidonic acid, and nonhydrolyzable nucleotide GTP-γ-S. Kuroki and Minakami[382] proposed that increased calcium influx stimulated by ATP in cytochalasin B-treated human neutrophils triggers superoxide production. Gyllenhammar[383] has studied the kinetics of superoxide production by LTB_4-stimulated human neutrophils. Based on the inhibitory effects of ethanol and butanol on FMLP-stimulated superoxide production by human neutrophils, it was concluded that the phospholipase D activation is linked to superoxide production in the human neutrophil.[384] Gresham et al.[385] have shown that the superoxide ion, hydrogen peroxide, and lactoferrin are needed for amplifying the Fc receptor-mediated phagocytic function of human PMNs and monocytes.

There are new works concerning the activation of PMN superoxide production by various stimuli, including a mixture of arachidonic acid and cytosolic proteins,[386] a monoclonal antibody raised against PMNs,[387] a combination of fatty acids and OAG,[388] guanine nucleotides,[389] 1,2-didecanoyl-3-*sn*-phosphatidate,[390] etc. It has been shown[391] that recombinant human TNF and recombinant human lymphotoxin enhanced superoxide release by nylon fiber-adherent PMNs. Woodman et al.[392] have studied the effects of recombinant human colony stimulating factors, the protein synthesis inhibitor cycloheximide, and dihydrocytochalasin B on superoxide production by FMLP-stimulated human neutrophils. Mege et al.[393] have found that GM-CSF stimulated superoxide production by FMLP- and PAF- (but not zymosan- and PMA-) activated human neutrophils. It was proposed that GM-CSF potentiates the rise in calcium level produced by PAF and FMLP. It was found that riminophenazines exhibited a strong priming effect on FMLP-activated human neutrophils.[394] Briheim et al.[395] compared the production of oxygen radicals from primed and nonprimed human PMNs.

Inhibitory effects on the production of oxygen radicals by activated leukocytes were shown for sulfasalazine and its derivatives,[396] exogenous 15-HETE,[397] antibiotic cerulenin,[398] chlorpromazine, mepacrine, and some other antipsychotropic and antihistamine drugs,[399] and prostanoids.[400] Lee et al.[401] proposed that the inhibitory effects of atropine, phentolamine, and propranol on superoxide production by human leukocytes are due to their inhibition of calcium influx and NADPH oxidase activity. Jacob[402] has found that calcium ionophores A 23187 and losalocid inhibited superoxide production by FMLP-stimulated rabbit neutrophils. Cotgreave et al.[403] have shown that the suppression by ebselen of oxygen radical production by human granulocytes is due to its inhibitory effect on NADPH oxidase and protein kinase C. Zimmerman et al.[404] have studied the effects of various calcium channel antagonists on superoxide production by human neutrophils. Henderson et al.[405] have found that phospholipase A_2 inhibitors suppressed superoxide production by NADPH oxidase.

Dileepan et al.[406] concluded that the inhibitory effect of mast cell granules on superoxide production by PMA-stimulated and nonstimulated human eosinophils apparently is due to the SOD activity of the granules. Sedgwick et al.[407] compared the effects of various stimuli on superoxide production by normal human eosinophils and neutrophils. It has been shown[408] that gamma-IFN and LPS overcome the inhibitory effects of glucocorticoids on priming superoxide release by PMA- and FMLP-activated human monocytes. Witz and Czerniecki[409] have shown that tumor promoters resulted in superoxide production by murine peritoneal exudate cells.

It was concluded[410] that β_1- and β_2-adrenoreceptors regulate the production of oxygen radicals by bovine pulmonary alveolar macrophages because dobutamine and isoproterenol inhibited the macrophage zymosan-stimulated, luminol-dependent CL. Hishinuma et al.[411] assumed that both protein kinase C and calmodulin are involved in superoxide production by mouse macrophages activated with opsonized zymosan. Ryer-Powder and Forman[412] used Mn(IV)-desferrioxamine for the measurement of the superoxide ion produced by adhering lung macrophages. DiGregorio et al.[413] developed a kinetic model, describing production of the superoxide ion by single pulmonary alveolar macrophages.

It has been shown[414] that quinones (benzoquinone, menadione, and doxorubicin) exhibit a double effect on latex- and polyelectrolyte-activated rat peritoneal macrophages: they stimulated superoxide production at low concentrations (possibly affecting the activating process of NADPH oxidase) and inhibited it at high concentrations (by way of scavenging the superoxide ion). Gudewicz[415] found that doxorubicin inhibited the antibody-dependent cellular cytotoxicity and superoxide production by PMA-stimulated rat peritoneal macrophages. Zeidler et al.[416] have shown that bleomycin enhanced superoxide production in the most active alveolar macrophage subpopulation. Stimulatory effects on superoxide production by macrophages were also shown for endotoxin (in the case of zymosan- and PMA-activated hepatic macrophages from endotoxin-treated rats[417] and Bacillus-Calmette-Guérin-primed mouse macrophages),[418] ethanol,[419] chrysotile-asbestos (but not for amphiboles such as crocidolite),[420] and murine neuroblastoma cells or released by them [Met^5] enkephalin.[421] In contrast to findings obtained in References 417 and 418, it has been shown[422] that endotoxin inhibited superoxide production by human blood-derived macrophages stimulated with zymosan or PMA. Czerniecki and Witx[423] proposed that tumor promoters activate superoxide production in murine peritoneal macrophages via stimulation of the formation of 1,2-diacylglycerol.

Maridoneau-Parini et al.[424] have studied the mechanism of inhibition by dexamethasone (a glucocorticoid) of superoxide production by A 23187- and PMA-activated guinea pig alveolar macrophages. The linoleic acid metabolites, 9-hydroxyoctadecadienoic acid and 9-hydroperoxyoctadienoic acid produced by nonstimulated macrophages inhibited lucigenin-dependent CL (but not cytochrome c reduction) induced by zymosan- and TPA-activated pulmonary guinea macrophages.[425] Similarly, the inhibitory effect of linoleic acid hydroperoxide on superoxide production by alveolar macrophages was observed by Forman and Kim.[426] McLeish et al.[427] found that prostaglandin E_2 (PGE_2) suppressed the production of hydrogen peroxide by elicited murine peritoneal macrophages, but had no effect on superoxide production. Chiara et al.[428] have shown that cyclosporin A inhibited phorbol ester stimulation of superoxide production in resident mouse peritoneal macrophages.

Gorog et al.[429] have found that the production of oxygen radicals by human endothelial cells sharply increased during phagocytosis of polystyrene microspheres, formaldehyde-fixed human platelets, and chylomicron-size lipid particles. Maly et al.[430,431] concluded that normal nontransformed human tonsillar B-lymphocytes and B-cells produce the superoxide ion by a system similar to that of phagocytic NADPH oxidase. Hancock et al.[432] have shown that lymphocytes of three B-lymphocyte cell lines are able to generate 0.35 to 0.40 nmol $O_2^{\bar{}}$ per minute per milligram of protein upon stimulation with PMA. Kiss et al.[433] proposed that the TPA-inhibition of phosphatidylserine synthesis in human promyelocytic leukemia HL-60 cells is mediated by oxygen radicals.

Nilsson et al.[434] have found that rat intestinal cells sharply increased production of oxygen radicals after *in vitro* ischemia and reperfusion. It has been proposed[435] that oxygen radicals are formed upon stimulation of rat peritoneal mast cells with protamine sulfate. Cultured rat Kupffer cells produced a superoxide ion upon stimulation with zymosan and phorbol ester, superoxide production being unchanged after removal of calcium from the medium or the addition of calcium antagonists.[436] Latocha et al.[437] proposed that superoxide production by rat Kupffer cells is mediated by Fc receptors. Murrell et al.[438] have shown that nonstimulated and PMA- or A 23187-stimulated human fibroblasts produced a superoxide ion, but its amount was 10 to 15 times less than that produced by phagocytes. The cyclooxygenase inhibitor aspirin suppressed superoxide release by fibroblasts.[439]

Upon stimulation with zymosan, phagocytic hemocytes of several snail species produced oxygen radicals detected by the SOD-inhibitable, luminol-dependent CL and NBT reduction.[440] Miyanoshita et al.[441] have shown that cAMP inhibited phorbol ester-stimulated production of oxygen radicals in rat glomeruli. Nakamura et al.[442] have shown that NADPH oxidase in the plasma membrane fraction from porcine thyroid is able to produce a superoxide ion detected via the formation of Compound III of diacetyldeuteroheme-substituted HRP. Sandhu et al.[443] concluded that the superoxide ion is generated by rat testes Leydig cells. Using the spin-trapping technique, superoxide ion was detected in cell-free extracts of SOD-deficient strains of *E. coli* in the presence of NADPH and pyocyanine.[444] It was proposed[445] that streptozotocin enhances superoxide production by xanthine oxidase in pancreatic β-cells. Woodroofe et al.[446] have shown that superoxide production increases in interferon-gamma-treated microglia isolated from adult rat brain. Alvarez et al.[447] detected spontaneous lipid peroxidation and superoxide production in human spermatozoa.

REFERENCES

1. **Lynch, R. E. and Fridovich, I.,** Permeability of the erythrocyte stroma by superoxide radical, *J. Biol. Chem.*, 253, 4697, 1978.
2. **Rumyantseva, G. V., Weiner, L. M., Molin, Yu, N., and Budker, V. G.,** Permeation of liposome membrane by superoxide radical, *FEBS Lett.*, 108, 477, 1979.
3. **Takahashi, M. and Asada, K.,** Superoxide anion permeability of phospholipid membranes and chloroplast thylakoids, *Arch. Biochem. Biophys.*, 226, 558, 1983.
4. **Gus'kova, R. A., Ivanov, I. I., Kol'tover, V. K., Akhobadze, V. V., and Rubin, A. B.,** Permeability of bilayer lipid membranes for superoxide ($O_2^{\cdot-}$) radicals, *Biochim. Biophys. Acta*, 778, 579, 1984.
5. **Forman, H. J. and Thomas, M. J.,** Oxidant production and bactericidal activity in phagocytes, *Annu. Rev. Physiol.*, 48, 669, 1986.
6. **Rossi, F.,** The $O_2^{\cdot-}$-forming NADPH oxidase of the phagocytes: nature, mechanisms of activation and function, *Biochim. Biophys. Acta*, 853, 65, 1986.
7. **Badway, J. A. and Karnovsky, M.,** Production of superoxide by phagocytic leukocytes: a paradigm for stimulus-response phenomena, *Curr. Top. Cell. Regul.*, 28, 183, 1986.
8. **Bellavite, P.,** The superoxide-forming enzymatic system of phagocytes, *Free Rad. Biol. Med.*, 4, 225, 1988.
9. **Babior, B. M., Kipnes, R. S., and Curnutte, J. T.,** Biological defence mechanism. The production by leukocytes of superoxide, a potential bactericidal agent, *J. Clin. Invest.*, 52, 741, 1973.
10. **Babior, B. M.,** Oxidants from phagocytes: agents of defence and destruction, *Blood*, 64, 959, 1984.
11. **Curnutte, J. T. and Babior, B. M.,** Biological defence mechanism. The effect of bacteria and serum on superoxide production by granulocytes, *J. Clin. Invest.*, 53, 1662, 1974.
12. **Weening, R. S., Wever, R., and Roos, D.,** Quantitative aspects of the production of phagocytizing human granulocytes, *J. Lab. Clin. Med.*, 85, 245, 1975.
13. **Curnutte, J. T. and Babior, B. M.,** Effect of anaerobiosis and inhibitors on $O_2^{\cdot-}$ production by human granulocytes, *Blood*, 45, 851, 1975.
14. **Drath, D. B. and Karnovsky, M. L.,** Superoxide production by phagocytic leukocytes, *J. Exp. Med.*, 141, 257, 1975.

15. **DeChatelet, L. R., Mulliken, D., and McCall, C. E.,** Generation of superoxide anion by various types of phagocytes, *J. Infect. Dis.*, 131, 443, 1975.
16. **Babior, B. M.,** Superoxide production by phagocytes. Another look at the effect of cytochrome c on oxygen uptake by stimulated neutrophils, *Biochem. Biophys. Res. Commun.*, 91, 222, 1979.
17. **West, M.-Y., Sinclair, D. S., Southwell-Keely, P. T.,** Production of superoxide by neturophils, *Experientia*, 39, 61, 1983.
18. **Segal, A. W. and Meshulam, T.,** Production of superoxide by neutrophils: a reappraisal, *FEBS Lett.*, 100, 27, 1979.
19. **Green, T. R., Schaeffer, R. E., and Makler, M. T.,** Significance of O_2 availability and cycling on the respiratory burst response of human PMN's exposed to cytochrome c and superoxide dismutase, *Biochem. Biophys. Res. Commun.*, 94, 1213, 1980.
20. **Kakinuma, K. and Minakami, S.,** Effect of fatty acids on superoxide radical generation in leukocytes, *Biochim. Biophys. Acta*, 538, 50, 1978.
21. **Markert, M. and Allaz, M. J.,** Continuous monitoring of oxygen consumption and superoxide production by particle-stimulated human polymorphonuclear leukocytes, *FEBS Lett.*, 113, 225, 1980.
22. **Amano, D., Kagosaki, Y., Usui, T., Yamamoto, S., and Hayaishi, O.,** Inhibitory effects of superoxide dismustases and various other proteins on nitroblue tetrazolium reduction by phagocytizing guinea pig polymorphonuclear leukocytes, *Biochem. Biophys. Res. Commun.*, 66, 272, 1975.
23. **Briggs, R. T., Robinson, J. M., Karnovsky, M. L., and Karnovsky, M. J.,** Superoxide production by polymorphonuclear leukocytes. A. Cytochemical approach, *Histochemistry*, 84, 371, 1986.
24. **Allen, R. C., Stjernholm, R. L., and Steele, R. H.,** Evidence for the generation of an electronic excitation state(s) in human polymorphonuclear leukocytes and its participation in bactericidal activity, *Biochem. Biophys. Res. Commun.*, 47, 679, 1972.
25. **Allen, R. C., Yevich, S. J., Orth, R. W., and Steele, R. H.,** The superoxide anion and singlet molecular oxygen: microbicidal activity of the polymorphonuclear leukocytes, *Biochem. Biophys. Res. Commun.*, 66, 909, 1974.
26. **McPhail, L. C., DeChatelet, L. R., and Johnston, R. B., Jr.,** Generation of chemiluminescence by a particulate fraction isolated from human neutrophils, *J. Clin. Invest.*, 63, 648, 1979.
27. **Rosen, H. and Klebanoff, S. J.,** Chemiluminescence and superoxide production by myeloperoxidase-deficient leukocytes, *J. Clin. Invest.*, 58, 50, 1976.
28. **Allen, R. C. and Loose, L. D.,** Phagocytic activation of a luminol dependent chemiluminescence in rabbit alveolar and peritoneal macrophages, *Biochem. Biophys. Res. Commun.*, 69, 245, 1976.
29. **Minkenberg, I. and Ferber, E.,** Lucigenin-dependent chemiluminescence as a new assay for NAD(P)H-oxidase activity in particulate fractions of human polymorphonuclear leukocytes, *J. Immunol. Methods*, 71, 61, 1984.
30. **Suematsu, M., Oshio, C., Miura, S., and Tsuchiya,** Real-time visualization of oxyradical burst from single neutrophil by using ultrasensitive video intensifier microscopy, *Biochem. Biophys. Res. Commun.*, 149, 1106, 1987.
31. **Dahlgren, C.,** Difference in extracellular radical release after chemotactic factor and calcium ionophore activation of the oxygen radical-generating system in human neutrophils, *Biochim. Biophys. Acta*, 930, 33, 1987.
32. **Gyllenhammer, H.,** Lucigenin chemiluminescence in the assessment of neutrophil superoxide, *J. Immunol. Methods*, 97, 209, 1987.
33. **DeSole, P., DeLeo, M. E., Scanzano, A., and Lippa, S.,** Comparative analysis of luminol- and lucigenin-amplified chemiluminescence of polymorphonuclear leukocytes stimulated by zymosan, in *Bioluminescence and Chemiluminescence*, Schoelmerich, J., Ed., John Wiley & Sons, Chichester, 1987, 149.
34. **Dahlgren, C.,** Effects on extra- and intracellularly localized, chemoattractant-induced, oxygen radical production in neutrophils following modulation of conditions for ligand-receptor interaction, *Inflammation*, 12, 335, 1988.
35. **Gyllenhammer, H.,** Mechanisms for luminol-augmented chemiluminescence from neutrophils induced by leukotriene B_4 and N-formyl-methionyl-leucyl-phenylalanine, *Photochem. Photobiol.*, 49, 217, 1989.
36. **Nishida, A., Kimura, H., Nakano, M., and Goto, T.,** A sensitive and specific chemiluminescence method for estimating the ability of human granulocytes and monocytes to generate superoxide anion, *Clin. Chim. Acta*, 179, 177, 1989.
37. **Green, M. R., Hill, H. A. O., Okolow-Zubkowska, M. J., and Segal, A. W.,** The production of hydroxyl and superoxide radicals by stimulated human neutrophils-measurements by EPR spectroscopy, *FEBS Lett.*, 100, 23, 1979.
38. **Bannister, J. V., Bellavite, P., Serra, M. C., Thornally, P. J., and Rossi, F.,** An EPR study of the production of superoxide radicals by neutrophil NADPH oxidase, *FEBS Lett.*, 145, 323, 1982.
39. **Docampo, R., Casellas, A. M., Madeira, E. D., Cardoni, R. L., Moreno, S. N. J., and Mason, R. P.,** Oxygen-derived radicals from *Trypanosoma cruzi*-stimulated human neutrophils, *FEBS Lett.*, 155, 25, 1983.

40. **Rosen, G. M., Finkelstein, E., and Rauckman, E. J.,** A method for the detection of superoxide in biological systems, *Arch. Biochem. Biophys.*, 215, 367, 1982.
41. **Makino, R., Tanaka, T., Iizuka, T., Ishimura, Y., and Kanegasaki, S.,** Stoichiometric conversion of oxygen to superoxide anion during the respiratory burst in neutrophils. Direct evidence by a new method for measurement of superoxide anion with diacetyldeuteroheme-substituted horse radish peroxidase, *J. Biol. Chem.*, 261, 11444, 1986.
42. **Jones, H. P., Grisham, M. B., Bose, S. K., Shannon, V. A., Schott, A., and McCord, J. M.,** Effect of allopurinol on neutrophil superoxide production, chemotaxis, or degranulation, *Biochem. Pharmacol.*, 34, 3673, 1985.
43. **Babior, B. M., Curnutte, J. T., and Kipnes, R. S.,** Pyridine nucleotide-dependent superoxide production by a cell-free system from human granulocytes, *J. Clin. Invest.*, 56, 1035, 1975.
44. **Gabig, T., Kipnes, R. S., and Babior, B. M.,** Solubilization of the O_2^- forming activity responsible for the respiratory burst in human neutrophils, *Biochemistry*, 17, 4784, 1978.
45. **Gabig, T. G.,** The NADPH-dependent $O_2^{\cdot}$-generating oxidase from human neutrophils. Identification of a flavoprotein component that is deficient in a patient with chronic granulomatous disease, *J. Biol. Chem.*, 258, 6352, 1983.
46. **Deward, B., Baggiolini, M., Curnutte, J. T., and Babior, B. M.,** Subcellular localization of the superoxide-forming enzyme in human neutrophils, *J. Clin. Invest.*, 63, 21, 1979.
47. **Clark, R. A., Leidal, K. G., Pearson, D. W., Nauseef, W. M.,** NADPH oxidase of human neutrophils. Subcellular localization and characterization of an arachidonate-activatable superoxide-generating system, *J. Biol. Chem.*, 262, 4065, 1987.
48. **Goldstein, I. M., Cerqueira, M., Lind, S., and Kaplan, H. B.,** Evidence that the superoxide-generating system of human leukocytes is associated with the cell surface, *J. Clin. Invest.*, 59, 249, 1977.
49. **Green, T. R., Schaefer, R. E., and Marker, M. T.,** Orientation of the NADPH dependent superoxide generating oxidoreductase on the outer membrane of human PMN's, *Biochem. Biophys. Res. Commun.*, 94, 262, 1980.
50. **Hallett, M. B. and Campbell, A. K.,** Two distinct mechanisms for stimulation of oxygen-radical production by polymorphonuclear leukocytes, *Biochem. J.*, 216, 459, 1983.
51. **Gabig, T. G. and Babior, B. M.,** The O_2^--forming oxidase responsible for the respiratory burst in human neutrophils. Properties of the solubilized enzyme, *J. Biol. Chem.*, 254, 9070, 1979.
52. **Bettetini, D., Garrouste, F., Remacle-Bonnet, M., Culouscou, J.-M., Marvaldi, J., and Pommier, G.,** Enhancement of production of superoxide anion by human polymorphonuclear leukocytes exposed to products of the HT-29 human colonic adenocarcinoma cell line, *J. Natl. Cancer Inst.*, 77, 1225, 1986.
53. **Devalou, M. L., Elliott, G. R., and Regelmann, W. E.,** Oxidative response of human neutrophils, monocytes, and alveolar macrophages induced by unopsonized surface-adherent *Staphylococcus aureus*, *Infect. Immun.*, 55, 2398, 1987.
54. **Goldstein, I. M., Roos, D., Kaplan, H. B., and Weissmann, G.,** Complement and immunoglobins stimulate superoxide production by human leukocytes independently of phagocytosis, *J. Clin. Invest.*, 56, 1155, 1975.
55. **Green, M. J., Hill, H. A. O., and Tew, D. G.,** The rate of oxygen consumption and superoxide anion formation by stimulated human neutrophils: the effect of particle concentration and size, *FEBS Lett.*, 216, 31, 1987.
56. **Murata, T., Sallivan, J. A., Sawyer, D. W., and Mandell, C. L.,** Influence of type and opsonization of ingested particle on intracellular free calcium distribution and superoxide production by human neutrophils, *Infect. Immun.*, 55, 1784, 1987.
57. **Lee, C. S. Han, E. S., and Lee, K. S.,** Effect of calcium antagonists on superoxide generation, NADPH oxidase activity and phagocytic activity in activated neutrophils, *Korean J. Pharmacol.*, 23, 33, 1987.
58. **Suzuki, H., Pabst, M. J., Johnston, R. B., Jr.,** Enhancement by Ca^{2+} or Mg^{2+} of catalytic activity of the superoxide-producing NADPH oxidase in membrane fractions of human neutrophils and monocytes, *J. Biol. Chem.*, 260, 3635, 1985.
59. **Korchak, H. M., Ljubich, L. B., Rich, A. M., and Weissmann, G.,** Activation of the neutrophil by calcium-mobilizing ligands. I. A chemotactic peptide and the lectin concanavalin A stimulate superoxide anion generation but elicit different calcium movements and phosphoinositide remodeling, *J. Biol. Chem.*, 263, 11090, 1988.
60. **Gomez-Cambronero, J., Molski, T. F. P., Becker, E. L., and Sha'afi, R. I.,** The diacylglycerol kinase inhibitor R 59022 potentiates superoxide production but not secretion induced by fMET-LEU-PHE: effects of leupeptin and the protein kinase c inhibitor H-7, *Biochem. Biophys. Res. Commun.*, 148, 38, 1987.
61. **Finkel, T. H., Rabst, M. J., Suzuki, H., Guthrie, L. A., Forehand, J. R., Phillips, W. A., and Johnston, R. B., Jr.,** Priming of neutrophils and macrophages for enhanced release of superoxide anion by the calcium ionophore ionomycin. Implications for regulation of the respiratory burst, *J. Biol Chem.*, 262, 12589, 1987.
62. **Tsujimoto, M., Yokota, S., Vilcek, J., and Weissman, G.,** Tumor necrosis factor provokes superoxide anion generation from neutrophils, *Biochem. Biophys. Res. Commun.*, 137, 1094, 1986.

63. **Berkow, R. L., Wang, D., Larrick, J. W., Dodson, R. W., and Howard, T. H.,** Enhancement of neutrophil superoxide production by preincubation with recombinant human tumor necrosis factor, *J. Immunol.*, 139, 3783, 1987.
64. **Ginsburg, I., Ward, P. A., and Varani, J.,** Lysophosphatides enhance superoxide responses of stimulated human neutrophils, *Inflammation*, 13, 163, 1989.
65. **Berkow, R. L. and Dodson, M. R.,** Biochemical mechanisms involved in the priming of neutrophils by tumor necrosis factor, *J. Leukocyte Biol.*, 44, 345, 1988.
66. **Guarnieri, C., Georgountzos, A., Caldarera, I., Flamigni, F., and Ligalue, A.,** Polyamines stimulate superoxide production in human neutrophils activated by N-fMet-Leu-Phe but not by phorbol myristate acetate, *Biochim. Biophys. Acta*, 930, 135, 1987.
67. **Dahn, L. J., Hewett, J. A., and Roth, R. A.,** Bile and bile salts potentiate superoxide anion release from activated rat peritoneal neutrophils, *Toxicol. Appl. Pharmacol.*, 95, 82, 1988.
68. **Kharazmi, A., Nielsen, H., and Bendtzen, K.,** Modulation of human neutrophil and monocyte chemotaxis and superoxide responses by recombinant TNF-alpha and GM-CSF, *Immunobiology*, 177, 363, 1988.
69. **Sumimoto, H., Takeshige, K., and Minakami, S.,** Superoxide production of human polymorphonuclear leukocytes stimulated by leukotriene B_4, *Biochim. Biophys. Acta*, 803, 271, 1984.
70. **Gay, J. C., Beckman, J. K., Brash, A. R., Oates, J. A., and Lukens, J. N.,** Enhancement of chemotactic factor-stimulated neutrophil oxidative metabolism by leukotriene B_4, *Blood*, 64, 780, 1984.
71. **Badwey, J. A., Robinson, J. M., Horn, W., Soberman, R. J., Karnovsky, M. J., and Karnovski, M. L.,** Synergistic stimulation of neutrophils. Possible involvement of 5-hydroxy-6,8,11,14-eicosatetraenolate in superoxide release, *J. Biol. Chem.*, 263, 2779, 1988.
72. **Matsumoto, T., Takeshige, K., and Minakami, S.,** Spontaneous induction of superoxide release and degranulation of neutrophils in isotopic potassium medium: the role of intracellular calcium, *J. Biochem.*, 99, 1591, 1986.
73. **Curnutte, J. T., Babior, B. M., and Karnovsky, M.,** Floride-mediated activation of the respiratory burst in human neutrophils, *J. Clin. Invest.*, 63, 637, 1979.
74. **Rider, L. and Niedel, J. E.,** Diacylglycerol accumulation and superoxide anion production in stimulated human neutrophils, *J. Biol. Chem.*, 262, 5603, 1987.
75. **Fujita, I., Takeshige, K., and Minakami, S.,** Inhibition of neutrophil superoxide formation by 1-(5-isoquinolinesulfonyl)-2-methylpiperazine (H-7), an inhibitor of protein kinase C, *Biochem. Pharmacol.*, 35, 4555, 1986.
76. **Green, T. R., Wu, D. E., and Wirtz, M. K.,** The O_2^- generating oxidoreductase of human neutrophils: evidence of an obligatory requirement for calcium and magnesium for expression of catalytic activity, *Biochem. Biophys. Res. Commun.*, 110, 973, 1983.
77. **Christiansen, N. O., Larsen, C. S., Juhl, H., and Esmann, N.,** Membrane-associated protein kinase C activity in superoxide-producing human polymorphonuclear leukocytes, *J. Leukocyte Biol.*, 44, 33, 1988.
78. **Tsusaki, B. E., Kanda, S., and Huang, L.,** Stimulation of superoxide release in neutrophils by 1-oleoyl-2-acetyl-glycerol incorporated into pH-sensitive liposomes, *Biochem. Biophys. Res. Commun.*, 136, 242, 1986.
79. **Smith, R. J., Sam, L. M., and Justen, J. M.,** Diacylglycerols modulate human polymorphonuclear neutrophil responsiveness: effects on intracellular calcium mobilization, granule exocytosis, and superoxide anion production, *J. Leukocyte Biol.*, 43, 411, 1988.
80. **Christiansen, N. O., Larsen, C. S., and Esmann, V.,** A study on the role of protein kinase C and intracellular calcium in the activation of superoxide generation, *Biochim. Biophys. Acta*, 971, 317, 1988.
81. **Sha'afi, R. I., Molski, T. F. P., Gomez-Cambronero, J., and Huang, C.-K.,** Dissociation of the 47-kilodalton protein phosphorylation from degranulation and superoxide production in neutrophils, *J. Leukocyte Biol.*, 43, 18, 1988.
82. **Dale, M. M. and Peufield, A.,** Comparison of the effects of indomethacin, RHC 80267 and R 59022 on superoxide production by 1,oleoyl-2,acetylglycerol and A 23187 in human neutrophils, *Br. J. Pharmacol.*, 92, 63, 1987.
83. **Nath, J. and Powledge, A.,** Temperature-dependent inhibition of Met-Leu-Phe-stimulated superoxide generation by C-I and H-7 in human neutrophils, *Biochem. Biophys. Res. Commun.*, 156, 1376, 1988.
84. **Yagawa, K., Hayashi, S., Nakanishi, M., Aso, H., Ogata, K., Maruyama, M., Ichinose, Y., and Shigematsu, N.,** Restoration of Fc receptor-mediated desensitization of superoxide generation in human PMN by PMA, *Int. J. Immunopharmacol.*, 9, 497, 1987.
85. **Suzuki, Y. and Furuta, H.,** Stimulation of pig neutrophil superoxide anion-producing system with thymol, *Inflammation*, 12, 575, 1988.
86. **Welch, W. D.,** Effect of halothane and N_2O on the oxidative activity of human neutrophils, *Anesthesiology*, 57, 172, 1982.
87. **Nakagawara, M., Takeshige, K., Takamatsu, J., Takahashi, S., Yoshitake, J., and Minakami, S.,** Inhibition of superoxide production and Ca^{2+} mobilization in human neutrophils by halothane, eufluraine, and isoflurane. *Anesthesiology*, 64, 4, 1986.

88. **Tsuchiya, M., Okimasu, E., Ueda, W., Hirakawa, M., and Utsumi, K.**, Halothane, and inhalation anesthetic, activates protein kinase C and superoxide generation by neutrophils, *FEBS Lett.*, 242, 101, 1988.
89. **Curnutte, J. T., Badwey, J. A., Robinson, J. M., Karnovsky, M. J., and Karnovsky, M. L.**, Studies on the mechanism of superoxide release from human neutrophils stimulated with arachidonate, *J. Biol. Chem.*, 259, 11851, 1984.
90. **Badwey, J. A., Curnutte, J. T., Robinson, J. M., Berde, C. B., Karnovsky, M. J., and Karnovsky, M. L.**, Effects of free fatty acids on release of superoxide and on change of shape by human neutrophils. Reversibility by albumin, *J. Biol. Chem.*, 259, 7870, 1984.
91. **Morimoto, Y. M., Sato, E., Nobori, K., Takahashi, R., and Utsumi, K.**, Effect of calcium ion on fatty acid-induced generation of superoxide in guinea pig neutrophils, *Cell Struct. Funct.*, 11, 143, 1986.
92. **Ohtsuka, T., Ozawa, M., Katayama, T., Okamura, N., and Ishibashi, S.**, Further evidence for the involvement of the phosphorylation of 46K protein(s) in the regulation of superoxide anion production in guinea pig polymorphonuclear leukocytes, *Arch. Biochem. Biophys.*, 260, 226, 1988.
93. **Tanaka, T., Kanegasaki, S., Makino, R., Iizuka, T., and Ishimura, Y.**, Saturated and trans-unsaturated fatty acids elicit high levels of superoxide generation in intact and cell-free preparations of neutrophils, *Biochem. Biophys. Res. Commun.*, 144, 606, 1987.
94. **Tanaka, T., Makino, R., Iizuka, T., Ishimura, Y., and Kanegasaki, S.**, Activation by saturated and monounsaturated fatty acids of the superoxide anion-generating system in a cell-free preparation from neutrophils, *J. Biol. Chem.*, 263, 13670, 1988.
95. **Morimoto, Y. M., Sato, E., Nobori, K., Takahashi, R., and Utsumi, K.**, Effect of calcium ion on fatty acid-induced generation of superoxide in guinea pig neutrophils, *Cell Struct. Funct.*, 11, 143, 1986.
96. **Badwey, J. A., Robinson, J. M., Curnutte, J. T., Karnovsky, M. J., and Karnovsky, M. L.**, Retinoids stimulate the release of superoxide by neutrophils and change their morphology, *J. Cell. Physiol.*, 127, 223, 1986.
97. **Cohen, H. J. and Chovaniec, M. E.**, Superoxide generation by digitonin-stimulated guinea pig granulocytes. A basis for a continuous assay for monitoring superoxide production and for the study of the activation of the generating system, *J. Clin. Invest.*, 61, 1081, 1978.
98. **Washida, N., Sagawa, A., Tamoto, K., and Koyama, J.**, Comparative studies on superoxide anion production by polymorphonuclear leukocytes stimulated with various agents, *Biochim. Biophys. Acta*, 631, 371, 1980.
99. **Fujita, I., Takeshige, K., and Minakami, S.**, Characterization of the NADPH-dependent superoxide production activated by sodium dodecyl sulfate in a cell-free system of pig neutrophils, *Biochim. Biophys. Acta*, 931, 41, 1987.
100. **Ginsburg, I., Borinski, R., Malamud, D., Struckweier, F., and Klimetzek, V.**, Chemiluminescence and superoxide generation by leukocytes stimulated by polyelectrolyte-opsonized bacteria: role of histones, polyarginine, polylysine, polyhistidine, cytochalasins, and inflammatory exudates as modulators of oxygen burst, *Inflammation*, 9, 245, 1985.
101. **Ginsburg, I., Borinski, R., and Pabst, M.**, NADPH and "cocktails" containing polyarginine reactivate superoxide generation in leukocytes lysed by membrane-damaging gents, *Inflammation*, 9, 341, 1985.
102. **Ginsburg, I., Borinski, R., and Pabst, M.**, Superoxide generation by human blood leukocytes under the effect of cytolytic agents, *Int. J. Tissue React.*, 7, 143, 1985.
103. **Nakagawara, A. and Minakami, S.**, Generation of superoxide anions by leukocytes treated with cytochalasin E, *Biochem. Biophys. Res. Commun.*, 64, 760, 1975.
104. **Nakagawara, A. and Minakami, S.**, Role of cytoskeletal elements in cytochalasin E-induced superoxide production by human polymorphonuclear leukocytes, *Biochim. Biophys. Acta*, 584, 143, 1979.
105. **Edwards, S. W. and Lloyd, D.**, The relationship between superoxide generation, cytochrome b and oxygen in activated neutrophils, *FEBS Lett.*, 227, 39, 1988.
106. **Tamura, M., Tamura, T., Tyagi, S. R., and Lambeth, J. D.**, The superoxide-generating respiratory burst oxidase of human neutrophil plasma membrane. Phosphatidylserine as an effector of the activated enzyme, *J. Biol. Chem.*, 263, 17621, 1988.
107. **Hewett, J. A. and Roth, R. A.**, Dieldrin activates rat neutrophils in vitro, *Toxicol. Appl. Pharmacol.*, 96, 269, 1988.
108. **Sharp, B. M., Tsukayama, D. T., Gekker, G., Keane, W. F., and Peterson, P. K.**, β-Endorphin stimulates human polymorphonuclear leukocyte superoxide production via a stereoselective opiate receptor, *J. Pharmacol. Exp. Ther.*, 242, 579, 1987.
109. **Nicotra, J., Orsini, A. J., and DeBari, V. A.**, Chemiluminescence response of the human polymorphonuclear neutrophil to lipopolysaccharides, *Cell Biophys.*, 7, 285, 1985.
110. **Kano, S., Iizuka, I., Ishimura, Y., Fujiki, H., and Sugimura, T.**, Stimulation of superoxide anion formation by non-TPA-type tumor promoters polytoxin and thapsigargin in porcine and human neutrophils, *Biochem. Biophys. Res. Commun.*, 143, 672, 1987.
111. **Maly, F. E., Kapp, A., and Weck, A. L., De.**, A novel interleukin stimulating free radical production by granulocytes, *Free Rad. Res. Commun.*, 3, 57, 1987.

112. **Matsumoto, S., Takei, M., Mariyama, M., and Imanishi, H.,** Augmentation of antibody-dependent cellular cytotoxicity of polymorphonuclear leukocytes by interferon-gamma: mechanism dependent on enhancement of Fc receptor expression and increased release of activated oxygen, *Chem. Pharm. Bull.*, 35, 1571, 1987.
113. **Kuhns, D. B., Kaplan, S. S., and Basford, R. E.,** Hexachlorocyclohexanes, potent stimuli of superoxide anion production and calcium release in human polymorphonuclear leukocytes, *Blood,* 68, 535, 1986.
114. **Smith, R. J., Bowman, B. J., and Iden, S. S.,** Stimulation of the human neutrophil superoxide anion-generating system with 1-O-hexadecyl/octadecyl-2-acetyl-*sn*-glycerol-3-phosphorylcholine, *Biochem. Pharmacol.*, 33, 973, 1984.
115. **Gabler, W. L., Creamer, H. R., and Bullock, W. W.,** Modulation of the kinetics of induced neutrophil superoxide generation by fluoride, *J. Dent. Res.*, 65, 1159, 1986.
116. **Monboisse, J. C., Bellou, G., Dufer, J., Randoux, A., and Borel, J. P.,** Collagen activates superoxide anion production by human polymorphonuclear neutrophils, *Biochem. J.*, 246, 599, 1987.
117. **Ballavite, P., Corson, F., Dusi, S., Grzeskowiak, M., Della-Blanca, V., and Rossi, F.,** Activation of NADPH-dependent superoxide production in plasma membrane extracts of pig neutrophils by phosphatidic acid, *J. Biol. Chem.*, 263, 8210, 1988.
118. **Anderson, R., Beyers, A. D., Savage, J. E., and Nel, A. E.,** Apparent involvement of phospholipase A_2, but not protein kinase C, in the pro-oxidative interactions of clofazimine with human phagocytes, *Biochem. Pharmacol.*, 37, 4635, 1988.
119. **Jay, M., Kojima, S., and Gillespie, M. N.,** Nicotine potentiates superoxide anion generation by human neutrophils, *Toxicol. Appl. Pharmacol.*, 86, 484, 1986.
120. **Gabler, W. L., Bullock, W. W., and Creamer, H. R.,** The influence of chlorhexidine on superoxide generation by induced human neutrophils, *J. Periodontal Res.*, 22, 445, 1987.
121. **Rebut-Bonneton, C., Bailly, S., and Pasquier, C.,** Superoxide anion production in glass-adherent polymorphonuclear leukocytes and its relationship to calcium movement, *J. Leukocyte Biol.*, 44, 402, 1988.
122. **Neumann, M., and Kownatzki, E.,** The effect of adherence on the generation of reactive oxygen species by human neutrophilic granulocytes, *Agents Actions,* 26, 183, 1989.
123. **Hoover, R. L., Robinson, J. M., and Karnovsky, M. J.,** Adhesion of polymorphonuclear leukocytes to endothelium enhances the efficiency of detoxification of oxygen-free radicals, *Am. J. Pathol.*, 126, 258, 1987.
124. **Bender, J. G. and Epps, D. E. V.,** Stimulus interactions in release of superoxide ion (O_2^-) from human neutrophils, *Inflammation,* 9, 67, 1985.
125. **Robinson, J. M., Badwey, J. A., Karnovsky, M. L., and Karnovsky, M. J.,** Superoxide release by neutrophils: synergistic effects of a phorbol ester and a calcium ionophore, *Biochem. Biophys. Res. Commun.*, 122, 734, 1984.
126. **Ohtsuka, T., Ozawa, M., Katayama, T., and Ishibashi, S.,** Synergism of phosphorylation of 46K protein(s) and arachidonate release in the induction of superoxide anion production in guinea pig polymorphonuclear leukocytes, *Arch. Biochem. Biophys.*, 262, 416, 1988.
127. **Christiansen, N. O.,** A time-course study on superoxide generation and protein kinase activation in human neutrophils, *FEBS Lett.* 239, 195, 1988.
128. **Reibman, J., Korchak, H. M., Vosshall, L. B., Haines, K. A., Rich, A. M., Weissmann, G., and Cristello, P.,** Changes in diacylglycerol labeling, cell shape, and protein phosphorylation distinguish "triggering" from "activation" of human neutrophils, *J. Biol. Chem.*, 263, 6322, 1988.
129. **Jandle, R. C., Andre-Schwartz, J., Borges-DuBois, L., Kipnes, R. S., McMurrich, B. J., and Babior, B. M.,** Termination of the respiratory burst in human neutrophils, *J. Clin. Invest.*, 61, 1176, 1978.
130. **Edwards, S. W. and Swan, T. F.,** Regulation of superoxide generation by myeloperoxidase during the respiratory burst of human neutrophils, *Biochem. J.*, 237, 601, 1986.
131. **Dahlgren, C. and Lock, R.,** The limitation of the human neutrophil chemiluminescence response by extracellular peroxidase is stimulus dependent: effect of added horseradish peroxidase on the response induced by both soluble and particulate stimuli, *J. Clin. Lab. Immunol.*, 26, 49, 1988.
132. **Henderson, L. M., Chappell, J. B., and Jones, O. T. G.,** Superoxide generation by the electrogenic NADPH oxidase of human neutrophils is limited by the movement of a compensating charge, *Biochem. J.*, 255, 285, 1988.
133. **Henderson, L. M., Chappell, J. B., and Jones, O. T. G.,** The superoxide-generating NADPH oxidase of human neutrophils is electrogenic and associated with a proton channel, *Biochem. J.*, 246, 325, 1987.
134. **Simchowitz, L.,** Intracellular pH modulates the generation of superoxide radicals by human neutrophils, *J. Clin. Invest.*, 76, 1079, 1985.
135. **Rossi, F., Della Bianca, V., and Davoli, A.,** A new way for inducing a respiratory burst in guinea pig neutrophils. Change in the Na^+, K^+ concentration of the medium, *FEBS Lett.*, 132, 273, 1981.
136. **Wong, K. and Chew, C.,** Slow exponential decay of rate of superoxide production in phorbol ester-activated human neutrophils, *Inflammation,* 9, 407, 1985.
137. **Gabig, T. G., Bearman, S. I., and Babior, B. M.,** Effect of oxygen tension and pH on the respiratory burst of human neutrophils, *Blood,* 53, 1133, 1979.

138. **Gyllenhammar, H.,** Effects of extracellular pH on neutrophil superoxide anion production and chemiluminescence augmented with luminol, lucigenin or DMNH, *J. Clin. Lab. Immunol.*, 28, 97, 1989.
139. **Bannister, J. V., Bellavite, P., Serra, M. C., Thornalley, P. J., and Rossi, F.,** An ESR study of the production of superoxide radicals by neutrophil NADPH oxidase, *FEBS Lett.*, 145, 323, 1982.
140. **Loschen, G.,** Superoxide formation accounts for almost all extra oxygen consumption upon stimulation of leukocytes with PMA, in *Superoxide and Superoxide Dismutase in Chemistry, Biology and Medicine*, Rotilio, G., Ed., Elsevier, Amsterdam, 1986, 373.
141. **Green, T. R. and Wu, D. E.,** The NADPH: O_2 reductase of human neutrophils. Stoichiometry of univalent and divalent reduction of O_2, *J. Biol. Chem.*, 261, 6010, 1986.
142. **Hoffstein, S. T., Gennaro, D. E., and Manzi, R. M.,** Neutrophils may directly synthesize both H_2O_2 and O_2^- since surface stimuli induce their release in stimulus-specific ratios, *Inflammation*, 9, 425, 1985.
143. **Kensler, T. W. and Trush, M. A.,** Inhibition of oxygen radical metabolism in phorbol ester-activated polymorphonuclear leukocytes by an antitumor promoting copper complex with superoxide dismutase-mimetic activity, *Biochem. Pharmacol.*, 32, 3485, 1983.
144. **Baader, W. J., Hatzelmann, A., and Ullrich, V.,** The suppression of granulocyte functions by lipophilic antioxidants, *Biochem. Pharmacol.*, 37, 1089, 1988.
145. **Kraut, E. H., Metz, E. N., and Sagone, A. L.,** *In vitro* effects of ascorbate on white cell metabolism and the chemiluminescence response, *J. Reticuloendothelial Soc.*, 27, 359, 1980.
146. **Hemila, H., Roberts, P., and Wikstroem, M.,** Activated polymorphonuclear leukocytes consume vitamin C, *FEBS Lett.*, 178, 25, 1984.
147. **Engle, W. A., Yoder, M. C., Baurley, J. L., and Yu, P.-L.,** Vitamin E decreases superoxide anion production by polymorphonuclear leukocytes, *Pediatr. Res.*, 23, 245, 1988.
148. **Holt, M. E., Ryall, M. E. T., and Campbell, A. K.,** Albumin inhibits human polymorphonuclear leukocyte luminol-dependent chemiluminescence: evidence for oxygen radical scavenging, *Br. J. Exp. Pathol.*, 65, 231, 1984.
149. **Gallin, J. I., Seligmann, B. E., Cramer, E. B., Schiffmann, E., and Fletcher, M. P.,** Effects of vitamin K on human neutrophil function, *J. Immunol.*, 128, 1399, 1982.
150. **Niwa, Y., Somiya, K., Miyachi, Y., Kanoh, T., and Sakane, T.,** Luminol-independent chemiluminescence by phagocytes is markedly enhanced by dexamethasone, not by other glucocorticosteroids, *Inflammation*, 11, 163, 1987.
151. **Ward, P. A., Thomas, T. W., McCulloch, K. K., and Johnson, K. J.,** Regulatory effects of adenosine and adenine nucleotides on oxygen radical responses of neutrophils, *Lab. Invest.*, 58, 438, 1988.
152. **Kuroki, M. and Minakami, S.,** Extracellular ATP triggers superoxide production in human neutrophils, *Biochem. Biophys. Res. Commun.*, 182, 377, 1989.
153. **Kuhn, D. B., Wright, D. G., Nath, J., Kaplan, S. S., and Basford, R. E.,** ATP induces transient elevations of calcium^{2+} in human neutrophils and primes these cells for enhanced superoxide anion generation, *Lab. Invest.*, 58, 448, 1988.
154. **Ward, P. A., Cunningham, T. W., Blair, B. A. M., and Johnson, K. J.,** Differing calcium requirements for regulatory effects of ATP, AT S and adenosine on superoxide radical anion responses of human neutrophils, *Biochem. Biophys. Res. Commun.*, 154, 746, 1988.
155. **Melloni, E., Pontremoli, S., Salamino, F., Sparatore, B., Michetti, M., Sacco, O., and Horecker, B. L.,** ATP induces the release of a neutral serine proteinase and enhances the production of superoxide anion in membranes from phorbol ester-activated neutrophils, *J. Biol. Chem.*, 261, 11437, 1986.
156. **Ward, P. A., Cunningham, T. W., McCulloch, K. K., Phan, S. H., Powell, J., and Johnson, K. J.,** Platelet enhancement of superoxide anion responses in stimulated human neutrophils. Identification of platelet factor as adenine nucleotide, *Lab. Invest.*, 58, 37, 1988.
157. **Schinetti, M. L. and Lazzarino, G.,** Inhibition of phorbol ester-stimulated chemiluminescence and superoxide production in human neutrophils by fructose 1,6-diphosphate, *Biochem. Pharmacol.*, 35, 1762, 1986.
158. **Cronstein, B. N., Rosenstein, E. D., Kramer, S. B., Weissmann, G., and Hirschborn, R.,** Adenosine; a physiologic modulator of superoxide anion generation by human neutrophils. Adenosine adds via an A_2 receptor on human neutrophils, *J. Immunol.*, 135, 1366, 1985.
159. **Cronstein, B. N., Kramer, S. B., Rosenstein, E. D., and Korchak, H.,** Occupancy of adenosine receptors raises cyclic AMP alone and in synergy with occupancy of chemoattractant receptors and inhibits membrane depolarization, *Biochem. J.*, 252, 709, 1988.
160. **Lad, P. M., Goldberg, B. J., Smiley, P. A., and Olson, C. V.,** Receptor-specific threshold effects of cyclic AMP are involved in the regulation of enzyme release and superoxide production from human neutrophils, *Biochim. Biophys. Acta*, 846, 286, 1985.
161. **McGarrity, S. T., Stephenson, A. H., Hyers, T. M., and Webster, R. O.,** Inhibition of neutrophil superoxide anion generation by platelet products: role of adenine nucleotides, *J. Leukocyte Biol.*, 44, 411, 1988.
162. **Fantone, J. C. and Kinnes, D. A.,** Prostagladin E_1 and prostaglandin I_2. Modulation of superoxide production by human neutrophils, *Biochem. Biophys. Res. Commun.*, 113, 506, 1983.

163. **Simpkins, C. O., Sione, S. T., Tate, E. A., and Johnson, M.,** The effect of enkerphalins and prostaglandins on superoxide release by neutrophils, *J. Surg. Res.*, 41, 645, 1986.
164. **Gryglewski, R. J., Szczeklik, A., and Wandzilak, M.,** The effect of six prostaglandins, prostacyclin and iloprost on generation of superoxide anions by human polymorphonuclear leukocytes stimulated by zymosan or formyl-methionyl-leucyl-phenylalanine, *Biochem. Pharmacol.*, 36, 4209, 1987.
165. **Witz, G., Lawrie, N. J., Amoruso, M. A., and Goldstein, B. D.,** Inhibition by reactive aldehydes of superoxide anion radical production from stimulated polymorphonuclear leukocytes and pulmonary alveolar macrophages. Effects of cellular sulfhydryl groups and NADPH oxidase activity, *Biochem. Pharmacol.*, 36, 721, 1987.
166. **Cohen, H. J. and Chovaniec, M. E.,** Superoxide generation by digitonin-stimulated guinea pig granulocytes. The effects of N-ethyl maleimide, divalent cations, and glycolic and mitochondrial inhibitors on the activation of the superoxide generating system, *J. Clin. Invest.*, 61, 1088, 1978.
167. **Fantozzi, R., Brunelleschi, S., Guiliattini, L., Blandina, P., Masini, E., Cavallo, G., and Mannaioni, P. F.,** Mast cell and neutrophil interactions—a role for superoxide anion and hystamine, *Agents Actions*, 16, 260, 1985.
168. **Nielson, C. P., Brenner, D., and Olson, R. D.,** Doxorubicin and doxorubicinol-induced alterations in human polymorphonuclear leukocyte oxygen metabolite generation, *J. Pharmacol. Exp. Ther.*, 238, 19, 1986.
169. **Schinetti, M. L., Rossini, D., and Bertelli, A.,** Interaction of anthracycline antibiotics with human neutrophils: superoxide production, free radical formation and intracellular penetration, *J. Cancer Res. Clin. Oncol.*, 113, 15, 1987.
170. **Mian, M., Brunelleschi, S., Tarli, S., Rubino, A., Benetti, D., Fantozzi, R., and Zilletti, L.,** Rhein: an anthraquinone that modulates superoxide anion production from human neutrophils, *J. Pharm. Pharmacol.*, 39, 845, 1987.
171. **Van der Anwera, P., Petrikkos, G., Husson, M., and Klastersky, J.,** Influence of various antibiotics on superoxide generation by normal human neutrophils, *Arch. Int. Physiol. Biochim.*, 94, S23, 1986.
172. **Goldstein, I. M., Roos, D., Weissman, G., and Kaplan, H. B.,** Effect of corticosteroids on human polymorphonuclear leukocyte function in vitro. Reduction of lysosomal enzyme release and superoxide production, *Inflammation*, 1, 305, 1976.
173. **Whitecomb, J. M. and Schwartz, A. G.,** Dehydroepiandrosterone and 16 α-bromo-epiandrosterone inhibit 12-*O*-tetradecanoylphorbol-13-acetate stimulation of superoxide radical production by human polymorphonuclear leukocytes, *Carcinogenesis*, 6, 333, 1985.
174. **Styrt, B., Rocklin, R. E., and Klempner, M. S.,** Inhibition of neutrophil superoxide production by fanetizole, *Inflammation*, 9, 233, 1985.
175. **Simpkins, C. O., Alailima, S. T., and Tate, E. A.,** Inhibition by naloxone of neutrophil superoxide release: a potentially useful antiinflammatory effect, *Circ. Shock*, 20, 181, 1986.
176. **Fantozzi, R., Brunellschi, S., Cremonesi, P., Pagella, P. G., Ciani, D., and Sportoletti, G. C.,** Drug modulation of superoxide anion production from human neutrophils, *Int. J. Tissue React.*, 7, 149, 1985.
177. **Neal, T. M., Vissers, M. C. M., and Winterbourn, C. C.,** Inhibition by nonsteroidal anti-inflammatory drugs of superoxide production and granule enzyme release by polymorphonuclear leukocytes stimulated with immune complexes of formyl-methionyl-leucyl-phenylalanine, *Biochem. Pharmacol.*, 36, 2511, 1987.
178. **Capsoni, F., Venegoni, E., Minonzio, F., Ougari, A. M., Maresca, V., and Zanussi, C.,** Inhibition of neutrophil oxidative mechanism by nimesulide, *Agents Actions*, 21, 121, 1987.
179. **Pasini, F. L., Ceccatelli, L., Capecchi, P. L., Orrico, A., Pasqui, A. L., and Perri, T.,** Benzodiazepines inhibit in vitro free radical formation from human neutrophils induced by FMLP and A 23187, *Immunopharmacol. Immunotoxicol.*, 9, 101, 1987.
180. **Finkelstein, A. E., Ladizesky, M., Borinsky, R., Kohn, E., and Ginsburg, I.,** Antiarthritic synergism of combined oral and parenteral chrysotherapy. II. Increased inhibition of activated leukocyte oxygen burst by combined gold action, *Inflammation*, 12, 383, 1988.
181. **Goldstein, I. M., Lind, S., Hoffstein, S., and Weissmann, G.,** Influence of local anesthetics upon human polymorphonuclear leukocyte function in vitro. Reduction of lysosomal enzyme release and superoxide anion production, *J. Exp. Med.*, 146, 483, 1977.
182. **Neal, T. M., Winterbourn, C. C., and Vissers, M. C. M.,** Inhibition of neutrophil degranulation and superoxide production by sulfasalazine. Comparison with 5-aminosalicylic acid, sulfapyridine and olsalazine, *Biochem. Pharmacol.*, 36, 2765, 1987.
183. **Taniguchi, K. and Takanaka, K.,** Inhibitory effects of various drugs on phorbol myristate acetate and n-formyl methionyl leucyl phenylalanine induced O_2^- production in polymorphonuclear leukocytes, *Biochem. Pharmacol.*, 33, 3165, 1984.
184. **Irita, K., Fujita, I., Takeshige, K., Minakami, S., and Yoshitake, J.,** Calcium channel antagonist induced inhibition of superoxide production in human neutrophils. Mechanisms independent of antagonizing calcium influx, *Biochem. Pharmacol.*, 35, 3465, 1986.
185. **Flament, J., Schandene, L., and Boeynaems, J.-M.,** Effect of the 5-lipoxygenase inhibitor piriprost on superoxide production by human neutrophils, *Prostaglandins Leukocytes*, 34, 175, 1988.

186. **Ichikawa, S., Omura, K., Katayama, T., Okamura, N., Ohtsuka, T., Ishibashi, S., and Masayasu, H.,** Inhibition of superoxide anion production in guinea pig polymorphonuclear leukocytes by a seleno-organic compound, Ebselen, *J. Pharmacobiol.-Dyn.*, 10, 595, 1987.
187. **Anderson, R.,** Erythromycin and roxithromycin potentiate human neutrophil locomotion in vitro by inhibition of leukoattractant-activated superoxide generation and autoxidation, *J. Infect. Dis.*, 159, 966, 1989.
188. **Beilke, M. A., Collins-Lech, C., and Sohnle, P. C.,** Effects of dimethyl sulfoxide on the oxidative function of human neutrophils, *J. Lab. Clin. Med.*, 110, 91, 1987.
189. **Hurst, N. P., Freuch, J. K., Corjatschko, L., and Betts, W. H.,** Studies on the mechanism of inhibition of chemotactic tripeptide stimulated human neutrophil polymorphonuclear leukocyte superoxide production by chloroquine and hydroxychloroquine, *Ann. Rheim. Dis.*, 46, 750, 1987.
190. **Hurst, N. P., Freuch, J. K., Gorjatschko, L., and Betts, W. H.,** Chloroquine and hydroxychloroquine inhibit multiple sites in metabolic pathways leading to neutrophil superoxide release, *J. Rheumatol.*, 15, 23, 1988.
191. **Nakamura, H., Uetani, Y., Komura, M., Takada, S., Sano, K., and Matsuo, T.,** Inhibitory action of bilirubin on superoxide production by poly-morphonuclear leukocytes, *Biol. Neonate*, 52, 273, 1987.
192. **Gennaro, R. and Romeo, D.,** The release of superoxide anion from granulocytes: effect of inhibitors of anion permeability, *Biochem. Biophys. Res. Commun.*, 88, 44, 1979.
193. **Younes, M. and Rolbke, A.,** Inhibition of granulocyte-mediated release of oxygen free radicals following glutathione depletion, *Toxicol. Lett.*, 41, 139, 1988.
194. **Mehta, J. L., Lawson, D., and Mehta, P.,** Modulation of human neutrophil superoxide production by the selective thromboxane synthetase inhibitor U 63557 A, *Life Sci.*, 43, 928, 1988.
195. **Muid, R. E., Twomey, B., and Dale, M. M.,** The effect of inhibition of both diacylglycerol metabolism and phospholipase A_2 activity on superoxide generation by human neutrophils, *FEBS Lett.*, 234, 235, 1988.
196. **Tawara, K., Fujisawa, S., and Nakai, K.,** Effect of NCO-700, an inhibitor of thiol protease, on reactive oxygen production by chemotactic peptide-stimulated rabbit peripheral granulocytes, *Experientia*, 44, 346, 1988.
197. **Salin, M. L. and McCord, J. M.,** Free radicals and inflammation. Protection of phagocytosing leukocytes by superoxide dismutase, *J. Clin. Invest.*, 56, 1319, 1975.
198. **Auclair, C., Torres, M., and Hakim, J.,** Involvement of hydroxyl radical in NAD(P)H oxidation and associated oxygen reduction by the granule fraction of human blood polymorphonuclear, *Biochem. Biophys. Res. Commun.*, 81, 1067, 1978.
199. **Tauber, A. I., Gabig, T. G., and Babior, B. M.,** Evidence for production of oxidizing radicals by the particulate O_2^--forming system from human neutrophils, *Blood*, 53, 666, 1979.
200. **Takanaka, K. and O'Brien, P. J.,** Generation of activated oxygen species by polymorphonuclear leukocytes, *FEBS Lett.*, 110, 283, 1980.
201. **Sagone, A. L., Decker, M. A., Wells, R. M., and DeMocko, C.,** A new method for the detection of hydroxyl radical production by phagocyting cells, *Biochim. Biophys. Acta*, 628, 90, 1980.
202. **Thomas, M. J., Shirley, P. S., Hedrick, C. C., and DeChatelet, L. R.,** Role of free radical processes in stimulated human polymorphonuclear leukocytes, *Biochemistry*, 25, 8042, 1986.
203. **Bannister, J. V., Bellavite, P., Davoli, A., Thornalley, P. J., and Rossi, F.,** The generation of hydroxyl radicals following superoxide production by neutrophil NADPH oxidase, *FEBS Lett.*, 150, 300, 1982.
204. **Britigan, B. E., Rosen, G. M., Chai, Y., and Cohen, M. S.,** Do human neutrophils make hydroxyl radical? Determination of free radicals generated by human neutrophils activated with a soluble or particulate stimulus using electron paramagnetic resonance spectroscopy, *J. Biol. Chem.*, 261, 4426, 1986.
205. **Britigan, B. E., Rosen, G. M., Thompson, B. Y., Chai, Y., and Cohen, M. S.,** Stimulated human neutrophils limit iron-catalyzed hydroxyl radical formation as detected by spin-trapping technique, *J. Biol. Chem.*, 261, 17026, 1986.
206. **Britigan, B. E., Cohen, M. S., and Rosen, G. M.,** Detection of the production of oxygen-centered free radicals by human neutrophils using spin trapping techniques: a critical perspective, *J. Leukocyte Biol.*, 41, 349, 1987.
207. **Rosen, G. M., Britigan, B. E., Cohen, M. S., Ellington, S. P., and Barber, M. J.,** Detection of phagocyte-drived free radicals with spin trapping techniques, *Biochim. Biophys. Acta*, 969, 236, 1988.
208. **Ueno, I., Kohno, M., Mitsuta, K., Mizuta, Y., and Kanegasaki, S.,** Reevaluation of the spin-trapped adduct formed from 5,5-dimethyl-1-pyrroline-1-oxide during the respiratory burst in neutrophils, *J. Biochem.*, 105, 905, 1989.
209. **Janzen, E. G., Jandrisits, L. T., and Barber, D. L.,** Studies on the origin of the hydroxyl spin adduct of DMPO produced from the stimulation of neutrophils by phorbol-12-myristate-13-acetate, *Free Rad. Res. Commun.*, 4, 115, 1987.
210. **Winterbourn, C. C.,** Myeloperoxidase as an effective inhibitor of hydroxyl radical production—implication for the oxidative reactions of neutrophils, *J. Clin. Invest.*, 78, 545, 1986.
211. **Umeda, T., Hara, T., Hayashida, M., and Niijima, T.,** Role of hydroxyl radical in neutrophil-mediated cytotoxicity, *Cell. Mol. Biol.*, 31, 229, 1985.

212. **Chiba, T., Nagai, Y., and Kakinuma, K.,** Cerebroside sulfuric ester (sulfatide) induces oxygen radical generation in guinea pig leukocytes, *Biochim. Biophys. Acta,* 930, 10, 1987.
213. **Kleinhaus, F. W. and Barefoot, S. T.,** Spin trap determination of free radical burst kinetics in stimulated neutrophils, *J. Biol. Chem.,* 262, 12452, 1987.
214. **Samuni, A., Black, C. D. V., Krishna, C. M., Malech, H. L., Bernstein, E. F., and Russo, A.,** Hydroxyl radical production by stimulated neutrophils reappraised, *J. Biol. Chem.,* 263, 13797, 1988.
215. **Meretey, K., Autal, M., Rozsayay, Z., Bohn, U., Elekes, E., and Geuti, G.,** PHA- and Con A-induced chemiluminescence of human blood mononuclear cells and granulocytes in luminol or lucigenin, *Inflammation,* 11, 417, 1987.
216. **DeChatelet, L. R., Shirley, P. S., McPhail, L., Huntley, C. C., Muss, H. B., and Bass, D. A.,** Oxidative metabolism of the human eosinophils, *Blood,* 50, 525, 1977.
217. **Shult, P. A., Graziano, F. M., Wallow, I. H., and Busse, W. E.,** Comparison of superoxide generation and luminol-dependent chemiluminescence with eosinophils and neutrophils from normal individuals, *J. Lab. Clin. Med.,* 106, 638, 1985.
218. **Petreccia, D. C., Nauseef, W. M., and Clark, R. A.,** Respiratory burst of normal human eosinophils, *J. Leukocyte Biol.,* 41, 283, 1987.
219. **Yoshie, O., Majima, T., and Saito, H.,** Membrane oxidative metabolism of human eosinophilic cell line Eol-1 in response to phorbol diester and formyl peptide: synergistic augmentation by interferon-γ and tumor necrosis factor, *J. Leukocyte Biol.,* 45, 10, 1989.
220. **Kroegel, C., Yukawa, T., Westwick, J., and Barnes, P. J.,** Evidence for two platelet activating factor receptors on eosinophils: dissociation between RAF-induced intracellular calcium mobilization degranulation and superoxide anion generation in eosinophils, *Biochem. Biophys. Res. Commun.,* 162, 511, 1989.
221. **Henderson, W. R. and Kaliner, M.,** Immunologic and nonimmunologic generation of superoxide from must cells and basophils, *J. Clin. Invest.,* 61, 187, 1978.
222. **Johnston, R. B., Lehmeyer, J. E., and Guthrie, L. A.,** Generation of superoxide anions and chemiluminescence by human monocytes during phagocytosis and on contact with surface bound immunoglobin G, *J. Exp. Med.,* 143, 1551, 1976.
223. **Weiss, S. J., King, G. W., and LoBuglio, A. F.,** Superoxide generation by human monocytes and macrophages, *Am J. Hematol.,* 4, 1, 1978.
224. **Reiss, M. and Roos, D.,** Differences in oxygen metabolism of phagocytosing monocytes and neutrophils, *J. Clin. Invest.,* 61, 480, 1978.
225. **French, J. K., Hurst, N. P., McColl, S. R., and Cleland, L.,** Effect of piroxicam on superoxide production by human blood mononuclear cells, *J. Rheumatol.,* 14, 1018, 1987.
226. **Rouis, M., Nigon, F., and Chapman, M. J.,** Platelet activating factor is a potent stimulant of the production of active oxygen species by human monocyte-derived macrophages, *Biochem. Biophys. Res. Commun.,* 156, 1293, 1988.
227. **Nielsen, H. and Valerius, N. H.,** Thalidomide enhances superoxide anion release from human polymorphonuclear and mononuclear leukocytes, *Acta Pathol. Microbiol. Imm. Scand.,* 94, 233, 1986.
228. **Sagone, A. L., King, G. W., and Metz, E. N.,** A comparison of the metabolic response to phagocytosis in human granulocytes and monocytes, *J. Clin. Invest.,* 57, 1352, 1976.
229. **Zeller, J. M., Caliendo, J., Lint, T. F., and Nelson, D. J.,** Changes in respiratory burst activity during human monocyte differentiation in suspension culture, *Inflammation,* 12, 585, 1988.
230. **Aasen, T. B., Bolan, B., Glette, J., Ulvik, R. J., and Schreiner, A.,** Lucigenin-dependent chemiluminescence in mononuclear phagocytes. Role of superoxide anion, *Scand. J. Clin. Lab. Invest.,* 47, 673, 1987.
231. **Falck, P.,** Bestimmung der Superoxid-Dismutase-Aktivität in Immunozellen mittels Lucigenin-vermittelter Chemiluminescenz, *Allerg. Immunol.,* 32, 199, 1986.
232. **Hurst, N. P., French, J. K., Bell, A. L., Nuki, G., O'Donnell, M. L., Betts, W. H., and Cleland, L. G.,** Different effects of mepacrine, chloroquine and hydroxychloroquine on superoxide anion generation, phospholipid methylation and arachidonic acid release by human blood monocytes, *Biochem. Pharmacol.,* 35, 3083, 1986.
233. **Leonard, E. J., Shenei, A., and Skeel, A.,** Dynamics of chemotactic peptide-induced superoxide generation by human monocytes, *Inflammation,* 11, 229, 1987.
234. **Scully, S. P., Segel, G. B., and Lichtman, M. A.,** Relationship of superoxide production to cytoplasmic free calcium in human monocytes, *J. Clin. Invest.,* 77, 1349, 1986.
235. **Costa-Casnellie, M. R., Segel, G. B., and Lichtman, M. A.,** Signal transduction in human monocytes: relationship between superoxide production and the level of kinase C in the membrane, *J. Cell. Physiol.,* 129, 336, 1986.
236. **Johansson, A. and Dahlgren, C.,** Characterization of the luminol-amplified light-generating reaction induced in human monocytes, *J. Leukocyte Biol.,* 45, 444, 1989.
237. **Schaffner, A. and Rellstal, P.,** γ-Interferon restores listericidal activity and concurrently enhances release of reactive oxygen metabolites in dexamethasone-treated human monocytes, *J. Clin. Invest..,* 82, 913, 1988.

238. **Davis, P. and Johnston, C.,** Effect of gold compounds on function of phagocytic cells. Comparative inhibition of activated polymorphonuclear leukocytes and monocytes from rheumatoid arthritis and control subjects, *Inflammation,* 10, 311, 1986.
239. **Hoffman, T. and Lizzio, E. F.,** Albumin in monocyte function assays. Differential stimulation of superoxide or arachidonate release by calcium ionophores, *J. Immunol. Methods,* 112, 9, 1988.
240. **Saito, H., Salmon, J. A., and Moncada, S.,** Influence of cholesterol feeding on the production of eicosanoids, tissue plasminogen activator and superoxide anion (O_2^-) by rabbit blood monocytes, *Atherosclerosis,* 61, 141, 1986.
241. **Geffner, J. R., Giorduno, M., Serebrinsky, G., and Isturiz, M.,** The role of reactive oxygen intermediates in nonspecific monocyte cytotoxicity induced by immune complexes, *Clin. Exp. Immunol.,* 87, 646, 1987.
242. **Heales, S. J. R., Blair, J. A., Meinschad, C., and Ziegler, I.,** Inhibition of monocyte luminol-dependent chemiluminescence by tetrahydrobiopterin, and the free radical oxidation of tetrahydrobiopterin, dihydrobiopterin, and dehydroneopterin, *Cell. Biochem. Funct.,* 6, 191, 1988.
243. **Britigan, B. E., Coffman, T. J., Adelberg, D. R., and Cohen, M. S.,** Mononuclear phagocytes have the potential for sustained hydroxyl radical production. Use of spin-trapping techniques to investigate mononuclear phagocyte free radical production, *J. Exp. Med.,* 168, 2367, 1988.
244. **Kemmerich, B., Rossing, T. H., and Pennington, J. E.,** Comparative oxidative microbicidal activity of human blood monocytes and alveolar macrophages and activation by recombinant gamma interferon, *Am. Rev. Respir. Dis.,* 136, 266, 1987.
245. **Oyanagui, Y.,** Inhibition of superoxide anion production in macrophages by anti-inflammatory drugs, *Biochem. Pharmacol.,* 25, 1473, 1976.
246. **Johnston, R. B., Godzik, C. A., and Conh, Z. A.,** Increased superoxide anion production by immunologically activated and chemically elicited macrophages, *J. Exp. Med.,* 148, 115, 1978.
247. **Stroke, S. H., Davis, W. B., and Sorber, W. A.,** Effect of phagocytosis on superoxide anion production and superoxide dismutase levels in BCG-activated and normal rabbit alveolar macrophages, *J. Reticuloendothelial. Soc.,* 24, 101, 1978.
248. **Miles, P. R., Castranova, V., and Lee, P.,** Reactive forms of oxygen and chemiluminescence in phagocyting rabbit alveolar macrophages, *Am. J. Physiol.,* 235, C103, 1978.
249. **Diaz, P., Jones, D. G., and Kay, A. B.,** Histamine-coated particles generate superoxide (O_2^-) and chemiluminescence in alveolar macrophages, *Nature,* 278, 454, 1979.
250. **Kayashima, K., Onoue, K., Nakagawara, A., and Minakami, S.,** Superoxide-anion-generating activities of macrophages as studied by using cytochalasin E and lectins as synergistic stimulants for superoxide release, *Microbiol. Immunol.,* 24, 449, 1980.
251. **Forman, H. J., Nelson, J., and Harrison, G.,** Hyperoxia alters effect of calcium on rat alveolar macrophage superoxide production, *J. Appl. Physiol.,* 60, 1300, 1986.
252. **Zeidler, R. B. and Conley, N. S.,** Superoxide generation by pig alveolar macrophages, *Compt. Biochem. Physiol.,* 85, Part B, 101, 1986.
253. **Leurs, R., Timmerman, H., and Bast, A.,** Inhibition of superoxide anion radical production by ebselen (PZ 51) and its sulfur analog (PZ 25) in guinea pig alveolar macrophages, *Biochem. Int.,* 18, 295, 1989.
254. **Ando, M., Suga, M., Sugimoto, M., and Tokuomi, H.,** Superoxide production in pulmonary alveolar macrophages and killing of BCG by the superoxide-generating system with or without catalase, *Infect. Immun.,* 24, 404, 1979.
255. **Hume, D. A., Gordon, S., Thornalley, P. J., and Bannister, J. V.,** The production of oxygen-centered radicals by Bacillus-Calmette-Guerin activated macrophages. An electron paramagnetic resonance study of the response to phorbol myristate acetate, *Biochim. Biophys. Acta,* 763, 245, 1983.
256. **Ito, M., Karmali, R., and Krim, M.,** Effect of interferon on chemiluminescence and hydroxyl radical production in murine macrophages stimulated by PMA, *Immunology,* 56, 533, 1985.
257. **Korkina, L. G., Suslova, T. B., Guyaeva, Zh. G., Zezin, A. B., Velichkovsky, B. T., and Kabanov, V. A.,** Chemiluminescence of peritoneal macrophages activated non-natural polyelectrolytes, *Dokl. Akad. Nauk SSSR,* 282, 206, 1985.
258. **Korkina, L. G., Korepanova, E. A., Velichkovsky, B. T., and Vladimirov, Yu. A.,** Membrane mechanism of macrophage activation. I. Peculiarities of cell activation by peptide polycations, protamine and polymyxin, *Biol. Membr.,* 3, 1250, 1986.
259. **Korkina, L. G., Suslova, T. B., Gulyaeva, Zh. G., Zezim, A. B., Velichkovsky, B. T., Vladimirov, Yu. A., and Kabanov, V. A.,** Membrane mechanism of macrophage activation. II. Stimulation of chemiluminescence in the suspension of peritoneal macrophages by synthetic polyelectrolytes, *Biol. Membr.,* 4, 1093, 1987.
260. **Velichkovsky, B. T., Vladimirov, Yu. A., Korkina, L. G., and Suslova, T. B.,** Physico-chemical mechanism of the interaction of phagocyting cells with fibrogenic dusts, *Vestn. Akad. Med. Nauk SSSR,* N 10, 45, 1982.
261. **Velichkovsky, B. T., Cheremishina, Z. P., Suslova, T. B., Korkina, L. G., and Olenev, V. I.,** Molecular mechanism of asbestos biological activity, *Gib. Tr.,* N 9, 5, 1986.

262. **Velichkovsky, B. T., Korkina, L. G., and Suslova, T. B.,** Determination of dust toxicity by the chemiluminescence technique, *Gig. Tr.*, N 5, 31, 1983.
263. **Velichkovsky, B. T., Korkina, L. G., Suslova, T. B., Cheremishina, Z. P., Deeva, I. B., and Kruglikov, G. G.,** Principal molecular mechanisms of cytotoxic action of fibrogenic dusts, *Vestn. Akad. Med. Nauk SSSR*, N1, 7, 1988.
264. **Storch, J., Ferber, E., and Munder, P. G.,** Influence of platelet activating factor and a nonmetabolizable analogue on superoxide production by bone marrow derived macrophages, *J. Leukocyte Biol.*, 44, 385, 1988.
265. **Sugioka, K., Nakano, M., Karashige, S., Akuzawa, Y., and Gobo, T.,** A chemiluminescent probe with a *Cypridina* luciferin analog, 2-methyl-6-phenyl-3,7-dihydroimidazo 1,2-a pyrazin-3-one, specific and sensitive for O_2^- production in phagocytizing macrophages, FEBS Lett., 197, 27, 1986.
266. **Hoffman, M. and Autor, A. P.,** Production of superoxide anion by an NADPH-oxidase from rat pulmonary macrophages, *FEBS Lett.*, 121, 352, 1980.
267. **Sakata, A., Ida, E., Tominaga, M., and Onoue, K.,** Arachidonic acid acts as an intracellular activator of NADPH-oxidase in Fc receptor-mediated superoxide generation in macrophages, *J. Immunol.*, 138, 4353, 1987.
268. **Tsunawaki, S. and Nathan, C. F.,** Release of arachidonate and reduction of oxygen. Independent metabolic bursts of the mouse peritoneal macrophage, *J. Biol. Chem.*, 261, 11563, 1986.
269. **Ida, E., Sakata, A., Tominaga, M., Yamasaki, H., and Onoue, K.,** Arachidonic acid release is closely related to the Fcγ receptor-mediated superoxide generation in macrophages, *Microbiol. Immunol.*, 32, 1127, 1988.
270. **Kitagawa, S. and Johnston, R. B., Jr.,** Relationship between membrane potential changes and superoxide-releasing capacity in resident and activated mouse peritoneal macrophages, *J. Immunol.*, 135, 3417, 1985.
271. **Hishinuma, K., Hosono, A., Mashiko, S., Inaba, H., and Kimura, S.,** Glycolytic inhibitor failed to inhibit superoxide generation in mouse macrophages: an investigation using a chemiluminescence probe specific for superoxide anion, *Agric. Biol. Chem.*, 53, 1189, 1989.
272. **Esterline, R. L. and Trush, M. A.,** Lucigenine chemiluminescence and its relationship to mitochondrial respiration in phagocytic cells, *Biochem. Biophys. Res. Commun.*, 159, 584, 1989.
273. **Speer, C. P., Gahr, M., and Pabst, M. J.,** Phagocytosis-associated oxidative metabolism in human milk macrophages, *Acta Paediatr. Scand.*, 75, 444, 1986.
274. **Sodhi, A. and Gupta, P.,** Increased release of hydrogen peroxide (H_2O_2) and superoxide anion ($O_2^{\cdot}$) by murine macrophages in vitro after cis-platin treatment, *Int. J. Immunopharmacol.*, 8, 709, 1986.
275. **Amano, F., Nishijima, M., Akagawa, K., and Akamatsu, Y.,** Enhancement of O_2^- generation and tumoricidal activity of murine macrophages by a monosaccharide precursor of *Escherichia coli* lipid A, *FEBS Lett.*, 192, 263, 1985.
276. **Miyagawa, N., Okamoto, Y., and Nakano, H.,** Effect of C-reactive protein on peritoneal macrophages. II. Human C-reactive protein activates peritoneal macrophages of guinea pigs to release superoxide anion in vitro, *Microbiol. Immunol.*, 32, 709, 1988.
277. **Altavilla, D., Berlinghieri, M. C., Seminara, S., Iannello, D., and Foca, A.,** Different effects of bacterial lipopolysaccharide on superoxide anion production by macrophages from normal and tumor-bearing rats, *Immunopharmacology*, 17, 99, 1989.
278. **Case, B. W., Ip, M. P. C., Padilla, M., and Kleinerman, J.,** Asbestos effects on superoxide production. An in vitro study of hamster alveolar macrophages, *Environ. Res.*, 39, 299, 1986.
279. **Hansen, K. and Mossman, B. T.,** Generation of superoxide (O_2^-) from alveolar macrophages exposed to asbestiform and nonfibrous particles, *Cancer Res.* 47, 1687, 1987.
280. **Wilhelm J., Vilim, V., and Brzak, P.,** Participation of superoxide in luminol-dependent chemiluminescence triggered by mineral dust in rabbit alveolar macrophages, *Immunol. Lett.*, 15, 329, 1987.
281. **Badger, A. M.,** Enhanced superoxide production by rat alveolar macrophages stimulated in vitro with biological response modifiers, *J. Leukocyte Biol.*, 40, 725, 1986.
282. **Drath, D. B.,** Modulation of pulmonary macrophage superoxide release and tumoricidal activity following activation by biological response modifiers, *Immunopharmacology*, 12, 117, 1986.
283. **Cassatella, M. A., Bianca, V. D., Berton, G., and Rossi, F.,** Activation by interferon-γ of human macrophage capability to produce toxic oxygen molecules is accompanied by decreased K_m of the superoxide generating NADPH oxidase, *Biochem. Biophys. Res. Commun.*, 132, 908, 1985.
284. **Edwards, C. K., III, Ghiasuddin, S. M., Schepper, J. M., Yunger, L. M., and Kelley, K. W.,** A newly defined property of somatotropin: priming of macrophages for production of superoxide anion, *Science*, 239, 769, 1988.
285. **Lison, D., Dubois, P., and Lauwerys, R.,** In vitro effect of mercury and vanadium on superoxide anion production and plasminogen activator of mouse peritoneal macrophages, *Toxicol. Lett.*, 40, 29, 1988.
286. **Conley, N. S., Yarbo, J. W., Ferrari, H. A., and Zeidler, R. B.,** Bleomycin increases superoxide anion generation by pig peripheral alveolar macrophages, *Mol. Pharmacol.*, 30, 48, 1986.

287. **Korkina, L. G., Cheremisina, Z. P., Suslova, T. B., Durnev, A. D., Seredenin, S. B., and Velichkovskii, B. T.,** Bleomycin and oxygen free radicals, *Stud. Biophys.*, 126, 105, 1988.
288. **DiGregorio, K. A., Cilento, E. V., and Lantz, R. C.,** Measurement of superoxide release from single pulmonary alveolar macrophages, *Am. J. Physiol.*, 252, C677, 1987.
289. **Colton, C. A. and Gilbert, D. L.,** Production of superoxide anions by a CNS macrophages, the microglia, *FEBS Lett.*, 223, 284, 1987.
290. **Warren, J. S., Kunkel, R. C., Johnson, K. J., and Ward, P. A.,** Comparative superoxide radical anion responses of lung macrophages and blood phagocytic cells in the rat. Possible relevance to IgA immune complex induced lung injury, *Lab. Invest.*, 57, 311, 1987.
291. **Winsel, K., Slapke, J., Unger, U., Grollmuss, H., and Renner, H.,** Influence of antioxidants on the production of reactive oxygen metabolites by stimulated alveolar macrophages, *Z. Erkr. Atmungsorgane*, 169, 250, 1987.
292. **Afanas'ev, I. B., Korkina, L. G., Briviba, K. K., and Velichkovskii, B. T.,** Rutin is scavenger of Superoxide ion and "crypto"-hydroxyl radicals, in *Proc. 4th Biennial Gen. Meeting of the Soc. for Free Radical Res.*, Hayaishi, O., Niki, E., Kondo, M., and Yoshikawa, T., Eds., Elsevier, Amsterdam, 1989, 243.
293. **Parnham, M. J. and Kindt, S.,** A novel biologically active seleno-organic compound-III. Effects of PZ 51 (ebselen) on glutathione peroxidase and secretory activity of mouse macrophages, *Biochem. Pharmacol.*, 33, 3247, 1984.
294. **Gudewicz, P. W.,** Effect of anticancer drugs on macrophage-mediated antibody-dependent cytotoxicity and secretion of reactive oxygen intermediates, *Cancer Lett.*, 42, 67, 1988.
295. **Arduini, A., Mancinelli, G., Belfiglio, M., DeJulia, J., Damonti, V., Storto, S., and Federici, G.,** NPGB-induced inhibition of superoxide anion production by normal Lewis rat macrophages, *Neurochem. Res.*, 14, 55, 1989.
296. **Tritsch, G. and Niswander, P.,** Modulation of macrophage superoxide release by purine metabolism, *Life Sci.*, 32, 1359, 1983.
297. **Hancock, J. T. and Jones, O. T. G.,** The inhibition by diphenylene iodonium and its analogs of superoxide generation by macrophages, *Biochem. J.*, 242, 103, 1987.
298. **Samamura, D., Sato, S., Suzuki, M., Nomura, K., Hanada, K., and Hashimoto, I.,** Effect of cepharonthine on superoxide anion (O_2^-) production by macrophages, *J. Dermatol.*, 15, 304, 1988.
299. **Dileepan, K. D., Simpson, K. M., and Stechschulte, D. J.,** Modulation of macrophage superoxide-induced cytochrome c reduction by mast cells, *J. Lab. Clin. Med.*, 113, 577, 1989.
300. **Smith, R. L. and Weidemann, M. J.,** Reactive oxygen production associated with arachidonic acid metabolism by peritoneal macrophages, *Biochem. Biophys. Res. Commun.*, 97, 973, 1980.
301. **Schopf, R. E., Lutz, B., and Rehder, M.,** Regulation of phagocyte chemiluminescence by arachidonate metabolism, in *Bioluminescence and Chemiluminescence*, Schoelmerich, J., Ed., John Wiley & Sons, Chichester, 1987, 173.
302. **Ginsburg, I., Fligiel, S. E. G., Ward, P. A., and Varani, J.,** Lipoteichoic acid-antilipoteichoic acid complexes induce superoxide generation by human neutrophils, *Inflammation*, 12, 525, 1988.
303. **Lewis, J. G., Hamilton, T., and Adams, D. O.,** The effect of macrophage development on the release of reactive oxygen intermediates and lipid oxidation products, and their ability to induce oxidative DNA damage in mammalian cells, *Carcinogenesis*, 7, 813, 1986.
304. **Cheung, K., Lark, J., Robinson, M. F., Pomery, P. J., and Hunter, S.,** The production of hydroxyl radical by human neutrophils stimulated by arachidonic acid-measurements by ESR spectroscopy, *Aust. J. Exp. Biol. Med. Sci.*, 64, 157, 1986.
305. **Sakane, F., Kojima, H., Takahashi, K., and Kojama, J.,** Porcine polymorphonuclear leukocyte NADPH-cytochrome c reductase generates superoxide in the presence of cytochrome b_{559} and phospholipid, *Biochem. Biophys. Res. Commun.*, 147, 71, 1987.
306. **Baud, L., Hagege, J., Sraer, J., Rondeau, E., Perez, J., and Ardaileau, R.,** Reactive oxygen production by cultured rat glomerular mesangial cells during phagocytosis is associated with stimulation of lipoxygenase activity, *J. Exp. Med.*, 158, 1836, 1983.
307. **Sedor, J. R., Carey, S. W., and Emancipator, S. N.,** Immune complexes bind to cultured rat glomerular mesangial cells to stimulate superoxide release. Evidence for an Fc Receptor, *J. Immunol.*, 138, 3751, 1987.
308. **Adler, S., Baker, P. J., Johnson, R. J., Ochi, R. F., Pritzl, P., and Couser, W. G.,** Complement membrane attack complex stimulates production of reactive oxygen metabolites by cultured rat mesangial cells, *J. Clin. Invest.*, 77, 762, 1986.
309. **Basci, A. and Shah, S. V.,** Trypsin- and chymotrypsin-induced chemiluminescence by isolated glomeruli, *Am. J. Physiol.*, 252, C611, 1987.
310. **Matsubara, T. and Ziff, M.,** Superoxide anion release by human endothelial cells: synergism between a phorbol ester and a calcium ionophore, *J. Cell. Physiol.*, 127, 207, 1986.
311. **Matsubara, T. and Ziff, M.,** Increased superoxide anion release from human endothelial cells in response to cytokenes, *J. Immunol.*, 137, 3295, 1986.

312. **Gorog, P., Pearson, J. D., and Kakkar, V. V.,** Generation of reactive oxygen metabolites by phagocytosing endothelial cells, *Atherosclerosis*, 72, 19, 1988.
313. **Schinetti, M. L., Sbarbati, R., and Scarlattini, M.,** Superoxide production by human umbilical vein endothelial cells in anoxia-reoxygenation model, *Cardiovasc. Res.*, 23, 76, 1989.
314. **Zweier, J. L., Kuppusamy, P., and Lutty, G. A.,** Measurement of endothelial free radical generation: evidence for a central mechanism of free radical injury in postischemic tissues, *Proc. Natl. Acad. Sci. U.S.A.*, 85, 4046, 1988.
315. **Newburger, P. E., Speier, C., Borregoard, N., Walsh, C. E., Whitin, J. C., and Simons, E. R.,** Development of the superoxide-generating system during differentiation of the HL-60 human promyelocytic leukemia cell line, *J. Biol. Chem.*, 259, 3771, 1984.
316. **Gaut, J. R. and Carchman, R. A.,** A correlation between phorbol diester-induced protein phosphorylation and superoxide anion generation in HL-60 cells during granulocytic maturation, *J. Biol. Chem.*, 262, 826, 1987.
317. **Peterson, D. A., Mehta, N., Butterfield, J., Husak, M., Christopher, M. M., Jagarlapudi, S., and Eaton, J. W.,** Polyunsaturated fatty acids stimulate superoxide formation in tumor cells: a mechanism for specific cytotoxicity and a model for tumor necrosis factor?, *Biochem. Biophys. Res. Commun.*, 155, 1033, 1988.
318. **Mazzone, A., Ricevuti, G., Rizzo, S. C., and Sacchi, S.,** The probable role of superoxide produced by blast cells in leukemic cutaneous spreading, *Int. J. Tissue React.*, 8, 493, 1986.
319. **Kapp, A., Wolff-Vorbeck, G., and Peter, H. H.,** Chemiluminescence response of human B-cell lines, *Free Rad. Res. Commun.*, 2, 337, 1987.
320. **Volkman, D. J., Buescher, E. S., Gallin, J. I., and Fauci, A. S.,** B Cell lines as models for inherited phagocytic diseases: abnormal superoxide generation in chronic granulomatous disease and giant granules in Chediak-Higashi syndrome, *J. Immunol.*, 133, 3006, 1984.
321. **Shibanuma, M., Kuroki, T., and Nose, K.,** Effects of the protein kinase C inhibitor H-7 and calmodulin antagonist W-7 on superoxide production in growing and resting human histocytic leukemia cells (U 937), *Biochem. Biophys. Res. Commun.*, 144, 1313, 1987.
322. **Balsinde, J. and Mollinedo, F.,** Specific activation by concanavalin A of the superoxide anion generation capacity during U 937 differentiation, *Biochem. Biophys. Res. Commun.*, 151, 802, 1988.
323. **Shibanuma, M., Kuroki, T., and Nose, K.,** Superoxide as a signal for increase in intracellular pH, *J. Cell. Physiol.*, 136, 379, 1988.
324. **Leroyer, V., Werner, L., Shaughnessy, S., Goddard, G. J., and Orr, F. W.,** Chemiluminescence and oxygen radical generation by Walker carcinoma cells following chemotactic stimulation, *Cancer Res.*, 47, 4771, 1987.
325. **Das, U. N., Begin, M. E., Ells, G., Huang, Y. S., and Horrobin, D. F.,** Polyunsaturated fatty acids augment free radical generation in tumor cells in vitro, *Biochem. Biophys. Res. Commun.*, 145, 15, 1987.
326. **Rieder, H., Ramadori, G., and Bueschenfelde, K. H. M.,** Guinea pig Kupfer cells can be activated in vitro to an enhanced superoxide response. I. Comparison with peritoneal macrophages, *J. Hepatol.*, 7, 338, 1988.
327. **Suthanthiran, M., Solomon, S. D., Williams, P. S., Rubin, A. L., Nobogrodsky, A., and Stenzel, K. H.,** Hydroxyl radical scavengers inhibit human natural killer cell activity, *Nature*, 307, 276, 1984.
328. **Bray, R. A. and Brahmi, Z.,** Role of lipoxygenation in human natural killer cell activation, *J. Immunol.*, 136, 1783, 1986.
329. **Duwe, A. K., Werkmeister, J., Roder, J. C., Lanzon, R., and Payne, U.,** Natural killer cell-mediated lysis involves an hydroxyl radical-dependent step, *J. Immunol.*, 134, 2637, 1985.
330. **Gibboney, J. J., Haak, R. A., Kleinhaus, F. W., and Brahmi, Z.,** Electron spin resonance spectroscopy does not reveal hydroxyl radical production in activated natural killer lymphocytes, *J. Leukocyte Biol.*, 44, 545, 1988.
331. **Chaudhri, G., Clark, I. A., Hunt, N. H., Cowden, W. B., and Ceredig, R.,** Effect of antioxidants on primary alloantigen-induced T cell activation and proliferation, *J. Immunol.*, 137, 2646, 1986.
332. **Chaudhri, G., Hunt, N. H., Clark, I. A., Ceredig, R.,** Antioxidants inhibit proliferation and cell surface expression of receptors for interleukin-2 and transferrin in T lymphocytes stimulated with phorbol myristate acetate and ionomycin, *Cell. Immunol.*, 115, 204, 1988.
333. **Melinn, M. and McLaughlin, H.,** Nitroblue tetrazolium reduction in lymphocytes, *J. Leukocyte Biol.*, 41, 325, 1987.
334. **Markus, A. J., Silk, S. T., Safier, L. B., and Ullman, H. L.,** Superoxide production and reducing activity of human platelets, *J. Clin. Invest.*, 59, 149, 1977.
335. **Allen, R. C.,** Chemiluminescence from eukaryotic and prokariotic cells: reducing potential and oxygen requirements, *Photochem. Photobiol.*, 30, 157, 1979.
336. **Shvinka, J. E., Toma, M. K., Galinino, N. I., Skards, I. V., and Viesturs, U. E.,** Production of superoxide radicals during bacterial respiration, *J. Gen. Microbiol.*, 113, 377, 1979.
337. **Fischer, S. M. and Adams, L. M.,** Suppression of tumor promoter-induced chemiluminescence in mouse epidermal cells by several inhibitors of arachidonic acid metabolism, *Cancer Res.*, 45, 3130, 1985.

338. **Heinecke, J. W., Baker, L., Rosen, H., and Chait, A.,** Superoxide-mediated modification of low density lipoprotein by arterial smooth muscle cells, *J. Clin. Invest.*, 77, 757, 1986.
339. **Kahl, R., Weimann, A., and Hildebrandt, A. G.,** Detection of active oxygen in rat hepatocyte suspensions with the chemiluminigenic probe lucigenin, *Biochem. Biophys. Res. Commun.*, 140, 468, 1986.
340. **Chan, P. H., Chen, S. F., and Yu, A. C. H.,** Induction of intracellular superoxide radical formation by arachidonic acid and by polyunsaturated fatty acids in primary astrocytic cultures, *J. Neurochem.*, 50, 1185, 1988.
341. **Aitken, R. J. and Clarkson, J. S.,** Cellular basis of defective sperm function and its association with the genesis of reactive oxygen species by human spermatozoa, *J. Reprod. Fertil.*, 81, 459, 1987.
342. **Takahashi, A., Totsune-Nakano, H., Nakano, M., Mashiko, S., Suzuli, N., Ohma, C., and Inaba, H.,** Generation of superoxide anion and tyrosine cation-mediated chemiluminescence during the fertilization of sea urchin eggs, *FEBS Lett.*, 246, 117, 1989.
343. **Asada, K., Takahashi, M.-A., Tanaka, K., and Nakano, Y.,** Formation of active oxygen and its fate in chloroplasts, in *Biochemical and Medical Aspects of Active Oxygen*, Tokyo, 1977, 45.
344. **Merzlyak, M. N.,** Activated oxygen and oxidative processes in membranes of plant cells, *Itogi Nauki Tekh.*, 6, 1989, 6, VINITI, Moscow.
345. **Asada, K. and Kiso, K.,** Initiation of aerobic oxidation of sulfite by illuminated spinach chloroplasts, *Eur. J. Biochem.*, 33, 253, 1973.
346. **Asada, K., Kuniaki, K., and Kyoto, Y.,** Univalent reduction of molecular oxygen by spinach chloroplasts on illumination, *J. Biol. Chem.*, 249, 2175, 1974.
347. **Allen, J. F. and Hall, D. O.,** Superoxide reduction as a mechanism of ascorbate-stimulated oxygen uptake by isolated chloroplasts, *Biochem. Biophys. Res. Commun.*, 52, 856, 1973.
348. **Allen, J. F. and Hall, D. O.,** The relationship of oxygen uptake to electron transport in photosystem I of isolated chloroplasts: the rate of superoxide and ascorbate, *Biochem. Biophys. Res. Commun.*, 58, 579, 1974.
349. **Epel, B. L., Neumann, J., and Fishman, R.,** The mechanism of the oxidation of ascorbate and Mn^{2+} by chloroplasts. The role of the radical superoxide, *Biochim. Biophys. Acta*, 325, 520, 1973.
350. **Elstner, E. F. and Konze, J. R.,** Determination of superoxide free radical ion and hydrogen peroxide as products of photosynthetic oxygen reduction, *Z. Naturforsch.*, 30c, 53, 1975.
351. **Nelson, N., Nelson, H., and Racker, E.,** Photoreaction of FMN-tripcine and its participation in photophosphorylation, *Photochem. Photobiol.*, 16, 481, 1972.
352. **Takahashi, M. and Asada, K.,** Superoxide production in aprotic interior of chloroplast thylakoids, *Arch. Biochem. Biophys.*, 267, 714, 1988.
353. **Harbour, J. R. and Bolton, J. R.,** Superoxide formation in spinach chloroplasts: electron spin resonance detection by spin trapping, *Biochem. Biophys. Res. Commun.*, 64, 803, 1975.
354. **Ginkel, G. van and Raison, J. K.,** Light-induced formation of O_2^- oxygen radicals in systems containing chlorophyll, *Photochem. Photobiol.*, 32, 793, 1980.
355. **Bowyer, J. R. and Camilleri, P.,** Spin-trap study of the reactions of ferredoxin with reduced oxygen species in pea chloroplasts, *Biochim. Biophys. Acta*, 808, 235, 1985.
356. **Mikhailik, O. M., Shevchenko, A. I., and Ostrovskays, L. K.,** Superoxide anion radical formation in lupine chloroplasts under normal and pathological conditions, *Biokhimiya*, 51, 1186, 1986.
357. **Mikhailik, O. M., Shevchenko, A. I., and Ostrovskaya, L. K.,** Formation of superoxide anion-radicals by chloroplasts, *Biofizika*, 32, 89, 1987.
358. **Boucher, F. and Gingras, G.,** The photogeneration of superoxide by isolated photoreaction center from *Rhodospirillium rubrum*, *Biochem. Biophys. Res. Commun.*, 67, 323, 1975.
359. **Upham, B. L. and Jahnke, L. S.,** Photooxidative reactions in chloroplast thylakoids. Evidence for a Fenton-type reaction promoted by superoxide or ascorbate, *Photosynth. Res.*, 8, 235, 1986.
360. **Gross, G. G., Janse, C., and Elstner, E. F.,** Involvement of malate, monophenols, and the superoxide radical in hydrogen peroxide formation by isolated cell walls from horseradish *(Armoracia lapathifolia Gilib)*, *Planta*, 136, 271, 1977.
361. **Elstner, E. F.,** Comparison of "inflammation" in pine needles and humans, in *Oxygen Radicals in Chemistry and Biology*, Bors, W., Saran, M., and Tait, D., Eds., Walter de Gruyter, Berlin, 1984, 969.
362. **Averyanov, A. A.,** Generation of superoxide radical by intact pea roots, *Fiziol. Rast. (Physiology of Plants)*, 32, 268, 1985.
363. **Aver'yanov, A. A., Lapikova, V. P., Umnov, A. M., and Dzhavakhiya, V. G.,** Generation of superoxide radical by rice leaves in relation to plant tolerance to blast infection, *Fiziol. Rast.*, 34, 373, 1987.
364. **Doke, N.,** Generation of superoxide anion by potato tuber protoplasts upon the hypersensitive response to hyphal wall components of *Phytophthora infestans* and specific inhibition of the reaction by suppressor of hypersensitivity, *Physiol. Plant. Pathol.*, 23, 359, 1983.
365. **Doke, N.,** NADPH-dependent O_2^- generation in membrane fractions isolated from wounded potato tubers inoculated with *Phytophthora infestans*, *Physiol. Plant. Pathol.*, 27, 311, 1985.
366. **Vianelli, A. and Macri, F.,** NAD(P)H oxidation elicits anion superoxide formation in radish plasmalemma vesicles, *Biochim. Biophys. Acta*, 980, 202, 1989.

367. **Lynch, D. V. and Thompson, J. E.,** Lipoxygenase-mediated production of superoxide anion in senescing plant tissue, *FEBS Lett.*, 173, 251, 1984.
368. **Doke, N.,** Involvement of superoxide anion generation in hypersensitive response of potato tuber tissues to infection with an incompatible race of *Phytophthora infestans* and to hyphal wall components, *Physiol. Plant. Pathol.*, 23, 345, 1983.
369. **Doke, N. and Ohashi, Y.,** Involvement of a superoxide-generating system in the induction of necrotic lesions on tobacco leaves infested with tobacco mosaic virus, *Physiol. Mol. Plant. Pathol.*, 32, 163, 1988.
370. **Zacheo, G. and Bleve-Zacheo, T.,** Involvement of superoxide dismutases and superoxide radicals in the susceptibility and resistance of tomato plants to *Meloidogyne incognita* attack, *Physiol. Mol. Plant. Pathol.*, 32, 313, 1988.
371. **Cakmak, I., Van de Wetering, D. A. M., Marschner, H., and Bienfait, H. F.,** Involvement of superoxide radical in extracellular ferric reduction by iron-deficient bean roots, *Plant Physiol.*, 85, 310, 1987.
372. **Chai, H. B. and Doke, N.,** Superoxide anion generation: a response of potato leaves to infection with *Phytophthora infestans, Phytopathology*, 77, 645, 1987.
373. **Dahlgren, C.,** Is lysosomal fusion required for the granulocyte chemiluminescence reaction?, *Free Rad. Biol. Med.*, 6, 399, 1989.
374. **Vilim, V. and Wilhelm, J. W.,** What do we measure by a luminol-dependent chemiluminescence of phagocytes?, *Free Rad. Biol. Med.*, 6, 623, 1989.
375. **Ellis, J. A., Cross, A. R., and Jones, O. T. G.,** Studies on the electron-transfer mechanism of the human neutrophil NADPH oxidase, *Biochem. J.*, 262, 575, 1989.
376. **Pilloud, M. C., Dousiere, J., and Vignais, P. V.,** Parameters of activation of the membrane-bound superoxide radical anion generating oxidase from bovine neutrophils in a cell-free system, *Biochem. Biophys. Res. Commun.*, 159, 783, 1989.
377. **Pou, S., Rosen, G. M., Britigan, B. E., and Cohen, M. S.,** Intracellular spin-trapping of oxygen-centered radicals generated by human neutrophils, *Biochim. Biophys. Acta*, 991, 459, 1989.
378. **Ohsaka, A., Saito, M., Suzuki, I., Miura, Y., Takaki, F., and Kitagawa, S.,** Phorbol myristate acetate potentiates superoxide release and membrane depolarization without affecting an increase in cytoplasmic free calcium in human granulocyte stimulated by chemotactic peptide, lectins and the calcium ionophore, *Biochim. Biophys. Acta*, 941, 19, 1988.
379. **Dahlgren, C.,** Polymorphoncular leukocyte chemiluminescence induced by formylmethionyl-leucyl-phenylalanine and phorbol myristate acetate: effects of catalase and superoxide dismutase, *Agents Actions*, 21, 104, 1987.
380. **Greenwald, R. A., Rush, S. W., Moak, S. A., and Weitz, Z.,** Conversion of superoxide generated by polymorphonuclear leukocytes to hydroxyl radical: a direct spectrophotometric detection system based on degradation of deoxyribose, *Free Rad. Biol. Med.*, 6, 385, 1989.
381. **Ligeti, E., Tardif, M., and Vignais, P.,** Activation of superoxide radical anion generating oxidase of bovine neutrophils in cell-free system. Interaction of a cytosolic factor with the plasma membrane and control by nucletides, *Biochemistry*, 28, 7116, 1989.
382. **Kuroki, M. and Minakami, S.,** Extracellular ATP triggers superoxide production in human neutrophils, *Biochem. Biophys. Res. Commun.*, 162, 377, 1989.
383. **Gyllenhammar, H.,** Correlation between neutrophil superoxide formation, luminol-augmented chemiluminescence and intracellular calcium levels upon stimulation with leukotriene B_4, formylpeptide and phorbolester, *Scand. J. Clin. Invest.*, 49, 317, 1989.
384. **Bonser, R. W., Thompson, N. T., Randall, R. W., and Garlaud, L. G.,** Phospholipase D activation is functionally linked to superoxide generation in the human neutrophil, *Biochem. J.*, 264, 612, 1989.
385. **Gresham, H. D., McGarr, J. A., Shackelford, P. G., and Brown, E. J.,** Studies on the molecular mechanisms of human Fc receptor-mediated phagocytosis. Amplification of ingestion is dependent on the generation of reactive oxygen metabolites and is deficient in polymorphonuclear leukocytes from patients with chronic granulomatous disease, *J. Clin. Invest.*, 82, 1192, 1988.
386. **Ligeti, E., Doussiere, J., and Vignais, P. V.,** Activation of the superoxide radical generation oxidase in plasma membrane from bovine polymorphonuclear neutrophils by arachidonic acid, a cytosolic factor of protein nature, and non-hydrolyzable analogs of GTP, *Biochemistry*, 27, 193, 1988.
387. **Ichinose, Y., Hara, N., Ohta, M., Motohiro, A., Kuda, T., Aso, H., and Yagawa, K.,** Phorbol myristate acetate modulates calcium ion-dependent superoxide anion generation induced by a monoclonal antibody raised against polymorphonuclear leukocytes, *Infect. Immun.*, 57, 2529, 1989.
388. **Ozawa, M., Ohtsuka, T., Okamura, N., and Ishibashi, S.,** Synergism between protein kinase C activator and fatty acids in stimulating superoxide anion production in guinea pig polymorphonuclear leukocytes, *Arch. Biochem. Biophys.*, 273, 491, 1989.
389. **Doussiere, J., Pilloud, M. C., and Vinas, P. V.,** Activation of bovine neutrophil oxidase in a cell free system. GTP-dependent formation of a complex between a cytosolic factor and a membrane protein, *Biochem. Biophys. Res. Commun.*, 152, 993, 1988.

390. **Ohtsuka, T., Ozawa, M., Okamura, N., and Ishibashi, S.,** Stimulatory effects of a short chain phosphatidate on superoxide anion production in guinea pig polymorphonuclear leukocytes, *J. Biochem.*, 106, 259, 1989.
391. **Kownatzki, E., Kapp, A., and Uhrich, S.,** Modulation of human neutrophilic granulocyte functions by recombinant human tumor necrosis factor and recombinant human lymphotoxin-promotion of adherence, inhibition of chemotactic migration and superoxide anion release from adherent cells, *Clin. Exp. Immunol.*, 74, 143, 1988.
392. **Woodman, R. C., Curnutte, J. T., and Babior, B. B.,** Evidence that de novo protein synthesis participates in a time-dependent augmentation of chemotactic peptide-induced respiratory burst in neutrophils. Effects of recombinant human colony stimulating factors and dihydrocytochalasin B, *Free Rad. Biol. Med.*, 5, 355, 1988.
393. **Mege, J. L., Gomez-Combronero, J., Molski, T. F. P., Becker, E. L., and Sha'afi, R. I.,** Effect of granulocyte-macrophage colony-stimulating factor on superoxide production in cytoplasts and intact human neutrophils: role of protein kinase and G-proteins, *J. Leukocyte Biol.*, 46, 161, 1989.
394. **Savage, J. E., Sullivan, J. F., Zeis, B. M., and Anderson, R.,** Investigation of the structural properties of dihydrophenazines which contribute to their pro-oxidative interaction with human phagocytes, *J. Antimicrob. Chemother.*, 23, 691, 1989.
395. **Briheim, G., Follin, P., Sandstedt, S., and Dahlgren, C.,** Relationship between intracellularly and extracellularly generated oxygen metabolites from primed polymorphonuclear leukocytes differs from that obtained from nonprimed cells, *Inflammation*, 13, 455, 1989.
396. **Carlin, G., Djursaeter, R., and Smedegaard, G.,** Inhibitory effects of sulfasalazine and related compounds on superoxide production by human polymorphonuclear leukocytes, *Pharmacol. Toxicol.*, 65, 121, 1989.
397. **Serhan, C. N. and Reardon, E.,** 15-Hydroxyeicosatetraenoic acid inhibits superoxide anion generation by human neutrophils: relationship to lipoxin production, *Free Rad. Res. Commun.*, 7, 341, 1989.
398. **Nakata, M., Tomita, T., Iizuka, T., and Kanegasaki, S.,** Inhibitory effect of antibiotic cerulenin on the respiratory burst in phagocytes. II. Inhibition by cerulenin of intracellular calcium mobilization in human neutrophils, *J. Antibiot.*, 42, 1178, 1989.
399. **Taniguchi, K., Urakami, M., and Takanaka, K.,** Effects of various drugs on superoxide generation, arachidonic acid release and phospholipase A_2 in polymorphonuclear leukocytes, *Jpn. J. Pharmacol.*, 46, 275, 1988.
400. **Stahlberg, H.-J., Loschen, G., and Flohé, L.,** Effects of prostacyclin analogs on cyclic adenosine monophosphate and superoxide production in human polymorphonuclear leukocytes stimulated by formylmethionyl-leucyl-phenylalanine, *Biol. Chem., Hoppe-Seyler*, 369, 329, 1988.
401. **Lee, C. S., Han, E. S., and Lee, K. S.,** Effects of atropine, phentolamine and propranolol on calcium uptake, superoxide generation and phagocytic activity in activated PMN leukocytes, *Korean J. Pharmacol.*, 24, 83, 1989.
402. **Jacob, J.,** Ca^{2+} ionophores inhibit superoxide generation by chemotactic peptide in rabbit neutrophils and the correlation with intracellular calcium, *Mol. Cell. Biochem.*, 84, 97, 1988.
403. **Cotgreave, I. A., Duddy, S. K., Kass, G. E. N., Thompson, D., and Moldeus, P.,** Studies on the anti-inflammatory activity of Ebselen. Ebselen interferes with granulocyte oxidative burst by dual inhibition of NADPH oxidase and protein kinase C?, *Biochem. Pharmacol.*, 38, 649, 1989.
404. **Zimmerman, J. J., Zuk, S. M., and Millard, J. R.,** In vitro modulation of human neutrophil superoxide anion generation by various calcium channel antagonists used in ischemia-reperfusion resuscitations, *Biochem. Pharmacol.*, 38, 3601, 1989.
405. **Henderson, L. M., Chappell, J. B., and Owen, O. T. G.,** Superoxide generation is inhibited by phospholipase A_2 inhibitors. Role for phospholipase A_2 in the activation of NADPH oxidase, *Biochem. J.*, 264, 249, 1989.
406. **Deleepan, K. N., Simpson, K. M., Lynch, S. R., and Stechschulte, D. J.,** Dismutation of eosinophil superoxide by mast cell granule superoxide dismutase, *Biochem. Arch.*, 5, 153, 1989.
407. **Sedgwick, J. B., Vrtis, R. F., Courley, M. F., and Busse, W. W.,** Stimulus-dependent differences in superoxide anion generation by normal human eosinophils and neutrophils, *J. Allerg. Clin. Immunol.*, 81, 876, 1988.
408. **Szefler, S. J., Norton, C. E., Ball, B., Gross, J. M., Aida, Y., and Pabst, M. J.,** IFN-γ and LPS overcome glucocorticoid inhibition of priming for superoxide release in human monocytes. Evidence that secretion of IL-1 and tumor necrosis factor-α is not essential for monocyte priming, *J. Immunol.*, 142, 3985, 1989.
409. **Witz, G. and Czerniecki, B. J.,** Tumor promoters differ in their ability to stimulate superoxide anion radical production by murine peritoneal exudate cells following in vivo administration, *Carcinogenesis*, 10, 807, 1989.
410. **Conlon, P. D., Ogunbiyi, P. O., Black, W. D., and Eyre, P.,** β-Adrenergic receptor function and oxygen radical production in bovine pulmonary alveolar macrophages, *Can. J. Physiol. Pharmacol.*, 66, 1538, 1988.

411. **Hishinuma, K., Hosono, A., Mashiko, S., Inaba, H., and Kimura, S.,** Identification of distinct pathways for superoxide generation in mouse macrophages: an investigation using a chemiluminescence probe specific for superoxide anion, *Agric. Biol. Chem.*, 53, 1453, 1989.
412. **Ryer-Powder, J. E. and Forman, H. J.,** Adhering lung macrophages produce superoxide demonstrated with desferal-Mn(IV), *Free Rad. Biol. Med.*, 6, 513, 1989.
413. **DiGregorio, K. A., Cilento, E. V., and Lantz, R. C.,** A kinetic model of superoxide production from single pulmonary alveolar macrophages, *Am. J. Physiol.*, 256, C405, 1989.
414. **Afanas'ev, I. B., Korkina, L. G., Suslova, T. B., and Soodaeva, S. K.,** Are quinones producers or scavengers of superoxide ion in cell?, *Arch. Biochem. Biophys.*, 281, 245, 1990.
415. **Gudewicz, P. W.,** Effect of anticancer drugs on macrophage-mediated antibody-dependent cytotoxicity and secretion of reactive oxygen intermediates, *Cancer Lett.*, 42, 67, 1988.
416. **Zeidler, R. B., Yarbro, J. W., and Conley, N. S.,** Bleomycin increases superoxide production in the most active alveolar macrophage subpopulation, *Int. J. Immunopharmacol.*, 9, 691, 1987.
417. **Arthur, M. J. P., Kowalski-Sounders, P., and Wright, R.,** Effect of endotoxin on release of reactive oxygen intermediates by rat hepatic macrophages, *Gastroenterology*, 95, 1588, 1988.
418. **Jackson, S. K., Stark, J. M., Rowlands, C. C., and Evans, J. C.,** Electron spin resonance detection of oxygen-centered radicals in murine macrophages stimulated with bacterial endotoxin, *Free Rad. Biol. Med.*, 7, 165, 1989.
419. **Dorio, R. J., Hoek, J. B., Rubin, E., and Forman, H. J.,** Ethanol modulation of rat alveolar macrophage superoxide production, *Biochem. Pharmacol.*, 37, 3528, 1988.
420. **Roney, P. L. and Holian, A.,** Possible mechanism of chrysotile asbestos-stimulated superoxide anion production in guinea pig alveolar macrophages, *Toxicol. Appl. Pharmacol.*, 100, 132, 1989.
421. **Willems, J., Leclercq, G., and Joniau, M.,** Enhanced oxygen metabolism of peritoneal macrophages in the presence of murine neuroblastoma cells is partly caused by enkephalins, *J. Neuroimmunol.*, 19, 269, 1988.
422. **Rellstab, P. and Schaffner, A.,** Endotoxin suppresses the generation of superoxide anion and hydrogen peroxide by resting and lymphokine-activated human blood-derived macrophages, *J. Immunol.*, 142, 2813, 1989.
423. **Czerniecki, B. J. and Witz, G.,** Tumor promoters stimulate the formation of 1,2-diacylglycerol in murine peritoneal macrophages: a possible mechanism for stimulating superoxide anion radical production, *Cancer Lett.*, 48, 29, 1989.
424. **Maridoneau-Parini, I., Errasfa, M., and Russo-Marie, F.,** Inhibition of superoxide generation by dexamethasone is mimicked by lipocortin I in alveolar macrophages, *J. Clin. Invest.*, 83, 1936, 1989.
425. **Engels, F., Henricks, P. A. J., van der Vliet, H., and Nijkamp, F. P.,** Modulatory activity of 9-hydroxy- and 9-hydroperoxy-octadecadienoic acid towards reactive oxygen species from guinea-pig pulmonary macrophages, *Eur. J. Biochem.*, 184, 197, 1989.
426. **Forman, H. J. and Kim, E.,** Inhibition by linoleic acid hydroperoxide of alveolar macrophage superoxide production: effects upon mitochondrial and plasma membrane potentials, *Arch. Biochem. Biophys.*, 274, 443, 1989.
427. **McLeish, K. R., Stelzer, G. T., and Wallace, J. H.,** Regulation of oxygen radical release from murine peritoneal macrophages by pharmacologic doses of PGE_2, *Free Rad. Biol. Med.*, 3, 15, 1987.
428. **Chiara, M. D., Bedoya, F., and Sobrino, F.,** Cyclosporin A inhibits phorbol ester-induced activation of superoxide production in resident mouse peritoneal macrophages, *Biochem. J.*, 264, 21, 1989.
429. **Gorog, P., Pearson, J. D., and Kakkar, V. V.,** Generation of reactive oxygen metabolites by phagocytosing endothelial cells, *Atherosclerosis*, 72, 19, 1988.
430. **Maly, F. E., Cross, A. R., Jones, O. T. G., Wolf-Vorbeck, G., Walter, C., Dahinden, C. A., and DeDeck, A. L.,** The superoxide generating system of B cell lines. Structural homology with the phagocytic oxidase and triggering via surface Ig, *J. Immunol.*, 140, 2334, 1988.
431. **Maly, F. E., Nakamura, M., Gauchat, J.-F., Urwyler, A., Walker, C., Dahinden, C. A., Cross, A. R., Jones, O. T. G., and DeWeck, A. L.,** Superoxide-dependent nitroblue tetrazolium reduction and expression of cytochrome b-245 components by human tonsillar B lymphocytes and B cell lines, *J. Immunol.*, 142, 1260, 1989.
432. **Hancock, J. T., Maly, F. E., and Jones, O. T. G.,** Properties of the superoxide-generating oxidase of B-lymphocyte cell lines. Determination of Michaelis parameters, *Biochem. J.*, 262, 373, 1989.
433. **Kiss, Z., Deli, E., and Kuo, J. F.,** Phorbol ester inhibits phosphatidylserine synthesis in human promyelocytic leukemia HL 60 cells. Possible involvement of free radicals and correlation with phosphorylation of nuclear protein 1b, *Biochem. J.*, 248, 649, 1987.
434. **Nilsson, U. A., Olsson, L.-I., Thor, H., Moldeus, P., and Bylund-Fellenius, A.-C.,** Detection of oxygen radicals during reperfusion of intestinal cells in vitro, *Free Rad. Biol. Med.*, 6, 251, 1989.
435. **Grupe, R., Ziska, T., Richter, A., Boehme, B., and Goeres, E.,** Lipoxygenase inhibitors and mast cells degranulation, *Biomed. Biochim. Acta*, 47, 955, 1988.

436. **Dieter, P., Schulze-Specking, A., and Decker, K.,** Calcium requirement of prostanoid but not of superoxide production by rat Kupffer cells, *Eur. J. Biochem.*, 177, 61, 1988.
437. **Latocha, G., Dieter, P., Schulze-Specking, A., and Decker, K.,** Fc receptors mediate prostaglandin and superoxide synthesis in culture rat Kupffer cells, *Biol. Chem. Hoppe-Seyler*, 370, 1055, 1989.
438. **Murrell, G. A. C., Francis, M. J. O., and Bromley, L.,** Fibroblasts release superoxide free radicals, *Biochem. Soc. Trans.*, 17, 483, 1989.
439. **Murrell, G. A. C., Francis, M. J. O., and Bromley, L.,** Cyclooxygenase and oxygen free radical-stimulated fibroblast proliferation, *Biochem. Soc. Trans.*, 17, 482, 1989.
440. **Dikkeboom, R., Van der Knaap, W. P. W., Van den Bovenkamp, W., Tijnagel, J. M. G. H., and Bayne, C. J.,** The production of toxic oxygen metabolites by hemocytes of different snail species, *Dev. Comp. Immunol.*, 12, 509, 1988.
441. **Miyanoshita, A., Takahashi, T., and Endou, H.,** Inhibitory effect of cyclic AMP on phorbol ester-stimulated production of reactive oxygen metabolites in rat glomeruli, *Biochem. Biophys. Res. Commun.*, 165, 519, 1989.
442. **Nakamura, Y., Ohtaki, S., Makino, R., Tanaka, T., and Ishimura, Y.,** Superoxide anion is the initial product in the hydrogen peroxide formation catalyzed by NADPH oxidase in porcine thyroid plasma membrane, *J. Biol. Chem.*, 264, 4759, 1989.
443. **Sandhu, S. S., Band, A., Abayasekera, D. R. E., Rice-Evans, C., and Cooke, B. A.,** A novel superoxide-generating system (Leydig cells), *Biochem. Soc. Trans.*, 17, 710, 1989.
444. **Schellhorn, H. E., Pou, S., Moody, C., and Hassan, H. M.,** An electron spin resonance study of oxyradical generation in superoxide dismutase- and catalase-deficient mutants of *Escherichia coli* K-12, *Arch. Biochem. Biophys.*, 271, 323, 1989.
445. **Nukatsuka, M., Sakurai, H., Yashimura, Y., Nishida, M., and Kawada, J.,** Enhancement by streptozotocin of superoxide anion radical generation by the xanthine oxidase system of pancreatic β-cells, *FEBS Lett.*, 239, 295, 1988.
466. **Woodroofe, M. N., Hayes, G. M., and Cuzner, M. L.,** Fc receptor density, MHC antigen expression and superoxide production are increased in interferon-gamma-treated microglia isolated from adult rat brain, *Immunology*, 68, 421, 1989.
447. **Alvarez, J. G., Touchstone, J. C., Blasco, L., and Storey, B. T.,** Spontaneous lipid peroxidation and production of hydrogen perioxide and superoxide in human spermatozoa: superoxide dismutase as major enzyme protectant against oxygen toxicity, *J. Androl.*, 8, 338, 1987.
448. **Phillips, W. A. and Hamilton, J. A.,** Phorbol ester-stimulated superoxide production by murine bone marrow-derived macrophages requires preexposure to cytokines, *J. Immunol.*, 142, 2445, 1989.

Chapter 5

THE INTERACTION OF OXYGEN RADICALS WITH NATURAL AND BIOLOGICALLY ACTIVE COMPOUNDS, CELLULAR COMPONENTS, AND CELLS

I. INTRODUCTION

As we see, there are many sources of oxygen radicals in cells that make them ubiquitous species. The role of oxygen radicals in physiological and pathological processes is now far from being completely understood and depends, in the first place, on the ability of oxygen radicals to interact with substrates, drugs, proteins, lipids, enzymes, DNA, and other cellular compounds. However, the effect of oxygen radicals is not confined to these processes; their action on such cell components as mitochondria, microsomes, plasmolemma, sarcoplasmic reticulum, etc., and on the cells themselves is also of great importance. It is common knowledge that oxygen radicals are able to damage cells and cellular components, being possibly a primary factor of many pathologies. However, it is now also clear that oxygen radicals may play the role of a benefactor in many biological processes. The participation of oxygen radicals in phagocytosis was already considered in Chapter 3. We will see further that oxygen radicals are also able to activate some enzymes, to promote iron transport, and to stimulate other important processes.

One of the most important questions is the identification of a genuine active species among various oxygen radicals participating in biological processes. We now can conceive of all possible types of oxygen radical attack and the relative roles of different radicals. The superoxide ion is a main precursor of all other active oxygen species, including hydrogen peroxide. In Volume I we discussed the relatively small reactivity of the superoxide ion in comparison with hydroxyl or even perhydroxyl radicals. However, there is convincing evidence that the superoxide ion is not only a precursor of other oxygen radicals, but is also a participant in DNA and cellular oxidative damage as well as in other processes (see below). In some enzymatic reactions, the superoxide ion plays an exclusive role.

The role of reactive hydroxyl radicals in damaging processes is especially important, the superoxide-driven Fenton reaction remaining the most probable route of their formation. Two mechanisms of the reactions of hydroxyl radicals should be considered: the interaction of substrates with "free" hydroxyl radicals generated in solution and the "site-specific" mechanism of reactions when hydroxyl radicals are formed as a result of the decomposition of hydrogen peroxide by transition metal ions absorbed on proteins, DNA, lipids, and other polymeric and monomeric compounds. At present, there are many findings supporting the idea that in the absence of specific chelators (among them, the most effective one is EDTA), a site-specific mechanism is the most important one for the reaction of hydroxyl radicals in biological systems.

Very important active oxygen species are oxy and perhydroxy complexes of transition metals ("crypto-hydroxyl" radicals, ferryl and perferryl ions). We already discussed the structure of crypto-hydroxyl radicals (Volume I, Chapter 9), which must be precursors of hydroxyl radicals in the Fenton reaction. It should be noted that although the perferryl ion FeO_2^{2+} and the ferryl ion FeO^{2+} are supposedly formed during the oxidation of ferrous ions by dioxygen, they may be identical to crypto-hydroxyl radicals. For example:

$$Fe^{2+} + H_2O_2 \rightarrow \underset{\text{"crypto-hydroxyl"}}{Fe^{2+}(H_2O_2)} \rightarrow \underset{\text{ferryl ion}}{FeO^{2+}} + H_2O \qquad (1)$$

(It should be noted that hypothetic perferryl and ferryl ions are free radicals because the interaction of ferrous ion with oxygen atoms leads to the appearance of an unpaired electron on the oxygen atom:

$$Fe^{2+} + O_2 \rightarrow FeO_2^{2+} \rightleftarrows (Fe^{3+}O_2^{\bar{\cdot}}) \tag{2}$$

The interaction of all mentioned oxygen species with substrates, cellular components, and cells will be considered in this chapter.

II. REACTIONS OF OXYGEN RADICALS WITH NATURAL SUBSTRATES AND BIOLOGICALLY ACTIVE COMPOUNDS

In Volume I (Chapter 10) we discussed the mechanism of the interaction of the superoxide ion with such important natural compounds as ascorbic acid, α-tocopherol, ubiquinones, catecholamines, flavonoids, flavins, metalloporphyrins, cobalamins, and chlorophylls. However, the superoxide ion is never formed alone in biological systems and frequently is only a source of much more active oxygen species. Therefore, the *in vitro* experiments with oxygen radicals generated by enzymes, xenobiotics, and phagocytes give, as a rule, quite different results from those obtained in "clean" chemical systems. Nonetheless, chemical studies are very useful to separate genuine effects of the superoxide ion from those of other more active oxygen species.

A. NADH

Land and Swallow[1] concluded that the superoxide ion slowly reacts with NADH via Reaction 3:

$$O_2^{\bar{\cdot}} + NADH \rightarrow HO_2^- + NAD\cdot \tag{3}$$

with $k_3 \ll 27\ M^{-1}\ s^{-1}$. A rate constant of the reaction of the perhydroxyl radical with NADH is much more.

$$HO_2\cdot + NADH \rightarrow H_2O_2 + NAD\cdot \tag{4}$$

$k_4 = (1.8 \pm 0.2) \cdot 10^5\ M^{-1}\ s^{-1}$, $E_a = 5.5 \pm 0.5$ kcal/mol.[2] However, Bielski and Chan[3-5] showed later that the rate of the reaction of the radiolytically and enzymatically generated superoxide ion with NADH is greatly enhanced in the presence of lactate dehydrogenase (LDH) due to the formation of the LDH-NADH complex. In addition, LDH transformed the NADH oxidation by the superoxide ion into a chain radical process for which the next mechanism was proposed:

$$LDH + NADH \rightleftarrows LDH\text{–}NADH \tag{5}$$

$$LDH\text{–}NADH + O_2^{\bar{\cdot}} + H^+ \rightarrow LDH\text{–}NAD\cdot + H_2O_2 \tag{6}$$

$$LDH\text{–}NADH + HO_2\cdot \rightarrow LDH\text{–}NAD\cdot + H_2O_2 \tag{7}$$

$$LDH\text{–}NAD\cdot + O_2 \rightarrow LDH\text{–}NAD^+ + O_2^{\bar{\cdot}} \tag{8}$$

$$LDH\text{–}NAD^+ \rightleftarrows LDH + NAD^+ \tag{9}$$

The rate constants for Reactions 6 and 7 are equal to $3.6 \cdot 10^4$ and $1.2 \cdot 10^6\ M^{-1}\ s^{-1}$, respectively.[5] It was found that the chain length depended on pH values, having a maximum at pH 7.2. Under pulse radiolysis conditions, an optimal chain length was equal to 18. It is of interest that using xanthine oxidase as a generator of the superoxide ion, a chain length of 73 can be achieved.[6] The oxidation of NADH was inhibited by superoxide dismutase (SOD) and sodium oxamate, a specific inhibitor of LDH.

Glyceraldehyde-3-phosphate dehydrogenase (GAPDH) also catalyzed the oxidation of NADH by oxygen radicals; however, in contrast to LDH, a maximal chain length of 76 was observed at pH 5.0. It was proposed that the superoxide ion is unreactive in the GAPDH-catalyzed oxidation of NADH, and the only oxygen radical participating in the reaction is $HO_2\cdot$. A rate constant for Reaction 10 was equal to $2.0 \cdot 10^7$ M^{-1} s^{-1}.

$$\text{GAPDH–NADH} + HO_2\cdot \rightarrow \text{GAPDH–NAD}\cdot + H_2O_2 \quad (10)$$

Since a direct hydrogen atom abstraction by superoxide ion (such as an early proposed Reaction 3) is impossible (Volume I, Chapter 4), one may suggest that the formation of an LDH-NADH complex must enhance the rate of NADH deprotonation by $O_2^{\overline{\cdot}}$ to make a deprotonation-oxidation mechanism more favorable (Volume I, Chapter 5). On the other hand, the formation of the complex must also enhance a rate for hydrogen abstraction by a perhydroxyl radical (Reaction 7). It is therefore possible that the catalysis of NADH oxidation by dehydrogenases increases simultaneously both deprotonation and hydrogen atom abstraction by the superoxide ion and perhydroxyl radical, respectively.

Vanadate and molybdate are also able to catalyze the oxidation of NADH and reduced nicotinamide mononucleotide (NMNH) by superoxide ion.[7,8] It was proposed that vanadate forms a peroxy complex with the superoxide ion which is a genuine oxidant of NADH and NMNH. Superoxide ion was produced by xanthine oxidase[7,8] or rat liver microsomes.[9] SOD, but not catalase, inhibited the vanadate-catalyzed oxidation of NADH and NMNH.

We already discussed (Chapter 2) the ability of peroxidases to generate superoxide ion, catalyzing the oxidation of NADH and other substrates. The superoxide ion formed reacts with substrates and by this initiates a chain oxidation. Thus Halliwell and DeRycker[10] described the oxidation of NADH, dihydroxyfumarate, and dithiothreitol catalyzed by horseradish peroxidase (HRP). The reactions were inhibited by SOD, indicating the participation of a superoxide ion.

B. CATECHOLS AND MELANINS

The oxidation of catechols by superoxide ion and dioxygen was already discussed (Volume I, Chapters 5 and 10). The best known example of such reactions is the oxidation of epinephrine by superoxide ion which was proposed[11] and now widely applied as a test of superoxide production. It has been shown[12,13] that rat liver microsomes stimulate the SOD-inhibitable oxidation of epinephrine into adrenochrome. The superoxide ion is able not only to oxidize, but also to bind catechols to biological polymers. Thus Dybing et al.[14] observed the oxidation and covalent binding of methyldopa by rat liver microsomes in the presence of an NADPH-generating system which was inhibited by SOD.

Melanins are polymers derived from enzymatic oxidation of catecholamines (dopa and cysteinyldopa). These pigments possibly play an important role in the protection of skin from ultraviolet (UV) radiation. Although melanins are able to produce superoxide ion during irradiation with UV light,[15] they are also able to react with $O_2^{\overline{\cdot}}$.[16] Using ESR spectroscopy, it was shown[16] that potassium superoxide reacts with melanins to form melanin free radicals with rate constants equal to $(1.2—1.6) \cdot 10^6$ M^{-1} s^{-1}. The reaction was inhibited by SOD. Similar results were obtained by studying the reactions of melanins with the superoxide ion produced by pulse-radiolysis:[17] a rate constant for the reaction of $O_2^{\overline{\cdot}}$ with dopa-melanin was equal to $5 \cdot 10^5$ M^{-1} s^{-1}. Unexpectedly, the authors were unable to measure a rate constant for the interaction of superoxide ion with 5-*S*-cysteinyldopa-melanin. Since melanins contain both *o*-quinone and *o*-hydroquinone moieties, $O_2^{\overline{\cdot}}$ is possibly able to react with them simultaneously as a reductant and an oxidant (Volume I).

C. THIOLS

The mechanism of the reactions of the superoxide ion with thiols was already considered by us in Volume I (Chapter 5). It was shown that a deprotonation-oxidation mechanism (Reactions 11 to 14) is actually the only pathway for this reaction:

$$RSH + O_2^{\overline{\cdot}} \rightarrow RS^- + HO_2\cdot \qquad (11)$$

$$HO_2\cdot + O_2^{\overline{\cdot}} \rightarrow HO_2^- + O_2 \qquad (12)$$

$$RS^- + O_2 \rightarrow RS\cdot + O_2^{\overline{\cdot}} \qquad (13)$$

$$2\ RS\cdot \rightarrow RSSR \qquad (14)$$

In cells, superoxide ion may react with both low-molecular thiols, such as glutathione and cysteine,[18-20] and proteins containing the sulfhydryl groups.[21] Thus Wefers and Sies[19] have shown that glutathione inhibited the SOD-inhibitable reduction of cytochrome c by xanthine oxidase and that SOD suppressed the formation of oxidized glutathione (GSSG) and glutathione sulfonate (GSO_3H). On these grounds, it was concluded that superoxide ion is able to oxidize glutathione in disulfide and sulfonate. (The formation of sulfonates in the reaction with superoxide ion was shown for many simple thiols in Volume I, Chapter 5). Ross et al.[20] confirmed that xanthine oxidase catalyzes the oxidation of glutathione into disulfide. However, from comparison of the inhibitory effects of SOD and catalase, it was concluded that the superoxide-mediated oxidation of glutathione is significant, but quantitatively less important than glutathione oxidation by hydrogen peroxide.

Using the competition technique, Asada and Kanematsu[18] determined the rate constants for the reaction of superoxide ion generated by xanthine oxidase with cysteine, reduced glutathione, homocysteine, and dithiothreitol (Table 1). Glutathione, cysteine, and other thiols are also oxidized by the superoxide ion produced by stimulated leukocytes.[22] These reactions were inhibited by SOD and catalase and apparently proceeded via a direct interaction of the superoxide ion with thiols without iron ion participation. The superoxide ion produced by xanthine oxidase reacted with the thiol-containing nucleic acid base 6-mercaptopurine with a rate constant of $1.4 \cdot 10^4\ M^{-1}\ s^{-1}$.[23]

The superoxide ion can play an important role in protein *S*-thiolation, i.e., the formation of mixed disulfides between glutathione and the protein sulfhydryl groups. During oxidative stress, *S*-thiolation affects such cellular activities as membrane-associated calcium pumping, NADPH production by the pentose-P pathway, etc. It was concluded[21] that the superoxide-mediated reaction of reduced glutathione with the protein sulfhydryl groups is a more rapid and more important pathway of *S*-thiolation than a slow interaction between oxidized glutathione and reduced protein. This conclusion was confirmed by the fact that creatine kinase and glycogen phosphorylase effectively inhibited cytochrome c reduction by xanthine oxidase. Moreover, these enzymes were more effective inhibitors than glutathione. Therefore, the superoxide ion must react preferentially with the protein sulfhydryl groups even in the presence of excess glutathione.

Another example of superoxide-mediated reactions of thiols is the interaction of thiols with catechols. Ito and Fujita[24] have found that oxygen radicals generated by xanthine oxidase catalyzed the conjugation of dopa and 5-*S*-cysteinyldopa with cysteine to form 5-*S*-cysteinyldopa and 2,5-dicysteinyldopa, respectively. The reaction was inhibited by SOD and was

TABLE 1
Rate Constants for the Reactions of Superoxide Ion with Native Substrates, Biologically Active Compounds, Enzymes, and Proteins

Substance	Reaction product	k ($M^{-1}s^{-1}$)	Ref.
NADH		27	1
Cysteine		$2.7 \cdot 10^6$	18
Homocysteine		$4.5 \cdot 10^5$	18
Glutathione (reduced)		$6.7 \cdot 10^5$	18
Dithiothreitol		$1.0 \cdot 10^6$	18
Melamin		$(1.2—1.6) \cdot 10^6$	16
Dopa-melamin		$5 \cdot 10^5$	17
Desferrioxamine		900	62
		$1.35 \cdot 10^6$	63
Cytochrome c(III)	Cytochrome c(II)	$(2.6 \pm 0.1) \cdot 10^5$	153
Acetylated cyt c(III)	Acetylated cyt c(II)	$3.5 \cdot 10^5$	156
HbO_2 (+ 2 H^+)	metHb + H_2O_2	$(4 \pm 1) \cdot 10^3$	164
metHb	HbO_2	$6 \cdot 10^3$	164
		$1.4 \cdot 10^3$	161
Collagen		$4.8 \cdot 10^6$	291
(Cd,Zn)metallothionein		$(5.8 \pm 0.5) \cdot 10^5$	298
Zn(II)metallothionein		$(4.0 \pm 0.4) \cdot 10^5$	298
Cd(II)metallothionein		$(5.6 \pm 0.6) \cdot 10^5$	298
Catalase Fe(III)	Catalase Fe(III)$O_2^{\cdot}$ (Comp. III)	$4.5 \cdot 10^4$	168
		$2.5 \cdot 10^5$	169
Catalase Fe(III)H_2O_2(Comp. I)	Catalase Fe(II)H_2O_2(Comp. II)	$5.0 \cdot 10^6$	170
HRP(c)Fe(III)	HRP9(c)Fe(III)$O_2^{\cdot}$ (Comp. III)	$6 \cdot 10^6$	178
HRP Fe(III)H_2O_2(Comp. I)	HRP Fe(II)H_2O_2(Comp. II)	$1.6 \cdot 10^6$	178
Equine myeloPer Fe(III)	myeloPer Fe(III)$O_2^{\cdot}$ (Comp. III)	$(2.1 \pm 0.2) \cdot 10^6$	176
Human myeloPer Fe(III)	myeloPer Fe(III)$O_2^{\cdot}$ (Comp. III)	$(1.1 \pm 0.3) \cdot 10^6$	176
Indoleamine 2,3-dioxygenase (IADO)Fe(III)	(IADO)Fe(III)$O_2^{\cdot}$	$1.1 \cdot 10^6$	185
Transketolase		10^6	242
Laccase		$2 \cdot 10^6$	243
Glucose oxidase		ca 10^6	231
Semiquinone of glucose oxidase		$8.5 \cdot 10^8$	231

not affected by the hydroxyl radical scavengers, mannitol and formate, indicating mediation by superoxide ion and not by hydroxyl radicals.

D. SUGARS, NUCLEOTIDES, AND NUCLEOSIDES

Sugars are very sensitive to damage by oxygen radicals, although the relative importance of various oxygen species in the reactions with sugars or the compounds containing the sugar moiety is not always certain. Recently, Mashino and Fridovich[25] studied the reaction of superoxide ion generated by xanthine oxidase with dihydroxyacetone, which they consider the simplest model sugar compound. It was found that xanthine oxidase stimulated dioxygen consumption by aqueous solutions of dihydroxyacetone. The oxidation was strongly inhibited by SOD and was only slightly affected by catalase. Mannitol did not inhibit the oxidation, and benzoate only weakly affected it. On these grounds, it was concluded that superoxide ion is able to oxidize dihydroxyacetone directly without participating hydroxyl radicals formed by the Fenton reaction. These authors proposed that superoxide ion reacted with the enolic form of dihydroxyacetone. It seems sound, but the proposed Reaction 15 is of course impossible (Volume I, Chapter 4):

$$HOCH_2C(OH)=CHOH + O_2^{\cdot} \longrightarrow HOCH_2C(OH)=CHO\cdot + HO_2^- \qquad (15)$$

As in other reactions of the superoxide ion with oxidizable compounds, this reaction must proceed by a deprotonation-oxidation mechanism:

$$HOCH_2C(OH){=}CHOH + O_2^{\bar{\cdot}} \rightarrow HOCH_2C(OH){=}CHO^- + HO_2\cdot \quad (16)$$

$$O_2^{\bar{\cdot}} + HO_2\cdot \rightarrow O_2 + HO_2^- \quad (12)$$

$$HOCH_2C(OH){=}CHO^- + O_2 \rightarrow HOCH_2C(OH){=}CHO\cdot + O_2^{\bar{\cdot}} \quad (17)$$

Irrespective of the mechanism, these data indicate the possibility of direct interaction of superoxide ion with sugars. However, as in the other cases, the most effective sugar-damaging agents are probably the oxygen radicals formed in the Fenton reaction. The reactions of these radicals were extensively studied with deoxyribose, a simple but very important sugar contained in DNA and deoxynucleotides. It was proposed[26] that the ferrous ion-dependent damage of deoxyribose which results in the formation of thiobarbituric acid-reactive products (TBA-reactive products) is mediated by hydroxyl radicals because it was inhibited by hydroxyl radical scavengers (mannitol, thiourea, and catechol). Later on, the importance of the site-specific mechanism of hydroxyl radical attack on deoxyribose in real biological systems and in the *in vitro* experiments carried out without effective chelators was shown.[27-29]

Gutteridge[30] has studied the formation of TAB-reactive products during the oxidative damage of deoxysugars (deoxyribose, deoxygalactose, and deoxyglucose and deoxynucleotides (deoxyadenosine and deoxyinosine) promoted by the Fenton and the superoxide-driven Fenton reaction. Based on the effects of hydroxyl radical scavengers, he concluded that the damage was mediated by free hydroxyl radicals only in the presence of EDTA; in the absence of EDTA, the reaction apparently proceeded by a site-specific mechanism, with hydroxyl radicals forming in the reaction of hydrogen peroxide with iron ions bound to substrates. Another explanation of the inability of hydroxyl radical scavengers to inhibit destruction of these compounds in the absence of EDTA is the formation of oxidants (such as perferryl or ferryl ions) other than hydroxyl radicals. The site-specific mechanism was also proposed for depolymerization and destruction of oligo- and polysaccharides (β-cyclodextrin, hyaluronate, pectin, amylose, and dextran) by the ascorbate-copper and hydrogen peroxide-copper systems.[31]

Yamane et al.[32] have studied the interaction of the superoxide ion with nucleotides and nucleosides in aprotic media. They found that nucleotides (adenosine-3′-monophosphate, adenosine-5′-monophosphate, and guanosine-3′-monophosphate) reacted with the superoxide ion, releasing free nucleobases (adenosine and guanosine) with a good yield (15 to 80%). Nucleosides reacted with a smaller yield (3 to 14%). It was concluded that the cleavage of the *N*-glycosidic bonds by the superoxide ion was enhanced by the formation of phosphate peroxy radicals. These findings seem to indicate the possibility of the cleavage of nucleotides and nucleosides by the superoxide ion in cells. However, the results obtained cannot simply be extrapolated to aqueous solution, since there the nucleophilic character of the superoxide ion becomes weaker.

One of the most important known reactions of oxygen radicals with sugars is their interaction with hyaluronic acid. Hyaluronic acid is a polymer consisting of repeating disaccharide units of D-glucuronic acid and *N*-acetylglucosamine. It is supposed that the destruction of hyaluronic acid by oxygen radicals takes place in synovial fluid from inflamed joints. For the first time, the effect of oxygen radicals on hyaluronic acid was shown by McCord,[33] who found that superoxide ions generated by xanthine oxidase decreased the viscosity of synovial fluid due to the depolymerization of hyaluronic acid. SOD and catalase inhibited depolymerization. Later, it was shown[34,35] that the destruction of hyaluronic acid is promoted by oxygen radicals produced not only by xanthine oxidase, but also by activated leukocytes.

The main active oxygen species in this process is apparently not the superoxide ion, but a hydroxyl radical formed in the superoxide-driven Fenton reaction, since catalase decreased and iron ions increased the rate of depolymerization.[33] Furthermore, depolymerization of hyaluronic acid was inhibited not only by antioxidative enzymes (SOD and catalase) and hydroxyl radical scavengers (mannitol and formate), but also by iron chelators (DTPA and bathophenanthroline).[36] Carlin and Djursaeter[37] have shown that in the absence of chelators, xanthine oxidase scarcely diminished the viscosity of hyaluronic acid, whereas the addition of EDTA or ferritin resulted in an immediate decrease in viscosity. A similar effect was observed after the addition of ferrous ions.[35] Recently, it was shown[38] that among activated neutrophils, only PMA-stimulated neutrophils are really able to promote the destruction of hyaluronic acid, whereas the stimulation of superoxide production by FMLP, Con A, and digitonin did not lead to its depolymerization. This difference was explained by the fact that in contrast to all other stimuli studied, PMA does not stimulate the release of myeloperoxidase during neutrophil activation. When the other stimuli were applied, the released myeloperoxidase apparently decomposed hydrogen peroxide and thereby suppressed the formation of hydroxyl radicals.

E. OTHER EXAMPLES OF THE INTERACTION OF OXYGEN RADICALS WITH BIOLOGICALLY ACTIVE COMPOUNDS

There are various compounds whose reactions with oxygen radicals have been studied due to their role in physiological and pathological processes. As is known (Volume I, Chapter 10), the superoxide ion is able to react with ubiquinones by a one-electron transfer mechanism. It was also found that ubiquinones inhibited the luminol- and lucigenin-dependent chemiluminescence (CL) induced by two oxygen radical-generating systems: xanthine oxidase and the NADPH oxidase of leukocytes.[39] Similarly, Nakamura et al.[40] have shown that the reduction of ubiquinone Q_5 by the NADPH oxidase from guinea pig macrophages is mediated by the superoxide ion.

Methylguanidine (MG) is a toxin which is present in an increased amount in uremia and possibly participates in carcinogenesis. It was proposed[41] that MG formation from creatinine (Reaction 18) is mediated by oxygen radicals, because this reaction was sharply accelerated by the superoxide ion generated by xanthine oxidase and by hydroxyl radicals formed in the Fenton reaction.

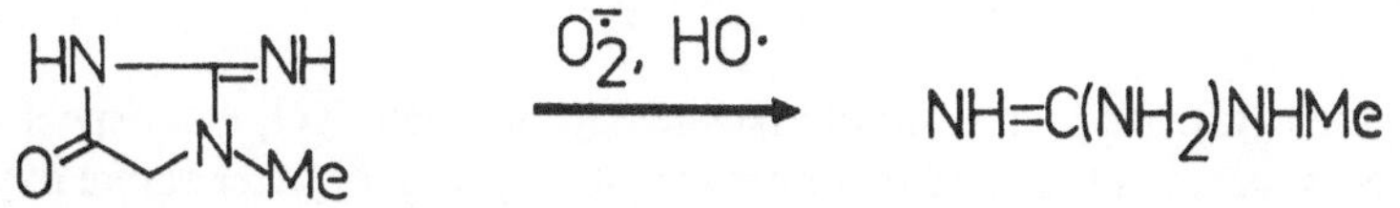

(18)

SOD and hydroxyl radical scavengers (sorbitol and ethanol) inhibited this process. Since the accelerating effect of hydroxyl radicals was more powerful, the superoxide ion can be simply a HO· precursor in the superoxide-drived Fenton reaction. It was later shown[42] that hydroxyl radical scavengers (DMSO, sorbitol, etc.), catalase, and glutathione inhibited the synthesis of MG in isolated rat hepatocytes, whereas ascorbic acid accelerated it. These findings support the possibility of the *in vivo* mediation of MG synthesis from creatinine by hydroxyl radicals.

Uemura et al.[43] have shown that oxygen radicals produced by the xanthine-xanthine oxidase-EDTA system promoted the formation of β-carboline (5-hydroxy-2,3-dihydrotryptoline) from 5-hydroxytryptamine and glycine. It was proposed that oxygen radicals oxidize glycine into glyoxylic acid, which then condenses with 5-hydroxytryptamine with subsequent oxidative decarboxylation of cyclic intermediate into a final product.

Winterbourn et al.[44] found that autoxidation of the pyrimidine aglycone divicine (2,6-diamino-4,5-dihydroxypyrimidine) was inhibited by SOD. Divicine stimulates a rapid depletion of glutathione in erythrocytes and is believed to be a toxic component of fava beans, which are able to promote favism, an acute hemolytic disease associated with glucose-6-phosphate dehydrogenase deficiency. It is supposed that the superoxide-mediated autoxidation of divicine induces the depletion of glutathione in erythrocytes. Recently, it was confirmed[45] that the superoxide ion mediates the autoxidation of three cytotoxic pyrimidines: dialuric acid, divicine, and isouramil.

It is known that phentermine (2-methyl-1-2-aminopropane, $MPPNH_2$) is oxidized by rat liver microsomes into 2-methyl-2-nitro-1-phenylpropane ($MPPNO_2$). This process consists of two stages: *N*-hydroxylation of $MPPNH_2$ into *N*-hydroxyphentermine (MPPNHOH), which is thought to be a two-electron oxidation catalyzed by cytochrome P-450 and the SOD-sensitive oxidation of MPPNHOH into $MPPNO_2$.[46,47] It was proposed that the superoxide ion mediates the transformation of MPPNHOH by Reaction 19:

$$\text{MPPNHOH} \xrightarrow{O_2^{\overline{\cdot}}} \text{MPPNO} \xrightarrow{O_2} \text{MPPNO}_2 \qquad (19)$$

Such a mechanism is supported by the fact that xanthine oxidase, another generator of the superoxide ion, also oxidizes MPPNHOH into $MPPNO_2$; this reaction is completely inhibited by SOD. In addition, Reaction 19 proceeds with KO_2 and NMe_4O_2 as sources of superoxide ion.[48] Hydroxyl radicals apparently do not participate in the enzymatic oxidation of MPPNHOH, because hydroxyl radical scavengers did not affect this process.

The mechanism of Reaction 19 remains obscure. Based on the experiments with KO_2 and NMe_4O_2, Fukuto et al.[48] concluded that the superoxide ion abstracts a hydrogen atom from the hydroxyl group of MPPNHOH (Reaction 20). However, this reaction is certainly impossible (Volume I, Chapter 4). The usual mechanism of initiation by deprotonation (Reaction 21) was rejected by these authors since there was no dioxygen formation by Reaction 12.

$$\text{MPPNHOH} + O_2^{\overline{\cdot}} \rightarrow \text{MPPNHO}\cdot + HO_2^- \qquad (20)$$

$$\text{MPPNHOH} + O_2^{\overline{\cdot}} \rightarrow \text{MPPNHO}^- + HO_2\cdot \qquad (21)$$

$$O_2^{\overline{\cdot}} + HO_2\cdot \rightarrow O_2 + HO_2^- \qquad (12)$$

These results are confusing, but it should be stressed that both KO_2 and especially NMe_4O_2 are not "clean" sources of $O_2^{\overline{\cdot}}$, and therefore the participation of other active oxygen species in the reaction studied cannot be excluded.

The superoxide ion produced by xanthine oxidase is able to reduce putidamonooxin, a component of 4-methoxybenzoate monooxygenase from *Pseudomonas putida*.[49]

Scislowski and Davis[50] have studied the reaction of the superoxide ion produced by xanthine oxidase, the dissolution of KO_2, and the oxidation of oxymyoglobin, or generated by isolated heart mitochondria with methionine. In all cases, the SOD-inhibitable production of methionine sulfoxide was observed. On these grounds, it was concluded that the superoxide ion may mediate the oxidation of methionine in biological systems, for example, during perfusation of skeletal muscles.

Younes[51] concluded that superoxide ion and hydrogen peroxide mediate the oxidation of thiobenzamide by microsomal flavin-containing monooxygenase because SOD and catalase inhibited this process.

The reactions of oxygen radicals with drugs are of special interest since, on one hand, they may be relevant to the transformation of prodrugs into genuine drugs and, on the other hand, they may be a route of drug detoxification. For example, Sagone and Husney[52] have shown that oxygen radicals produced by xanthine oxidase and zymosan-stimulated granulocytes promoted decarboxylation and hydroxylation of benzoate, salicyclic acid, and acetylsalicylic acid (aspirin). SOD, catalase, and hydroxyl radical scavengers (benzoate, phenol, and dimethylthiourea) inhibited these processes. It was proposed that these reactions are mediated by hydroxyl radicals and the anti-inflammatory properties of salicylates may be relevant to their scavenging ability.

Benzoate is also decarboxylated by rat liver microsomes.[53] This reaction was inhibited by hydroxyl radical scavengers (mannitol, dimethyl sulfoxide (DMSO), ethanol, etc.) and was accelerated by azide (an inhibitor of catalase) and EDTA. These findings also indicate the mediation of benzoate decarboxylation by hydroxyl radicals. Recently, it was concluded[54] that the transformation of 5-aminosalicylate, a metabolite of sulfasalazine (a drug applied for the treatment of human ulcerative colitis), by leukocytes into salicylate, gentisate, and other compounds is promoted by hydroxyl radicals.

Hydroxyl radicals are thought to participate in the oxidation of ethanol by liver microsomes and a reconstituted microsomal ethanol-oxidizing system,[55,56] since hydroxyl radical scavengers inhibited the oxidation of ethanol. SOD did not affect the microsomal oxidation of ethanol, but inhibited the xanthine oxidase-dependent ethanol oxidation. Thus hydroxyl radicals initiating ethanol oxidation may be formed with or without superoxide participating.

There seems to be two pathways for ethanol oxidation by microsomes: the P-450 cytochrome-dependent pathway and the ethanol oxidation catalyzed by NADPH cytochrome P-450 reductase.[57] The last is mediated by hydroxyl radicals because it increases in the presence of EDTA and is inhibited by DMSO or mannitol.[58] (Besides microsomes, the EDTA-catalyzed oxidation of ethanol can be promoted by xanthine oxidase.) The formation of carbon-centered free radicals of ethanol, 2-propanol, and 2-butanol during their microsomal oxidation has been shown by means of the spin-trapping technique.[59]

Shaw et al.[60] have shown that the superoxide ion generated from ethanol metabolism (during the oxidation of ethanol by alcohol dehydrogenase into acetaldehyde in the presence of xanthine oxidase) promoted the cleavage of folates, especially 5-methyltetrahydrofolate. The reaction was inhibited by SOD and desferrioxamine and stimulated by $FeCl_2$ and ferritin. Therefore, the occurrence of the superoxide-driven Fenton reaction seems to be possible. It was proposed that the *in vivo* superoxide production from the metabolism of ethanol in the presence of xanthine oxidase may stimulate folate deficiency in the alcoholic.

The iron chelator desferrioxamine (Desferal®) produced by *Streptomyces pilosus* is widely applied for the inhibition of iron-catalyzed free radical destructive processes in *in vitro* and *in vivo* systems. However, this compound may also react with oxygen radicals, which makes doubtful its use as the *in vivo* chelator. Indeed, the competition experiments with nitroblue tetrazolium (NBT) and cytochrome c showed that the superoxide ion produced by xanthine oxidase reacted with desferrioxamine.[61,62] Unfortunately, the determination of a rate constant for the reaction of superoxide ion with desferrioxamine by pulse-radiolysis[63] and the competition technique gave different results: $1.35 \cdot 10^6\ M^{-1}\ s^{-1}$ (Reference 63) and $900\ M^{-1}\ s^{-1}$ (Reference 62). The origin of this discrepancy is unknown, although one should note that the competition technique frequently gives results too low for the reactions of superoxide ion in comparison with pulse-radiolysis (Volume I, Chapter 7).

Recently, it was concluded[64] that a relatively stable nitroxide radical containing the $CH_2N(O)CO$ moiety is formed in the reaction of superoxide ion with desferrioxamine:

$$O_2^{\cdot -} + DesNOH + H^+ \rightarrow DesNO\cdot + H_2O_2 \tag{22}$$

This radical was able to suppress the activity of alcohol dehydrogenase, its inhibitory effect being reversed by thiol free radical scavengers (glutathione, cysteine, and methionine). The formation of a toxic nitroxide radical by Reaction 22 may be of importance for using desferrioxamine for treatment of hemochromatosis and iron poisoning.

III. THE INTERACTION OF OXYGEN RADICALS WITH LIPIDS

It has long been known that the attack of reactive free radicals on lipids initiates their peroxidation, forming the unstable products, lipid hydroperoxides, which are readily decomposed further to form the products of deep destruction. Lipid peroxidation is probably the most studied free radical biological process. There are many books and reviews concerning the investigation of lipid peroxidation *in vitro* and *in vivo* (see, for example, References 65 and 66). Earlier, lipid peroxidation was always considered as a free radical damaging process, but it is now known that enzymatic peroxidation of unsaturated acids is a metabolic route to prostaglandins and leukotrienes.

The traditional mechanism of lipid peroxidation was taken from chemical studies of the chain mechanism of hydrocarbon oxidation; it includes the stages of initiation, propagation, and termination:

$$\text{Initiation} \quad R_i\cdot + LH \rightarrow R_iH + L\cdot \tag{23}$$

$$L\cdot + O_2 \rightarrow LOO\cdot \tag{24}$$

$$\text{Propagation} \quad LOO\cdot + LH \rightarrow LOOH + L\cdot \tag{25}$$

$$\text{Termination} \quad 2\,LOO\cdot \rightarrow LOOL + O_2 \tag{26}$$

Here, R_i is an initiator radical, LH is a lipid molecule. Although thermal decomposition of lipid hydroperoxides LOOH is practically impossible, catalytic LOOH decomposition in the presence of transition metal ions or complexes is a very probable process. Therefore, Reaction 27 should be added.

$$LOOH + M^{n+} \rightarrow LO\cdot + HO^- + M^{(n+1)+} \tag{27}$$

Strictly speaking, there is no reliable evidence that a chain mechanism of lipid peroxidation is possible in biological systems. However, we need not discuss here all the problems of lipid peroxidation: this exceeds the limits of this book, as oxygen radicals participate only in the initiation step of this process. The nature and mechanism of the formation of an initiator radical is one of the most important problems of lipid peroxidation. This problem is directly connected with the mechanism of the interaction of oxygen radicals with lipids (and therefore will be discussed in this chapter), although the other, oxygen radical-independent mechanisms of initiation of lipid peroxidation are also possible. We will be concerned here with *in vitro* studies of lipid peroxidation, because *in vivo* experiments are, as a rule, very closely bound to the investigation of free radical pathological processes.

A. EXPERIMENTAL METHODS OF STUDYING LIPID PEROXIDATION

There are many experimental methods of studying lipid peroxidation based on measuring different characteristics of this process. Recently, Gutteridge[67] gave a useful, concise review of these methods. The most popular and well-known method is the detection of malondialdehyde (MDA), which is one of the final products of lipid oxidative degradation. The MDA content is usually measured via the formation of colored products (so-called TBA-reactive products) in the reaction with thiobarbituric acid (TBA). Actually, MDA is not the only product from the decomposition of lipid peroxides which is able to form a colored product with TBA, since similar reactions are known for the decomposition products from sugars and other biological polymers. However, it was recently shown[68] with the aid of high-performance liquid chromatography (HPLC) and optical spectroscopy that the TBA-reactive product formed in the reaction of hydroxyl radicals with deoxyribose is identical to that prepared from authentic MDA.

Other methods based on measuring the products of lipid peroxidation are the detection of lipid hydroperoxides as diene conjugates or their identification by HPLC and the detection of the production of hydrocarbon gases (mostly ethane and pentane). Although these three methods are based on measuring the products with completely different structures, there is, as a rule, a good correlation between them, indicating that MDA, lipid hydroperoxides, and hydrocarbon gases originated from the same free radical process. The difference between the production of hydrocarbon gases and two other methods appears only at different dioxygen pressures:[69,70] while MDA and lipid hydroperoxide formation increased with increasing dioxygen concentrations, a reverse dependence was observed for the formation of ethane and pentane. Kostrucha and Kappus[70] assumed that this difference is explained by the different origins of products, proposing that ethane and pentane are formed as a result of the decomposition of monohydroperoxides and that MDA originates from endoperoxy hydroperoxides. However, it is obvious that if hydrocarbons are formed from alkyl radicals (Reactions 28 and 29), a decrease in dioxygen concentration will enhance their production due to a decrease in the rate of formation of lipid hydroperoxides (Reaction 30).

$$R\cdot + LH \longrightarrow RH + L\cdot \qquad (28)$$

$$R\cdot + R\cdot \longrightarrow RR \text{ or } [RH + R(-H)] \qquad (29)$$

$$L\cdot + O_2 \longrightarrow LOO\cdot \xrightarrow{RH} LOOH + R\cdot \qquad (30)$$

Due to this, the observed opposite effects of dioxygen pressure on the yields of hydrocarbon gases and MDA or hydroperoxides may, in our opinion, be explained by the competition between Reactions 28 and 29 and Reaction 30.

As in studies of the chain oxidation of hydrocarbons, the measurement of dioxygen consumption may be applied for the kinetic investigation of lipid peroxidation. In addition, two important experimental techniques, spin trapping and CL, are also successfully used for studying lipid peroxidation. Rosen and Rauckman[71] were able to identify $O_2^{\overline{\cdot}}$ and HO· as DMPO-OOH and DMPO-OH spin adducts in liver microsomes. The formation of at least a part of DMPO-OH in the reaction of hydroxyl radicals with DMPO and not as a result of DMPO-OOH decomposition was confirmed by the fact that in the presence of ethanol, a small amount of DMPO-C_2H_4OH spin adduct was obtained. The spin-trapping technique was also used for the identification of hydroxyl radicals in other lipid peroxidation systems.[72-75]

Weimann et al.[76] have shown that the NADPH-dependent lipid peroxidation of rat liver microsomes stimulated the luminol- and lucigenin-dependent CL, which was inhibited by SOD, catalase, hydroxyl radical scavengers, and propyl gallate. The inhibitory effects of antioxidative enzymes and free radical scavengers confirm the participation of oxygen radicals in the stimulation of CL, although it does not mean that the same active species initiate both CL and lipid peroxidation. It has been shown[77,78] that there is a correlation between CL and MDA formation in the iron-dependent peroxidation of rat liver microsomes. However, Noll et al.[79] found that a time displacement exists between dioxygen uptake and MDA formation, on the one hand, and low-level CL, on the other hand, during NADPH-dependent microsomal lipid peroxidation at various partial dioxygen pressures. Therefore, low-level CL and MDA formation apparently characterize different pathways of lipid peroxidation.

B. SUPEROXIDE-MEDIATED INITIATION OF LIPID PEROXIDATION

Since the superoxide ion itself is not able to abstract a hydrogen atom even from the most active organic compounds (Volume I, Chapter 4), it cannot initiate lipid peroxidation directly by Reaction 23. Therefore, the superoxide-drive Fenton reaction in which very active hydroxyl (or crypto-hydroxyl) radicals are formed should be the most probable pathway for the initiation of lipid peroxidation:

$$O_2^{\overline{\cdot}} + Fe^{3+} \rightarrow O_2 + Fe^{2+} \quad (31)$$

$$Fe^{2+} + H_2O_2 \rightarrow Fe^{3+} + HO\cdot + HO^- \quad (32)$$

In 1973 to 1974, it was shown that xanthine oxidase promoted the lipid peroxidation of liver microsomes[80,81] and lysosomal membranes.[82] Fong et al.[82] concluded that lipid peroxidation was initiated by hydroxyl radicals formed in Reaction 32 because catalase and hydroxyl radical scavengers inhibited and SOD stimulated it (presumably, increasing the production of hydrogen peroxide). Contrary to this, SOD inhibited MDA formation when microsomal peroxidation[80] and the peroxidation of heavy mitochondrial fractions[83] were initiated by xanthine oxidase in the presence of EDTA. Kellogg and Fridovich[84] have found that both SOD and catalase inhibited liposome oxidation and erythrocyte lysis promoted by xanthine oxidase, whereas hydroxyl radical scavengers have very little effect. Similarly, SOD and catalase inhibited the peroxidation of arachidonic acid initiated by xanthine oxidase.[85] The reaction was apparently catalyzed by endogenous iron because EDTA and DTPA affected it.

Thomas et al.[86] have shown that the xanthine oxidase-stimulated peroxidation of linoleic acid dispersed in micelles was almost completely inhibited by SOD and that hydrogen peroxide and $FeCl_3$ had completely no effect on peroxidation. These authors suggested that the superoxide ion and lipid hydroperoxides were initiators of this process. Girotti and Thomas[87] concluded that the inhibitory action of antioxidative enzymes and free radical scavengers on the xanthine oxidase-initiated lipid peroxidation in erythrocyte membranes depended on the physical state of the membrane lipids. Lipid peroxidation of intact membranes was inhibited by SOD, catalase, and EDTA and promoted by mannitol. It was proposed that the absence of the inhibitory action of mannitol is explained by the site-specific mechanism of the formation of hydroxyl radicals. (The stimulatory effect of mannitol was explained by its protection of xanthine oxidase against inactivation.) However, when erythrocyte ghosts were dispersed with Triton® X-100, mannitol became strongly inhibitory. Therefore, in the last case lipid peroxidation was apparently initiated by free hydroxyl radicals.

Tien et al.[88] have shown that xanthine oxidase stimulated the peroxidation of phospholipid liposomes and detergent-dispersed linoleate only in the presence of iron ions. The peroxidation of phospholipid liposomes was stimulated by Fe(ADP), while the peroxidation of linoleate required EDTA. In contrast to the EDTA-stimulated peroxidation of detergent-dispersed linoleate, which was inhibited by catalase and hydroxyl radical scavengers, the peroxidation of liposomes in the presence of ADP was not sensitive to them. It was concluded that hydroxyl radicals participate in lipid peroxidation catalyzed by Fe(EDTA), whereas other active oxygen species initiate Fe(ADP)-dependent lipid peroxidation.

It was found[89] that catalase manifested a double effect on xanthine oxidase-initiated lipid peroxidation: it decreased the formation of conjugated dienes, but stimulated the formation of TBA-reactive substances during the oxidation of linoleic acid. It was proposed that the last effect is explained by the decomposition of lipid hydroperoxides by the iron-heme moiety of catalase. It is thought that the oxidation of linoleic acid was initiated by hydroxyl radicals.

Xanthine oxidase promoted the SOD-inhibitable lipid peroxidation of erythrocyte ghosts.[90]

Reaction was accelerated by oxyhemoglobin, which apparently changes the mechanism of peroxidation, as in its presence catalase inhibited and SOD did not affect the process. Similarly, xanthine oxidase stimulated peroxidation and induced the luminol-dependent CL after incubation with arachidonic acid.[91] A system consisting of xanthine oxidase, hypoxanthine, and the Fe^{3+}(ADP) complex stimulated both the lipid peroxidation of cardiac membrane phospholipids (as liposomes) and the decomposition of 15-hydroperoxyeicosatetraenoic acid.[92] It has also been shown[93] that acetaldehyde can be applied as a substrate in the xanthine oxidase-dependent peroxidation of hepatic lipid membranes instead of xanthine or hypoxanthine. It was proposed that this process may be a cause of alcohol-induced liver injury, since the oxidation of ethanol into acetaldehyde catalyzed by alcohol dehydrogenase may stimulate the production of oxygen radicals by xanthine oxidase.

It is known that lipid peroxidation depends on the composition of lipid membranes (mainly, on the concentration of unsaturated fatty acids and the initial level of hydroperoxides). Correspondingly, Girotti et al.[94] have found that the xanthine oxidase-promoted lipid peroxidation of erythrocyte ghosts was greatly accelerated after preliminary limited photoperoxidation. Peroxidation was inhibited by SOD and EDTA. It was concluded that the iron-catalyzed decomposition of lipid hydroperoxides by superoxide ion was an origin of both lipid peroxidation and lysis of erythrocyte ghosts.

Superoxide-mediated lipid peroxidation may also be initiated by compounds which produce superoxide ion during autoxidation. In Chapter 3 we already considered the superoxide-mediated lipid peroxidation stimulated by anthracycline antibiotics and some other xenobiotics. Similarly, dihydroxyfumarate (DHF) was found to be able to stimulate lipid peroxidation of cardiac sarcoplasmic reticulum and sarcolemmal membranes in the presence of Fe^{3+}(ADP).[95,96] Lipid peroxidation of sarcoplasmic reticulum was inhibited by SOD, catalase, and hydroxyl radical scavengers. The formation of hydroxyl radicals was confirmed by the spin-trapping technique. It was concluded that superoxide ion formed during the autoxidation of DHF reduced Fe^{3+}(ADP) and stimulated by this the Fenton reaction in which genuine initiating species (hydroxyl radical or ferryl ion) were formed.

Since oxygen radicals are released by activated phagocytes (Chapter 4), one may expect that phagocytes can initiate lipid peroxidation. Indeed, in 1974 it was shown[97] that leukocytes stimulate the peroxidation of linolenic acid. Since leukocytes from patients with chronic granulomatous disease were inactive, lipid peroxidation must be promoted by superoxide-generating NADPH oxidase. Later on, it was shown[98,99] that human polymorphonuclear leucocytes (PMNs) promoted lipid peroxidation of linolenic acid micelles and phospholipid liposomes. In both cases, iron ions and iron complexes accelerated lipid peroxidation. There are some unexplained peculiarities in the action of PMNs on linolenic acid micelles,[98] such as the inability of stimuli (PMA) to enhance lipid peroxidation and of SOD and catalase to inhibit it. However, the peroxidation of phospholipid liposomes by PMNs was greatly stimulated by PMA and was inhibited by SOD. Due to this NADPH oxidase must be responsible for the promotion of lipid peroxidation by leukocytes.

C. OXYGEN RADICALS IN NADPH-DEPENDENT LIPID PEROXIDATION

The NADPH-dependent microsomal lipid peroxidation is catalyzed by NADPH cytochrome P-450 reductase, an enzyme, placed in the same microsomal membrane. The occurrence of lipid peroxidation in microsomes in the presence of Fe^{3+}(ADP) was first shown in 1963 by Hochstein and Ernster[100] who proposed that this process is mediated by perferryl ion FeO_2^{2+}. It was later shown that in addition to NADPH cytochrome P-450 reductase, cytochrome P-450 also participates in the initiation of microsomal lipid peroxidation[101] and that NADPH can be replaced by NADH in this process.[102]

Contradictory results were obtained during study of the role of oxygen radicals and iron complexes in NADPH-dependent microsomal peroxidation. Fong et al.[82] have shown that this process was inhibited by catalase and hydroxyl radical scavengers, whereas SOD stimulated it presumably by increasing the formation of hydrogen peroxide. On these grounds it was concluded that NADPH-dependent lipid peroxidation must be initiated by hydroxyl radicals. Indeed, hydroxyl radicals were detected in this system by trapping them with DMPO and PBN.[103] However, Rosen and Rauckman[104] have shown, also using the spin-trapping technique, that a main radical formed in microsomal lipid peroxidation is the superoxide ion, whereas hydroxyl radicals are formed only in a small amount. Nonetheless, the inability of SOD to inhibit lipid peroxidation in rat liver microsomes was confirmed by other workers.[105]

There are similar contradictions concerning the effects of various free radical inhibitors on NADPH-dependent lipid peroxidation in other works. For example, peroxidation of methyl linoleate catalyzed by NADPH cytochrome P-450 reductase in the presence of ferric ions and EDTA was inhibited by SOD, catalase, ethanol, and mannitol,[106] while SOD, catalase, and hydroxyl radical scavengers had little or no effect on the NADPH-dependent peroxidation of rat liver microsomes.[107,77] Klimek[108] has shown that NADPH-dependent lipid peroxidation in human placental mitochondria was inhibited by SOD, but was not affected by hydroxyl radical scavengers. These discrepancies will be further discussed in Section E.

Morehouse et al.[109] compared the mechanisms of the xanthine oxidase-dependent and NADPH-cytochrome P-450 reductase-dependent microsomal lipid peroxidations. It was found that SOD completely inhibited the xanthine oxidase-dependent peroxidation, but had no effect on NADPH-dependent peroxidation. It was also shown that superoxide production measured by the reduction of acetylated cytochrome c was only about 1.5% of total microsomal NADPH-dependent reduction activity. Thus the superoxide-dependent contribution to NADPH-dependent microsomal lipid peroxidation is of little importance.

Another example of enzymatic lipid peroxidation is the NADH- and NADPH-dependent mitochondrial lipid peroxidation. It has been shown[110-112] that MDA is formed in bovine heart submitochondrial particles in the presence of NADH or NADPH, ADP, and ferric ions. Fe^{3+}(ADP) is presumably reduced by NADH dehydrogenase. The occurrence of peroxidation depended on the level of reduced ubiquinone; thus the addition of antimycin A resulted in the reduction of most of the endogenous ubiquinone and the inhibition of MDA formation.

D. PARTICIPATION OF OXYGEN RADICALS IN OTHER PEROXIDATION SYSTEMS

The superoxide ion is not an obligatory reductant of ferric ions or ferric complexes in lipid peroxidation. Sometimes, it is even sufficient to use ferrous ions without any reductants to promote lipid peroxidation. Thus Tien et al.[88] have shown that the Fenton reagent (Fe^{2+} + H_2O_2) efficiently promotes the peroxidation of phospholipid liposomes. Similarly, the Fenton reagent initiated the peroxidation of linolenic acid in micelles.[113] In this case, the rate of peroxidation depended on the type of detergent used; therefore, the site-specific mechanism of initiation by hydroxyl radicals was proposed. Ferrous ions initiated the peroxidation of bovine brain phospholipid liposomes; their effect was similar to that of a combination of ferric ions and xanthine oxidase.[114] Since peroxidation was not affected by catalase and hydroxyl radical scavengers, it was proposed that it was mediated by perferryl or ferryl ions. These works are probably only a small fraction of the papers concerning the initiation of lipid peroxidation by ferrous ions and complexes, hydrogen peroxide, or their combination.

When ferrous ions alone were insufficient for the promotion of lipid peroxidation, ascorbate, cysteine, and other reductants (in addition to the superoxide ion) were applied. Ascorbate, as is known, may manifest both prooxidant and antioxidant effects on lipid peroxidation. For example, ascorbate enhanced lipid peroxidation of lecithin liposomes at low iron concentrations, but inhibited it at high iron ion concentrations.[115] Usually, the inhibitory action of ascorbate is explained by its reaction with lipid peroxy radicals. However, a new explanation of this effect was recently proposed[116] (see below).

Another reductant of ferric ions is cysteine. It has been shown[117] that the combination of $FeSO_4$ and cysteine is an effective initiator of lipid peroxidation in rat liver microsomes. Since peroxidation was inhibited by high concentrations of benzoate and thiourea, chelators (DTPA and desferrioxamine), and phenolic antioxidants, the participation of hydroxyl radicals was proposed. The ferric ion-ascorbate and ferric ion-cysteine systems also initiated lipid peroxidation of muscle microsomes.[118] This process was slightly inhibited by hydroxyl radical scavengers, was not inhibited by SOD and catalase, and was strongly inhibited by butylated hydroxytoluene (BHT) and EDTA. On these grounds, the authors proposed the site-specific mechanism of initiation by hydroxyl radicals.

There are many other lipid peroxidation systems in which oxygen radicals may or may not be formed. Thus it has been shown[119] that carbon tetrachloride promotes lipid peroxidation of rat liver fractions. However, Kornbrust and Mavis[120] concluded that the CCl_4-dependent lipid peroxidation of rat liver microsomes was independent of iron ions (and oxygen radicals, respectively) and was initiated by the reaction of carbon tetrachloride with cytochrome P-450, giving a trichloromethyl radical. Keller et al.[121] proposed that vanadyl stimulates the peroxidation of linolenic acid via the destruction of lipid hydroperoxides to form hydroxyl radicals identified by the spin-trapping technique.

Oxidation of arachidonic acid by prostaglandin synthase and of linoleic acid by lipoxygenase is a normal metabolic route for these unsaturated acids. It has been shown[122,123] that these reactions proceed to form carbon-centered free radicals of substrates. The MDA formation during the oxidation of arachidonic acid by prostaglandin synthase was not inhibited by SOD, catalase, or formate[124] therefore, the participation of oxygen radicals remains questionable. However, it is also unknown how these enzymes catalyze the abstraction of hydrogen atoms from the substrates. It is possible that these reactions proceed via a site-specific mechanism, with hydroxyl radicals generated at the iron centers of enzymes.

E. THE ROLE OF OXYGEN RADICALS AND IRON IN LIPID PEROXIDATION

Apparently, we can now be certain that oxygen radicals participate in the initiation of lipid peroxidation only together with iron (or other transition metal) ions and complexes. Considering various initiating systems, we already mentioned the possibility of the formation of hydroxyl radicals, "crypto-hydroxyl" radicals, perferryl and ferryl ions, etc. Now we will discuss in detail the role of oxygen radicals and iron in the initiation of lipid peroxidation.

At the beginning, the superoxide-driven Fenton reaction appeared to be the most probable route to active oxygen radicals. Indeed, many authors[71-75,82,88-90] indicated an important role of hydroxyl radicals in lipid peroxidation. However, there are many contradictions in the effects of antioxidative enzymes (SOD and catalase), hydroxyl radical scavengers, and other inhibitors on various iron-dependent lipid peroxidation systems. For example, it was pointed out[80,125,126] that heavy contamination of microsomes with catalase and the occurrence of NADPH-dependent lipid peroxidation in Tris® buffer (which is an effective hydroxyl radical scavenger) do not agree with the participation of hydroxyl radicals in this peroxidation system.

It was then proposed that hydroxyl radicals initiate lipid peroxidation in the absence of chelators such as ADP. For example, lipid peroxidation initiated by ferrous ions and hydrogen peroxide in unbuffered solution was inhibited by catalase, Tris, and mannitol, indicating the importance of hydroxyl radical formation. In the presence of ADP, the Fenton Reaction 32 is thought to become impossible and lipid peroxidation is initiated by other active species such as ferryl or perferryl ions:[126]

$$Fe^{2+} + H_2O_2 \rightarrow HO\cdot + HO^- + Fe^{3+} \tag{32}$$

$$(ADP)Fe^{3+} + O_2^{\cdot-} \rightarrow (ADP)FeO_2^{2+} \tag{33}$$

$$(ADP)Fe^{2+} + O_2 \rightarrow (ADP)FeO_2^{2+} \tag{34}$$

$$(ADP)FeO_2^{2+} + 2\ e^- + 2\ H^+ \rightarrow (ADP)FeO^{2+} + H_2O \tag{35}$$

$$(ADP)FeO^{2+} + LH \rightarrow (ADP)Fe^{3+} + LOOH \tag{36}$$

EDTA also strongly affects the mechanism of iron-dependent lipid peroxidation. Bast and Steeghs[127] have found that EDTA increased hydroxyl radical formation in NADPH-dependent microsomal lipid peroxidation, but did not enhance the MDA formation. Tien et al.[88] have shown that EDTA inhibited peroxidation of phospholipid liposomes when its concentration exceeded the concentration of ferrous ions. Later on, it was concluded[128] that EDTA manifests a double effect on iron-stimulated phospholipid peroxidation: EDTA enhanced lipid peroxidation when $[EDTA] \leq [Fe^{2+}]$ and inhibited it when $[EDTA] > [Fe^{2+}]$. It was assumed that the stimulatory effect of EDTA was due to the production of hydroxyl radicals by the Fenton reaction.

It is frequently proposed that the absence of an inhibitory effect of hydroxyl radical scavengers on iron-catalyzed lipid peroxidation is a consequent of a site-specific mechanism of HO· forming. In this case an inhibitory effect of some chelators such as EDTA is logically explained by removing bound iron ions from the membrane. Vile and Winterbourn[129] recently showed that the inhibitory effect of EDTA and citrate on lipid peroxidation of ox brain phospholipid liposomes (initiated by xanthine oxidase in the presence of ferric ions) and of rat liver microsomes (initiated by NADPH and $FeCl_3$) can indeed be explained by extracting membrane-bound iron. However, the site-specific mechanism cannot explain the promotion of lipid peroxidation by ADP or ATP as all iron ions in the case of ATP and about one half in the case of ADP were unbound. On these grounds, the participation of the iron-dioxygen complexes (perhaps perferryl) in the initiation of lipid peroxidation should be accepted. Marton et al.[130] concluded that the NADPH-dependent lipid peroxidation of rat brain microsomes is initiated by the oxidized Fe(II)EDTA complex in the presence of iron ions and EDTA.

Recent results[131] support the idea that the formation of hydroxyl radicals is not important in NADPH-dependent microsomal lipid peroxidation. It was found that EDTA inhibited lipid peroxidation in the presence of ADP and citrate and simultaneously stimulated the formation of hydroxyl radicals (measured by the formation of ethylene from 2-keto-4-thiomethylbutyric acid and of formaldehyde from DMSO). In addition, catalase and hydroxyl radical scavengers inhibited the generation of hydroxyl radicals and did not affect lipid peroxidation.

The above results show that, excluding some special cases (such as the use of EDTA as a chelator), free hydroxyl radicals are apparently unimportant in iron-dependent enzymatic and nonenzymatic lipid peroxidation. At the same time, hydroxyl radicals formed by the site-specific mechanism may promote lipid peroxidation. However, the most important initiating agents are probably the active iron-oxygen species formed in the Fenton reaction ("crypto-hydroxyl" radicals) or during oxygenation of ferrous ions (perferryl or ferryl ions). (We already discussed an uncertainty in the structure of crypto-hydroxyl radicals. For example, it is possible that ferryl ion is a "crypto-hydroxyl" radical formed by Reaction 1.)

Another important factor of successful initiation of iron-dependent lipid peroxidation is the simultaneous presence of Fe^{2+} and Fe^{3+} ions. In 1983, Bucher et al.[132] found that the presence of ferric ions is an obligatory condition for the initiation of lipid peroxidation by chelated ferrous ions. Later Braughler et al.[133] showed that a combination of Fe^{2+}, Fe^{3+}, and H_2O_2 instantly initiated the formation of conjugated dienes, lipid hydroperoxides, and TBA-reactive products upon incubation with rat brain synaptosomes. During incubation, little or no hydroxyl radicals were formed, although about 90% H_2O_2 was consumed in the first seconds. Therefore, hydroxyl radicals certainly did not participate in lipid peroxidation. It was also found that the (1:1) Fe^{2+}/Fe^{3+} ratio was needed for obtaining a maximal rate of lipid peroxidation. If Fe^{2+} alone was used for initiation, a lag period was observed which was required for autoxidation of ferrous ions to yield the necessary Fe^{2+}/Fe^{3+} ratio. It was concluded that the main role of different free radical-generating systems used for initiation of lipid peroxidation is to create an optimal Fe^{2+}/Fe^{3+} ratio.

Minotti and Aust[134] have found that catalase, mannitol, and benzoate could either inhibit or stimulate lipid peroxidation of phospholipid liposomes initiated by the Fenton reagent. At the same time, these scavengers efficiently inhibited the formation of hydroxyl radicals detected as DMPO spin adducts in this system. It was therefore concluded that mannitol, benzoate, and catalase affected lipid peroxidation by changing the redox state of iron ions (and producing by this both stimulatory and inhibitory effects, depending on the movement to or from the optimal Fe^{2+}/Fe^{3+} ratio) and not as hydroxyl radical scavengers. A lag period was also observed when Fe^{2+} ions were applied for the initiation of lipid peroxidation in rat liver microsomes.[135] The lag period decreased in the presence of Fe^{3+} ions and increased in the presence of butylated hydroxytoluene. Hydroxyl radical scavengers, SOD, and catalase did not inhibit lipid peroxidation. These authors believe that the perferryl ion is the most likely candidate for the initiation of ferrous ion-dependent lipid peroxidation.

Minotti and Aust[136] assumed that the rate-determining step in the ferrous ion-citrate-dependent peroxidation of microsomal phospholipid liposomes is the autoxidation of the Fe^{2+}-citrate complex. Again, it was concluded that stimulatory and inhibitory effects of SOD and catalase were a consequence of the effect of these enzymes on the Fe^{2+}/Fe^{3+} ratio. Mannitol had no effect on ferrous ion-citrate-dependent lipid peroxidation, indicating that hydroxyl radicals did not participate in initiation. Similarly, Miller and Aust[116] recently proposed that a double effect of ascorbate on lipid peroxidation is explained not by the competition between the reduction of ferric ions and the interaction with lipid peroxy radicals, but by its effect on the Fe^{2+}/Fe^{3+} ratio. These authors found that microsomal lipid peroxidation initiated by iron chelates in the presence of ascorbate was not inhibited by SOD, catalase, mannitol, and benzoate. Because of this, the participation of oxygen radicals in the initiation of lipid peroxidation was excluded.

Although the aforementioned findings show the importance of the simultaneous presence of Fe^{2+} and Fe^{3+} ions in the reaction mixture, the mechanism of the accelerating action of Fe^{3+} ions on lipid peroxidation remains unknown. It was suggested[137] that an initiator can be the $Fe^{3+}O_2Fe^{2+}$ complex; however, there are no data confirming such a proposal. In recent work[102] the perferryl ion was again considered the most probable initiating agent of enzymatic microsomal lipid peroxidation. A role of chelators (ADP, EDTA, etc.) is thought to be to form such a complex, which can be reduced by NADPH and then oxidized by dioxygen to form the perferryl ion.

F. NATURAL IRON-CONTAINING CATALYSTS OF LIPID PEROXIDATION

In the works discussed above, the Fe(ADP) and Fe(EDTA) complexes were mainly used as catalysts of iron-dependent lipid peroxidation. However, it is very important to ascertain whether or not native iron-containing proteins stimulate lipid peroxidation. This question is clearly bound to the probability of the *in vivo* peroxidation processes. In Chapter 2 we already considered the ferritin-induced production of hydroxyl radicals in the presence of reductants. One may expect that the release of iron ions from ferritin by reductants can also stimulate lipid peroxidation. Indeed, Gutteridge et al.[138] have shown that iron-loading ferritin enhanced the rate of ascorbate-dependent lipid peroxidation. Later on, it was found that a combination of xanthine oxidase, ADP, and ferritin promoted the peroxidation of phospholipid liposomes[139] and that ferritin catalyzed the NADPH-dependent peroxidation of rat liver microsomes.[140] It was concluded that superoxide ion generated by xanthine oxidase reduced ferric ions contained in ferritin to ferrous ions which then initiated lipid peroxidation. Peroxidation was not inhibited by mannitol and was stimulated by catalase. It is possible that both ferrous and ferric ions participate in the initiation of lipid peroxidation.

In addition to the superoxide ion and ascorbate, the iron release from ferritin may be promoted by other reductants. Thus it was recently shown[141] that a combination of glutathione and alloxan stimulates the ferritin-dependent peroxidation of phospholipid liposomes. It was proposed that glutathione reduced alloxan to a free radical which then reduced ferric ions of ferritin.

Another potential native iron-containing catalyst of lipid peroxidation is hemoglobin. It has been shown[142] that hydrogen peroxide and hydroperoxides induced the release of iron ions from hemoglobin and promoted the peroxidation of linolenic acid and the degradation of deoxyribose. Both processes are thought to be mediated by hydroxyl radicals. Contrary to this, Kanner and Harel[143] concluded that the lipid peroxidation of sarcosomes initiated by methemoglobin or metmyoglobin in the presence of hydrogen peroxide is mediated by the oxygenated heme complexes.

IV. INTERACTION OF OXYGEN RADICALS WITH PROTEINS

The mechanism of the interaction of oxygen radicals with proteins depends on their structure and functions. On the one hand, oxygen radicals are able to damage proteins, promoting their inactivation and denaturation, whereas on the other hand they may mediate the biological functions of proteins. The last is mainly concerned with the enzymes, since some of them are able to catalyze superoxide-mediated reactions. Here, we will attempt to discuss all known processes occurring between oxygen radicals and proteins.

A. HEME-CONTAINING PROTEINS

This important group of proteins includes cytochromes, hemoglobins, myoglobins, and various enzymes (catalase, peroxidases, etc.) For all of them, a main pathway for the interaction with the superoxide ion is the reduction of heme iron or the formation of iron-dioxygen complexes. These processes may be of physiological importance or may be the origins of protein inactivation. In addition, hemoproteins may also be inactivated by oxygen radical attack on other protein fragments.

1. Cytochromes

Beginning with the pioneering work by McCord and Fridovich,[144] who showed that the reduction of cytochrome c by xanthine oxidase is mediated by superoxide ion, the reaction of superoxide ion with cytochrome c was studied in many works. Cytochrome c is a component of the respiratory chain located on the outer surface of the inner mitochondrial membrane. The reduction of cytochrome c by superoxide ion may be of physiological significance, since under certain disturbances $O_2^{\bar{}}$ is formed during the oxidation of various mitochondrial electron carriers (Chapter 2). However, the great interest in this reaction was not due to its potential physiological significance, but was a consequence of its wide use as a specific assay for the superoxide ion. Unfortunately, this method of superoxide detection may underestimate superoxide production due to the rapid spontaneous dismutation of $O_2^{\bar{}}$ and the oxidation of ferrocytochrome c by hydrogen peroxide.[145] However, it was recently concluded[146] that the concentrations of hydrogen peroxide formed in the experiments with xanthine oxidase probably are not sufficiently large to diminish a real rate of superoxide production.

$$O_2^{\overline{\cdot}} + \text{cyt c(III)} \rightleftarrows O_2 + \text{cyt c(II)} \quad (37)$$

Reaction 37 is a one-electron transfer process and may proceed by an outer-sphere or an inner-sphere mechanism. The differences between the one-electron reduction potentials of dioxygen and cytochrome c is about 0.4 V; therefore, the equilibrium for Reaction 37 must be completely shifted to the right, and the oxidation of ferrocytochrome c (cyt c[II]) by dioxygen should be negligible. The experimental findings confirm this conclusion. A rate constant for Reaction 37 was first determined by a stop-flow method[147] and was found to be equal to $1.6 \cdot 10^5 \, M^{-1} \, s^{-1}$ at pH 8.4. In subsequent works[148-154] the pulse-radiolysis technique was applied (Table 1). It was found[149,151] that a k_{37} value depends on pH, decreasing from $(1\text{—}2) \cdot 10^6 \, M^{-1} \, s^{-1}$ at pH 5 to 7 to zero at pH 11. Butler et al.[151] proposed that this dependence is explained by the existence of two pK_a values of 7.45 and 9.2, respectively. However, it was later found[152,153] that the change in the reaction rate at neutral pH is due to the presence in the incubation medium of copper impurities (which are responsible for the rapid decay of the superoxide ion by dismutation) and not due to cytochrome c having a pK_a value of 7.45. The copper impurities are apparently also responsible for overestimating k_{37} values at neutral pH; thus in the presence of EDTA (which binds copper ions to form inactive complexes), $k_{37} = (2.6 \pm 0.1) \cdot 10^5 \, M^{-1} \, s^{-1}$ at pH 7.8.[153] In accord with theoretical and experimental estimates, a rate constant for the reverse Reaction −37 does not exceed $4.5 \cdot 10^{-2} \, M^{-1} \, s^{-1}$ (Reference 150).

All workers agree that Reaction 37 is an outer-sphere process, as no intermediates were found during the cytochrome c reduction. However, a k_{37} value is clearly too small for a typical outer-sphere, exothermic, one-electron transfer reaction. It is obvious from comparison with a rate constant for Reaction 38, which is equal to $1.0 \cdot 10^9 \, M^{-1} \, s^{-1}$ at pH 6.2 and $6.3 \cdot 10^8 \, M^{-1} \, s^{-1}$ at pH 8.7.[150]

$$CO_2^{\overline{\cdot}} + \text{cyt c(III)} \rightarrow CO_2 + \text{cyt c(II)} \quad (38)$$

Certainly, CO_2 has a more negative reduction potential than O_2, and therefore k_{38} must be more than k_{37}. However, Reaction 37 remains an exothermic process, and it is well known that the rate constants for exothermic one-electron transfer reactions seldom differ one from another by more than an order of magnitude.

Various explanations of the slowness of Reaction 37 were offered. Thus Koppenol et al.[152] discuss three pathways of cytochrome c reduction by the superoxide ion: binding $O_2^{\overline{\cdot}}$ to the surface of the cytochrome c molecule and transferring an electron through the protein moiety via aromatic residues; binding $O_2^{\overline{\cdot}}$ to the exposed heme edge and transferring an electron via the porphyrin ring; and transferring an electron from $O_2^{\overline{\cdot}}$ to the heme iron via electron tunneling. It is believed that electron transfer from a superoxide ion to the exposed heme edge is the most probable pathway.

Seki et al.[150] proposed that an origin of the relative slowness of Reaction 37 is the involvement of water molecules in the transition state. However, it seems that the simplest explanation of this slowness is an inner-sphere mechanism of electron transfer between the superoxide ion and ferricytochrome c, since the inability to detect an intermediate may be simply due to its short lifetime.

The interaction of cytochrome c with the mitochondrial membrane may affect its redox properties and the rate of the reaction with the superoxide ion. Heijman et al.[154] used the sodium dodecyl sulfate-cytochrome c system as a model of mitochondrial membrane-bound cytochrome c. It was found that the detergent decreased the rate of Reaction 37, supposedly inducing conformation changes in the cytochrome c molecule.

The use of cytochrome c for the detection of superoxide ion in many biological systems such as microsomes is complicated due to its reduction by other reductants and enzymes. Therefore, it has been proposed to apply modified cytochromes c, which possess a greater specificity for superoxide detection.[155] Rates for the reduction of modified cytochromes c by superoxide ion are smaller than those for native cytochromes. Thus Ilan et al.[156] found that the superoxide ion reduces acetylated cytochrome c with a rate constant of $3.5 \cdot 10^5$ M^{-1} s^{-1} at pH 7.0, but does not at all reduce fully succinylated cytochrome c and carboxymethyl methionine-65,80 cytochrome c. The difference was related to the change in the reduction potentials of modified cytochromes. Kuthan et al.[157] also proposed that the rate of Reaction 37 for succinylated cytochrome c is diminished due to decreasing its net charge and reduction potential. However, despite a decrease in the rates of reduction of modified cytochromes c by the superoxide ion, these proteins are successfully used for the detection of the superoxide ion in microsomes since the rates of their reduction by microsomal cytochrome P-450 reductase are even much smaller. For example, there was a 30-fold and 190-fold increase in the specificity of acetylated and succinylated cytochromes c, respectively, when these proteins were applied for detecting superoxide production by NADPH cytochrome P-450 reductase.[158]

Recently, Morel et al.[159] proposed the use of sulfonated phenyl isothiocyanate cytochrome c for the detection of the superoxide ion in biological systems, since it is thought that this protein is a more selective scavenger of $O_2^{\overline{\cdot}}$ than earlier applied cytochrome c modifications. In contrast to the superoxide ion, the perhydroxyl radical HOO· is not able to reduce ferricytochrome c, but oxidizes ferrocytochrome c with a rate constant of $(0.5—5) \cdot 10^5$ M^{-1} s^{-1} (Reference 150). It is interesting that $O_2^{\overline{\cdot}}$ and even $CO_2^{\overline{\cdot}}$ are not able to reduce either substrate-free or substrate-bound bacterial cytochrome P-450.[160]

2. Hemoglobins

As we showed in Chapter 1, hemoglobins are able to react with dioxygen to form superoxide ion. At the same time, like porphyrins (Volume I, Chapter 10), both Fe^{3+} hemoglobin (methemoglobin [metHb]) and oxyhemoglobin (HbO_2) are also able to react with the superoxide ion:

$$O_2^{\overline{\cdot}} + HbO_2 \xrightarrow{2\,H^+} H_2O_2 + metHb + O_2 \qquad (39)$$

$$O_2^{\overline{\cdot}} + metHb \longrightarrow HbO_2 \qquad (40)$$

It has been shown [161-163] that both reactions proceeded with the superoxide ion generated radiolytically, during the oxidation of photoreduced riboflavin, and by xanthine oxidase. Rate constants for Reactions 39 and 40 were determined by the competition technique[164] and pulse-radiolysis[161] (Table 1).

Reactions 39 and 40 demonstrate that the superoxide ion is able to oxidize oxyhemoglobin and to reduce methemoglobin. However, the third reaction channel is also possible for the interaction of $O_2^{\overline{\cdot}}$ with hemoglobins: the oxidation of thiol groups. It has been shown[163] that the superoxide ion scarcely reacts with the thiol groups of normal hemoglobin, but oxidizes the thiol groups of some unstable hemoglobins. Oxidation of human hemoglobins by the superoxide ion may also be followed by intermolecular polymerization and cross-linking of subunits.[165]

3. Catalase and Peroxidases

Catalase and peroxidases are probably the most important heme-containing enzymes. Catalase decomposes hydrogen peroxide into water and dioxygen, while peroxidases catalyze the interaction of hydrogen peroxide and hydroperoxides with various substrates. During enzymatic cycles, these hemoproteins form several intermediates which are oxy and peroxy complexes: compound I, which is the EnzFe(III)H_2O_2 complex; compound II, which is the reduced compound I EnzFe(II)H_2O_2; and compound III, which is the superoxo complex of native enzyme, EnzFe(III)$O_2^{\bar{\cdot}}$. Compound III is an inactive form of these enzymes and may be formed by the oxidation of the reduced enzyme with dioxygen (Reaction 41) or by the interaction of an enzyme with superoxide ion (Reaction 42):

$$\text{EnzFe(II)} + O_2 \rightleftarrows \text{EnzFe(III)}O_2^{\bar{\cdot}} \quad (41)$$

$$\text{EnzFe(III)} + O_2^{\bar{\cdot}} \rightleftarrows \text{EnzFe(III)}O_2^{\bar{\cdot}} \quad (42)$$

Thus catalase and peroxidases react with the superoxide ion similarly to hemoglobin and myoglobin by an inner-sphere mechanism, forming oxygenated complexes, and not by an outer-sphere mechanism, as cytochrome c does.

Since compound III is an inactive form of catalase and peroxidases, one may expect that the reaction of superoxide ion with native enzymes will lead to their inactivation. Indeed, as early as 1976, Rister and Baehner[166] showed that the superoxide ion inactivated catalase. Later on, Kono and Fridovich[167] confirmed this conclusion. The rapid inhibition of catalase due to the formation of compound III was prevented and reversed by SOD. However, these authors found that the rapid reversible inhibition of catalase was accompanied by slow, irreversible inactivation which can be prevented, but not reversed by SOD. It was proposed that this slow process represents the transformation of catalytically active compound I into compound II by Reaction 43:

$$O_2^{\bar{\cdot}} + \text{compound I} \rightarrow O_2 + \text{compound II} \quad (43)$$

$$O_2^{\bar{\cdot}} + \text{CatFe(III)} \rightarrow \text{CatFe(III)}O_2^{\bar{\cdot}} \text{ (compound III)} \quad (44)$$

$$H_2O_2 + \text{CatFe(III)} \rightarrow \text{CatFe(III)}H_2O_2 \text{ (compound I)} \quad (45)$$

It was found[168] that a rate constant for Reaction 44 increased with decreasing pH (from $4.5 \cdot 10^4 \, M^{-1} \, s^{-1}$ at pH 9 to $4.6 \cdot 10^6 \, M^{-1} \, s^{-1}$ at pH 5). It is possibly explained by a more rapid reaction of catalase with perhydroxyl radical HOO· in comparison with $O_2^{\bar{\cdot}}$. Based on this proposal, a value of $4.5 \cdot 10^4 \, M^{-1} \, s^{-1}$ was accepted for k_{44}. However, using the pulse-radiolysis technique, Kobayashi et al.[169] determined $k_{44} = 2.5 \cdot 10^5 \, M^{-1} \, s^{-1}$ at pH 7.4. These authors concluded that the SOD concentration in erythrocytes is more than ten times greater than is required to prevent the catalase inactivation by Reaction 44. Gebicka et al.[170] determined the rate constants for Reactions 45 and 43 as $2.0 \cdot 10^7$ and $5.0 \cdot 10^6 \, M^{-1} \, s^{-1}$ (pH 7.5), respectively. It should be noted that catalase can also be inhibited by ascorbate, supposedly due to the formation of compound I in the reaction of catalase with hydrogen peroxide originating from the reduction of dioxygen by ascorbate.[171]

As with catalase, the superoxide ion forms compound III with myeloperoxidase,[172-177] HRP[175,178] and lactoperoxidase.[175] Bielski and Gebicki[179] found that the rate of the reaction of superoxide ion with HRP is substantially smaller than that of Reaction 46. This conclusion was recently confirmed for the HRP isoenzyme c.[178]

$$\text{PerFe(III)} + H_2O_2 \rightarrow \text{PerFe(III)}H_2O_2 \quad (46)$$

$$\text{PerFe(III)}H_2O_2 + Cl^- \rightarrow \text{PerFe(III)} + HOCl + HO^- \quad (47)$$

Much attention was attached to studying the interaction of the superoxide ion with myeloperoxidase from leukocytes since this enzyme catalyzes the oxidation of chloride into hypochlorite (via Reactions 46 and 47). This process is thought to be important for the bactericidal action of leukocytes because the leukocyte activation is accompanied by the release of superoxide ion (Chapter 4). Therefore, one may expect that the interaction of superoxide ion with myeloperoxidase may be of physiological importance. In 1972 Odajima and Yamazaki[172] showed that the superoxide ion obtained by electrochemical reduction of dioxygen in DMF converted myeloperoxidase to compound III. Then, it was found[174] that myeloperoxidase inhibited the reduction of cytochrome c by xanthine oxidase, i.e., myeloperoxidase is a more effective scavenger of the superoxide ion than cytochrome c. Indeed, a rate constant for Reaction 42 is equal to $(2.1 \pm 0.2) \cdot 10^6\ M^{-1}\ s^{-1}$ in the case of equine myeloperoxidase and $(1.1 \pm 0.3) \cdot 10^6\ M^{-1}\ s^{-1}$ in the case of human myeloperoxidase,[176] i.e., an order of magnitude greater than a rate constant of the reduction of cytochrome c by the superoxide ion (Table 1). Even SOD did not completely inhibit the formation of compound III from myeloperoxidase at concentrations as high as 3 μM.[174]

Metodiewa and Dunford[175] have studied the formation of compound III in the reactions of superoxide ion generated by xanthine oxidase with myeloperoxidase, HRP, and lactoperoxidase. It was concluded that Reaction 42 proceeds by a two-step mechanism, but the structure of a possible intermediate remains obscure. The role of compound III in the processes catalyzed by peroxidases is also unclear. As was pointed out, compound III is usually considered an inactive form of peroxidases and catalase; however for myeloperoxidase, it was proposed that compound III can react with superoxide ion to form compound I[174] or with hydrogen peroxide to form compound II, which is then converted by superoxide ion to myeloperoxidase.[177] In this way, compound III is involved in a peroxidative cycle (Reactions 46 and 47). Unfortunately, it is very difficult or in some cases even impossible to comprehend a mechanism for these hypothetical processes. Even if superoxide ion is able to reduce compound III, the reaction product cannot be compound I, since for this we need one electron and two protons more (Reaction 48):

$$[\text{PerFe(III)}O_2^{\cdot -} \rightleftarrows \text{PerFe(II)}O_2] + O_2^{\cdot -} \rightarrow \text{PerFe(II)}O_2^{\cdot -} + O_2 \qquad (48)$$

Compound III

H_2O_2 can formally substitute O_2 in compound III (Reaction 49) and $O_2^{\cdot -}$ can also formally reduce bound H_2O_2 in compound II (Reaction 50, a hypothetical hybrid of concerted intra-intermolecular Fenton reaction).

$$[\text{PerFe(III)}O_2^{\cdot -} \rightleftarrows \text{PerFe(II)}O_2] + H_2O_2 \rightarrow \text{PerFe(II)}H_2O_2 + O_2 \qquad (49)$$

Compound III Compound II

$$\text{PerFe(II)}H_2O_2 + O_2^{\cdot -} \rightarrow \text{PerFe(III)} + 2\ HO^- + O_2 \qquad (50)$$

However, from a chemical point of view, Reactions 49 and 50 are very doubtful.

The superoxide ion and perhydroxyl radical can also react with compound I of peroxidases. Thus for HRP, rate constants for Reactions 51 and 52 are equal to $1.6 \cdot 10^6$ and $2.2 \cdot 10^8\ M^{-1}\ s^{-1}$, respectively.[179]

$$\text{PerFe(III)}H_2O_2 + O_2^{\cdot -} \rightarrow \text{PerFe(II)}H_2O_2 + O_2 \qquad (51)$$

$$\text{PerFe(III)}H_2O_2 + HO_2^{\cdot} \rightarrow \text{PerFe(II)}H_2O_2 + O_2 + H^+ \qquad (52)$$

Jenzer et al.[180] have found that lactoperoxidase is irreversibly inactivated by excess hydrogen peroxide. It was concluded that inactivation was mediated by site-specific hydroxyl radicals, since hydroxyl radical scavengers and DMPO, a spin trap, did not prevent it. Hydroxyl radicals are thought to be formed in the Fenton-like reaction with ferrous lactoperoxidase as a catalyst.

4. Indoleamine 2,3-Dioxygenase

In 1971, Hirata and Hayaishi[181] have shown that SOD inhibited the transformation of D-tryptophan into *N*-formylkynurenine catalyzed by rabbit intestinal indoleamine 2,3-dioxygenase. At the same time, the addition of superoxide solution in DMF accelerated this process. On these grounds, it was concluded that superoxide ion is involved in the reactions of indoleamine 2,3-dioxygenase. Activation of this enzyme, containing protoheme IX as a prosthetic group, proceeds by its reduction to the ferrous state. It was found that methylene blue and the superoxide ion are effective reductants of indoleamine 2,3-dioxygenase. At the same time, it was supposed that the superoxide ion participates in the formation of the catalytically active oxygenated heme complex of the enzyme which transfers oxygen atoms to the products. Thus Hayaishi et al.[182] have shown that about 60% of the oxygen atoms incorporated into *N*-formyl-L-kynurenine were derived from the superoxide ion (added to incubation mixture as $K^{18}O_2$ in DMF) and about 40% were derived from dioxygen. It should, however, be noted that these authors erroneously accepted that the exchange reaction between $^{18}O_2^{\overline{\cdot}}$ and $^{16}O_2$ or $^{16}O_2^{\overline{\cdot}}$ and $^{18}O_2$ can be neglected (see Volume I, Chapter 7).

The formation of the $Fe^{3+}O_2^{\overline{\cdot}}$ heme complex was also shown spectrophotometrically in the reaction of KO_2 with a native ferric form of enzyme.[183] It was concluded that this complex was catalytically active as its decomposition was accelerated by active substrates such as tryptophan, 5-hydroxytryptophan, tryptamine, and serotonin and was not affected by the nonmetabolizable substrate analogs indoleacetic acid or skatole. It was also proposed[184] that an actual catalytic oxygenated complex is the tertiary complex substrate-enzyme-$O_2^{\overline{\cdot}}$. Indeed, the oxygenated enzyme complex prepared by the reaction of the superoxide ion with the ferric enzyme in the presence of tryptophan or by the reaction of the ferrous enzyme with dioxygen produced the product (*N*-formyl-L-kynurenine), oxidizing into the ferric form.[185] The participation of the superoxide ion as a cosubstrate in the reaction of indoleamine 2,3-dioxygenase was supported by the findings that the intracellular indoleamine 2,3-dioxygenase activity in rabbit intestinal cells increased when the substrates of xanthine oxidase (a generator of the superoxide ion) and diethyldithiocarbamate (the SOD inhibitor) were added and was abolished by allopurinol (a specific inhibitor of xanthine oxidase).[186]

Recently, Sono[187] studied the roles of the superoxide ion and an artificial substrate, methylene dye, in the reductive activation of indoleamine 2,3-dioxygenase. It was confirmed that the activation of the enzyme by xanthine oxidase plus hypoxanthine was completely inhibited by SOD, indicating that under these conditions $O_2^{\overline{\cdot}}$ was the sole reductant. Contrary to this, the activation of indoleamine 2,3-dioxygenase by ascorbate plus methylene blue was not affected by SOD and therefore was independent of the superoxide ion. The superoxide ion, together with dihydroflavin mononucleotide ($FMNH_2$), probably participates in the decyclization of L-tryptophan catalyzed by indoleamine 2,3-dioxygenase isolated from murine epididymis because SOD inhibited the enzyme activity by 50%.[188]

It has been shown[189] that the superoxide ion is able to activate a very close analog of indoleamine 2,3-dioxygenase, tryptophan 2,3-dioxygenase. However, in spite of their obvious similarity (both enzymes are hemoproteins and oxidize tryptophan into *N*-formylkynurenine), the reductive activation of tryptophan 2,3-dioxygenase by the superoxide ion is thought not to be physiologically meaningful and is not due to the heme reduction.

B. SUPEROXIDE DISMUTASE (SOD)

There are plenty of data in the literature concerning the interaction of oxygen radicals with enzymes which do not contain hemes. The most important among them is undoubtedly SOD, an enzyme responsible for destroying the superoxide ion by its dismutation. The role

TABLE 2
Rate Constants for Enzymatic Dismutation of Superoxide Ion

Enzyme	pH	k (M^{-1} s^{-1})	Ref.
Bovine CuSOD	7.8	$1.75 \cdot 10^9$	197
Bovine Cu(II)SOD	7.5	$(1.2 \pm 0.2) \cdot 10^9$	194
Bovine Cu(I)SOD		$(2.2 \pm 0.4) \cdot 10^9$	194
Bovine CuSOD	6—10.2	$(1.6—3.4) \cdot 10^9$	193
Bovine CuSOD	9	$2.3 \cdot 10^9$	198
Bovine Cu(II)SOD	7	$(1.4 \pm 0.2) \cdot 10^9$	173
Bovine Cu(I)SOD	7	$(1.9 \pm 0.6) \cdot 10^9$	173
Bovine CuSOD	9—9.9	$(2.37 \pm 0.19) \cdot 10^9$	201
Bovine CuSOD	7.8	$(1.3 \pm 0.1) \cdot 10^9$	196
Human CuSOD	5.7—10.5	$1.5 \cdot 10^9$	194
Escherichia coli FeSOD	6—10.2	$(0.4—1.9) \cdot 10^9$	193
E. coli FeSOD	7.4	$(3 \pm 1) \cdot 10^8$	208
C. vinosum FeSOD	6—10.2	$(0.8—4.4) \cdot 10^9$	450
Photobacterium leiognathi Fe(III)SOD	8.0	$5.2 \cdot 10^8$	207
P. leiognathi Fe(II)SOD	8.0	$5.5 \cdot 10^8$	207
E. coli MnSOD	6—10.2	$(0.3—2) \cdot 10^9$	193
E. coli MnSOD	7.9	$(1.5 \pm 0.15) \cdot 10^9$	205
E. coli Mn^-SOD $(+ 2 H^+)$	7.9	$(1.6 \pm 0.6) \cdot 10^9$	205
E. coli Mn^-SOD $(+ 2 H^+)$	7.9	$(1.6 \pm 0.25) \cdot 10^8$	205
MnSOD from chicken liver mitochondria	6—10.2	$(0.8—3.6) \cdot 10^9$	193
Bacillus stearothermophilus MnSOD		$5.6 \cdot 10^8$	206
B. stearothermophilus Mn^-SOD	$(+ 2 H^+)$	$5.6 \cdot 10^8$	206
B. stearothermophilus Mn^-SOD		$4.7 \cdot 10^7$	206
Paracoccus denitrificaus MnSOD	7.9	$3.33 \cdot 10^8$	209
D-Galactose oxidase		$3 \cdot 10^6$	228

of SOD in the detoxification of oxygen radicals in biological systems is of enormous importance. Now, we have special books dedicated to studying this enzyme (for example, see Reference 190) and five Conferences on Superoxide and Superoxide Dismutase (the last was held in Jerusalem, Israel, September 17 to 22, 1989). Historically, the identification of erythrocuprein as the superoxide-dismuting enzyme by McCord and Fridovich[191] in 1969 is rightly considered the beginning of a new era in the study of free radical processes in biological systems. Here, we will consider only the mechanism of the interaction of oxygen radicals with SOD.

SODs are enzymes contained in the active centers transition metals which catalyze the dismutation of the superoxide ion. SODs were isolated from a variety of sources, including human and bovine erythrocytes, yeast, *Escherichia coli*, chicken liver, etc.[192] Enzymes from the eucaryotes contain copper as a transition metal, whereas enzymes from the procaryotes contain manganese and iron. CuSOD also contains zinc, which, however, does not participate in the catalytic cycle. At the beginning, an outer-sphere, one-electron transfer mechanism was proposed for the dismutation of the superoxide ion by CuSOD:

$$Cu(II)SOD + O_2^{\overline{\cdot}} \rightarrow Cu(I)SOD + O_2 \tag{53}$$

$$Cu(I)SOD + O_2^{\overline{\cdot}} + 2 H^+ \rightarrow Cu(II)SOD + H_2O_2 \tag{54}$$

The rate constants for Reactions 53 and 54 were determined by competition with cytochrome c,[193] pulse-radiolysis,[173,194-196] flash photolysis,[197] polarography,[198] and the stopped-flow technique.[199] It was found that k_{53} and k_{54} values are practically equal, are independent of pH within a wide range, and are close to a diffusion limit (Table 2). In accord with the equality of k_{53} and k_{54} values, the Cu(II)/Cu(I) ratio is equal to one in the steady state of

CuSOD.[200] There is no difference in the rate constants for bovine and human SOD; for both enzymes, there was also no evidence for the formation of a Michaelis complex. All these findings confirm the outer-sphere mechanism of SOD catalysis.

It is known that CuSOD is a dimer of identical subunits, containing one Cu and one Zn. Fielden et al.[201] concluded that the reduction of one Cu by the superoxide ion makes the second Cu transiently unreactive. Because of this, only a half copper of CuSOD participates in the catalytic cycle.

Studying the interaction of superoxide ion with CuSOD by ESR spectroscopy at low temperatures confirmed the absence of the formation of long-lived complexes in both Reactions 53 and 54.[202] However, using high concentrations of the superoxide ion produced at the dropping mercury electrode in the presence of triphenylphosphine, Rigo et al.[203] showed a saturation character of the dismutation reaction catalyzed by CuSOD. Later on, the saturation kinetics were observed by Fee and Bull[199] for CuSOD in the stopped-flow experiments, also with high concentrations (about 5 m*M*) of superoxide ion at 5.5°C. These findings apparently indicate the possibility of the formation of weak intermediate complexes in Reactions 53 and 54 at very high (nonphysiological) concentrations of the superoxide ion and low (also nonphysiological) temperatures.

Viglino et al.[204] determined a rate constant for the oxidation of Cu(I)SOD by dioxygen (a reverse Reaction 53) which is equal to 5.4 M^{-1} s^{-1} (pH 8.4). A k_{-53} value increased with increasing pH, indicating that only the deprotonated form of Cu(I)SOD reacted with dioxygen.

Catalysis of superoxide dismutation by MnSOD from *Escherichia coli*[205] and *Bacillus stearother*[206] is a biphasic process. Such behavior of MnSOD is supposedly a consequence of the participation in the catalytic cycle of Mn(III), Mn(II), and Mn(I) states. It was assumed[205] that catalysis by Mn(III)/Mn(II) occurs with a rate constant of $1.5 \cdot 10^9$ M^{-1} s^{-1}, whereas a slower catalysis by Mn(II)/Mn(I) occurs with a rate of $1.6 \cdot 10^8$ M^{-1} s^{-1}. In contrast to MnSOD from *E. coli*, there was no evidence for a biphasic process in the case of FeSOD from *Photobacterium leiognathi*.[207] For all these enzymes, no evidence of saturation kinetics was obtained.

In contrast to bovine and human CuSOD, bacterial MnSOD, and FeSOD from *P. leiognathi*, Fee et al.[208] have shown that FeSOD from *E. coli* catalyzed the superoxide dismutation by a mechanism which exhibited saturation kinetics. Such behavior was interpreted as support for an inner-sphere mechanism by which the superoxide ion forms an intermediate complex with the enzyme:

$$O_2^{\overline{\cdot}} + \text{Fe(III)SOD} \rightleftarrows (O_2^{\overline{\cdot}})\text{Fe(III)SOD} \rightarrow O_2 + \text{Fe(II)SOD} \qquad (55)$$

Similarly, saturation kinetics was observed in the dismutation of the superoxide ion catalyzed by MnSOD isolated from *Paracoccus denitrificans*.[209] However, in this case it was found that the dismutation rate constant increased with an increase in pH. This finding may indicate the fact that the perhydroxyl radical and not the superoxide ion takes part in the reoxidation of MnSOD.

It has been shown[210,211] that the rates of superoxide dismutation catalyzed by CuSOD depend on ionic strength. McAdams[211] pointed out that the observed dependence of rate constants on high ionic strength should be considered only as an empiric one since a theoretical Bronsted-Bjerrum equation is not valid under these conditions. It was also shown that at low ionic strength where the Bronted-Bjerrum equation is valid, the rate constants were independent of ionic strength.

Cudd and Fridovich[212] proposed that a decrease in CuSOD activity with increasing ionic strength is explained by electrostatic interactions between the incoming superoxide ion and a cluster of cationic groups at the active center of the enzyme. It was also proposed that a pair of lysine residues placed close to the copper are sources of the electrostatic interactions. Indeed, acylation of lysine residues eliminated this electrostatic effect. Similarly, enhancing ionic strength resulted in a decrease in FeSOD and MnSOD activities.[213] In this case, acetylation of lysine residues also reversed this relationship. On these grounds, it was assumed[214] that during the reaction with CuSOD, the superoxide ion enters a positively charged channel in proximity to the active site of the enzyme which is formed by side chains of Agr-141, Lys-120, and Lys-134 at the distance of 0.5, 1.2, and 1.3 nm from the copper, respectively. Semsei and Zs.-Nagy[215] found that the dependence of SOD activity on ionic strength follows an exponential and not a linear curve. They proposed that both cations and anions must affect the SOD activity. Recently, a dependence of SOD activity on pH and ionic strength was confirmed for CuSOD isolated from ox, sheep, pig, and yeast.[216]

It is evident that the reoxidation of reduced SOD by the superoxide ion (Reaction 54 for CuSOD) cannot be an elementary step because $O_2^{\overline{\cdot}}$ must be protonized before it will be able to participate in one-electron transfer. In earlier work,[217] it was shown that the SOD activity is the same in H_2O and D_2O; therefore, it was concluded that a proton does not come to $O_2^{\overline{\cdot}}$ from solution. Richardson et al.[218] proposed that a possible origin of a proton is the acidic form of the bridging imidazolate of histidine-61 residue:

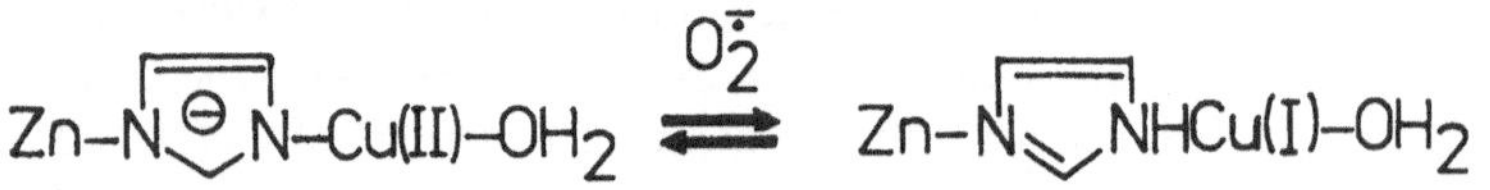

(56)

Supporting this idea, McAdams et al.[219] showed that the catalysis of superoxide dismutation by Co(II)-substituted CuSOD was accompanied by spectral changes which corresponded to rapid protonation and deprotonation of the Cu-facing N atom.

On the other hand, Fee and Ward[220] proposed that the superoxide ion receives a proton from a water molecule coordinated to the cuprous ion of the protein. In subsequent works, Fee and Bull[199,221] concluded that for both CuSOD and FeSOD a rate-limiting step of Reaction 57 is the breakdown of the peroxo complex:

$$M^{(n-1)+}SOD + O_2^{\overline{\cdot}} \xrightarrow{H_2O} HOO\text{–}M^{(n-1)+}SOD + HO^-$$

$$\xrightarrow{H_2O} M^{n+}SOD + H_2O_2 + 2\ HO^- \quad (57)$$

It is thought that in the case of CuSOD, the bridging imidazolate is not essential for the transfer of a proton to the superoxide ion. The catalytic process was accelerated in the presence of general acids and was slowed in D_2O.

However, there is another important point which was stressed by Argese et al.[222] These authors pointed out that the equality of the rate constants for Reactions 53 and 54 makes it difficult to believe that these rates correspond to chemical reactions involving Cu(II) and Cu(I) ions due to their different physicochemical characteristics. They therefore concluded that both reactions are diffusion-limited ones and are controlled by the rate of access of the superoxide ion to the active site, i.e, by the electrostatic interactions of $O_2^{\overline{\cdot}}$ with the charged groups of the enzyme.

In contrast to the superoxide ion, other active oxygen radicals promote the inactivation of SOD. In 1973, Fielden et al.[223] showed that hydrogen peroxide subsequently reduced and then inactivated SOD. It was proposed[224] that the inactivation of CuSOD by hydrogen peroxide (which proceeded with a rate constant of 3.1 $M^{-1}\ s^{-1}$ at 20 to 25°C) is due to the action on the enzyme of hydroxyl radicals formed by the Fenton-like reaction between Cu(I)SOD and H_2O_2. Hydroxyl radicals supposedly destroy a histidine residue close to the metal in SOD. Hodgson and Fridovich[217] determined the rate constant for the inactivation of CuSOD by hydrogen peroxide at pH 10 and 25°C as 6.7 M^{-1} s^{-1}. They found that inactivation was inhibited by formate, azide, and urate, but not benzoate or ethanol. The formation of active oxygen species $Cu(II)O^{\cdot-} \rightleftarrows Cu(I)O$ in the reaction of Cu(I)SOD with H_2O_2 was proposed. A similar mechanism was recently proposed for the inactivation of FeSOD from *E. coli* by hydrogen peroxide.[225]

Earlier, Blech and Borders[226] concluded that the perhydroxyl anion HOO^- and not H_2O_2 itself reacts with CuSOD, leading to its inactivation. This conclusion was recently confirmed by Viglino et al.,[227] who found that hydrogen peroxide participates in Reaction 58 only as the perhydroxyl anion.

$$Cu(II)SOD + HOO^- \rightleftarrows Cu(I)SOD + O_2^{\cdot-} + H^+ \tag{58}$$

$$Cu(I)SOD + H_2O_2 \rightarrow Cu(II)SOD + HO\cdot + HO^- \tag{59}$$

C. OTHER ENZYMES POSSESSING DISMUTASE ACTIVITY

SODs are unique enzymes, catalyzing the dismutation of the superoxide ion with maximal possible efficiency (the rate constants for Reactions 53 and 54 are close or equal to a diffusion limit). However, some other enzymes may also manifest (although significantly smaller) dismutase activity. Thus Cleveland and Davis[228] concluded that D-galactose oxidase from *Polyporus circinatus*, a Cu(II) protein, catalyzing the oxidation of D-galactose into galactohexodialdose, is able to dismute the superoxide ion with a rate constant of about $3 \cdot 10^6$ M^{-1} s^{-1}. Markossian et al.[229] have shown that cytochrome c oxidase inhibited the reduction of *p*-nitrotetrazolium chloride by NADH and the oxidation of epinephrine. It was concluded that cytochrome c oxidase has the dismutase activity which is equal approximately to 1 to 3% of that for CuSOD. Recently, Naqui and Chance[230] showed that the SOD activity of cytochrome oxidase is smaller (about 0.3%). However, the SOD activity of so-called "pulsed" cytochrome c oxidase (formed during reoxidation of the enzyme in the absence of hydrogen peroxide) is much more and equal to about 2.5%. Cuperus et al.[174] concluded that myeloperoxidase manifests SOD activity by subsequent formation of compounds III and I.

D. OXYGEN RADICALS IN ACTIVATION AND INACTIVATION OF ENZYMES

For most, but not all, enzymes which do not contain transition metal ions, interaction with oxygen radicals resulted in their inhibition and inactivation. We will consider here all the described reactions with various enzymes (including also some heme-containing enzymes for which the interaction with oxygen radicals takes place without the participation of transition metals).

It was shown in Volume I (Chapter 10) that the superoxide ion is able to react with flavins, forming flavin hydroperoxides. Similarly, it was found[231] that the superoxide ion reacted with glucose oxidase (GO), a flavoprotein, and its semiquinone:

$$O_2^{\cdot-} + GO \rightarrow (GO)^{\cdot-} + O_2 \tag{60}$$

$$O_2^{\cdot-} + (GO)^{\cdot-} \rightarrow GO(OOH) \tag{61}$$

The rate constants for Reactions 60 and 61 determined by the pulse-radiolysis technique are equal to about 10^6 M^{-1} s^{-1} and $8.5 \cdot 10^8$ M^{-1} s^{-1}, respectively. Henry et al.[232] have shown that the superoxide ion may participate as a reducing agent in the hydroxylation of tyramine catalyzed by dopamine-β-hydroxylase. However, the superoxide ion turns out to be a significantly less effective reductant than ascorbate, and therefore the reduction of dopamine-β-hydroxylase by O_2^{-} is unlikely to have physiological relevance. Similarly, it was proposed[233] that the superoxide ion participates in the reactions of propyl 4-hydroxylase and lysyl hydroxylase, which catalyze the hydroxylation of propyl and lysyl residues in peptide linkages. It was shown that although SOD did not inhibit these reactions, SOD-active copper chelates such as Cu(acetylsalicylate)$_2$ did, indicating in the authors' opinion the participation of the superoxide ion in the catalytic process.

It has been found[234] that microsomal glutathione *S*-transferase, an enzyme catalyzing the interaction of glutathione with xenobiotics, is activated by a combination of norepinephrine and the NADPH-generating system. This process was partially inhibited by SOD and catalase. On these grounds, it was proposed that oxygen radicals produced during the oxidation of norepinephrine oxidized the protein sulfhydryl group into the disulfide. Thus oxidative stress or the oxidation of xenobiotics may enhance the activity of glutathione *S*-transferase, which can be considered the regulatory function of this enzyme.

Mittal and Murad[235] proposed that hydroxyl radicals formed during the decomposition of hydrogen peroxide activate guanylate cyclase, an enzyme, catalyzing the formation of cyclic GMP from GTP. This proposal was based on the fact that SOD activated this enzyme apparently due to an increase in the rate of superoxide dismuting into hydrogen peroxide. The mechanism of activation of guanylate cyclase by oxygen radicals is unknown.

Schallreuter and co-workers[236-238] proposed that thioredoxin reductase, a ubiquitous electron-transfer enzyme contained in many tissues, catalyzes the reduction of the superoxide ion to hydrogen peroxide at the surface of the epidermis. This conclusion was based on the ability of thioredoxin reductase to reduce the stable nitroxide radical containing the quaternary ammonium moiety, which is unable to penetrate the cellular plasma membranes and therefore is reduced at the outer cell surface. It was proposed that thioredoxin reductase participates in the defense of keratinocytes, melanocytes, and melanoma cells against oxygen radicals.

Thioredoxin reductase contains two redox centers: FAD/FADH$_2$ and disulfide/thiol. It was assumed that the superoxide ion oxidizes both FADH$_2$[238] and the thiol group[236] to form hydrogen peroxide. However, these conclusions seem not to be properly substantiated. First of all, it is difficult to agree with the suggestion that the ability of thioredoxin reductase to reduce a stable neutral nitroxide radical can be interpreted as an argument for the possibility of a similar reaction with the superoxide ion. In contrast to the nitroxide radical, the superoxide ion is a negatively charged species, and its interaction with the enzyme (if possible) must proceed via another mechanism. In Volume I we showed that the superoxide ion cannot be directly reduced by thiols or flavins, neither can it abstract a hydrogen atom from these compounds; such reactions may proceed only by deprotonation-oxidation or a concerted mechanism. Certainly, such mechanisms may be realized in the reaction of O_2^{-} with thioredoxin reductase. However, unequivocal evidence of the interaction of the superoxide ion with this enzyme is needed.

It has been proposed[239] that the cleavage of catechols with the incorporation of dioxygen catalyzed by catechol dioxygenases may be mediated by the superoxide ion. This conclusion was based on the inhibitory effect of the low-molecular synthetic analog of SOD, copper salicylate, on the reactions catalyzed by protocatechuate 3,4-dioxygenase and 3,4-dihydroxyphenylacetate 2,3-dioxygenase (SOD did not inhibit these processes). Bateman et al.[240] assumed that enzymatic peptide amidation may be mediated by hydroxyl or site-specific hydroxyl radicals because the model ascorbate-copper system closely mimicked the characteristics of enzymatic peptide amidation.

The superoxide ion formed by photosystem I of chloroplasts probably takes part in the photorespiratory glycolate synthesis catalyzed by transketolase. Indeed, it has been shown[241,242] that a combination of transketolase and xanthine oxidase promotes glycolate formation during the oxidative split of fructose 6-phosphate. This reaction was inhibited by SOD and tiron, whereas the hydroxyl radical scavenger mannitol was ineffective. A rate constant for the reaction of the superoxide ion with transketolase determined by the competition method was equal to $1.0 \cdot 10^6\ M^{-1}\ s^{-1}$.

Guissani et al.[243] have studied the interaction of radiolytically generated $O_2^{\overline{\cdot}}$ and HO· with *Polyporus versicolor* laccase, a copper oxidase. Both radicals formed very unstable complexes with the enzyme. In the case of the superoxide ion, the rate constant for the reaction with laccase was about $2 \cdot 10^6\ M^{-1}\ s^{-1}$. It was concluded that laccase may be considered an efficient oxygen radical scavenger.

In contrast to these examples describing the activation of enzymes by oxygen radicals, a majority of the reactions of oxygen radicals with enzymes leads to their inactivation. In 1976, Rister and Bachner[166] concluded that the superoxide ion is able to inactivate catalase, glutathione peroxidase, and NADPH-cytochrome c reductase. The inactivation was prevented by SOD. However, Searle and Willson[244] found that radiolytically generated $O_2^{\overline{\cdot}}$ only slightly inhibited glutathione peroxidase, whereas hydroxyl radicals manifested a more prominent inactivating effect. By contrast, Blum and Fridovich[245] have shown that enzymatically produced superoxide ion inactivated an active, reduced form of glutathione peroxidase. SOD completely inhibited inactivation, whereas catalase, hydroxyl radical scavengers, and chelators did not. It was proposed that the superoxide ion in its protonized form (as HOO·) abstracts a hydrogen atom from the selenol group of an enzyme:

$$\text{EnzSeH} + O_2^{\overline{\cdot}} + H^+ \rightarrow \text{EnzSe}\cdot + H_2O_2 \qquad (62)$$

$$\text{EnzSe}\cdot + O_2 \rightarrow \text{EnzSeOO}\cdot \xrightarrow{\text{RH}} \text{EnzSeOOH} + R\cdot \qquad (63)$$

However, it may be equally accepted (see Volume I, Chapter 5) that the interaction of the superoxide ion with the selenol group proceeds via a deprotonation-oxidation mechanism:

$$\text{EnzSeH} + O_2^{\overline{\cdot}} \rightarrow \text{EnzSe}^- + HO_2\cdot \qquad (64)$$

$$O_2^{\overline{\cdot}} + HO_2\cdot \rightarrow O_2 + HO_2^- \qquad (65)$$

$$\text{EnzSe}^- + O_2 \rightarrow \text{EnzSe}\cdot + O_2^{\overline{\cdot}} \qquad (66)$$

Henry et al.[246] have shown that the superoxide ion generated by xanthine oxidase or prepared by the dissolution of KO_2 in DMSO irreversibly inactivated hydrogenases from *E. coli* and *Clostridium pasteurianum* (the enzymes catalyzing the reversible activation of molecular hydrogen). The inactivation was inhibited by SOD and catalase, but was not affected by hydroxyl radical scavengers (benzoate, formate, mannitol, tocopherol, etc.).

Oxygen radicals generated by a combination of xanthine, xanthine oxidase, ferric chloride, ascorbic acid, and ADP induced a significant reduction in the activity of iodothyronine 5′-monodeiodinase in rat liver microsomes and homogenates.[247] Simultaneously, there was an increase in MDA formation. The reduction of iodothyronine (T_4 and rT_3) 5′-monodeiodinating activity (MA) in liver tissue was blocked by SOD, tocopherol, and thiourea. It was supposed that lipid peroxidation can be an origin of MA reduction, although there may also be two independent free radical processes: lipid peroxidation and free radical reaction, directly affecting iodothyronine 5′-monodeiodinase. It was also proposed that hydroxyl radicals are apparently the most important oxygen radicals, participating in MA inhibition. The oxygen radical-mediated MA inhibition may be relevant to the development in patients of nonthyroidal illness.

Xanthine oxidase promoted the SOD-inhibitable inactivation of 3 α-hydroxysteroid dehydrogenase from *Pseudomonas testosteroni*.[248] It was proposed that the modification of histidine and cysteine residues of the enzyme by the superoxide ion can be an origin of inactivation.

Hydroxyl or site-specific hydroxyl radicals supposedly participate in the inactivation of acetylcholinesterase[249] and phosphoglucomutase[250] by the iron ions-ascorbate and ferritin-ascorbate systems. Similarly, it has been shown[251] that hydroxyl radicals inactivated α_1-antiproteinase, a major circulating inhibitor of serine proteases. Aruoma and Halliwell[252] confirmed the inactivating effect of oxygen radicals generated by pulse-radiolysis, the Fenton reaction, or xanthine oxidase in the presence of iron ions on this enzyme. It was also found that depending on the type of oxygen radicals, uric acid possessed both protective and inactivating action.

Kuo et al.[253] have shown that α,β-dihydroxyisovalerate dehydratase activity in *E. coli* cells sharply decreased upon exposure of the cells to paraquat and plumbagin. As inactivation of this enzyme was prevented by SOD, it was concluded that the inhibitory action of paraquat and plumbagin was mediated by superoxide ion.

Elliot et al.[254] have studied the action of oxygen radicals on lipoamide dehydrogenase, an enzyme, catalyzing the oxidation of the dehydrolipoamide residue of lipoate acyltransferase to the lipoamide residue. They found that although this enzyme contains free sulfhydryl groups, it is not inactivated by the superoxide ion. On the other hand, hydroxyl radicals were strong inactivating agents that promoted irreversible damage to the FAD moiety of the enzyme.

Oxygen radicals (supposedly the site-specific hydroxyl radicals) formed by the dithiothreitol plus iron ions or ascorbate plus copper ions systems under aerobic conditions inactivated yeast glutamine synthetase and *E. coli* adenylyltransferase.[255] In contrast, the inactivation of bovine serum albumin (BSA) and *Aspergillusniger* glucose oxidase was a very slow process. Kang et al.[256] have shown that the Fenton-like Cu^{2+} plus H_2O_2 system promoted the polymerization of lysozyme which was accompanied upon prolonged incubation by the polymer degradation. It was found that during this process the amino acid residues were oxidized.

Oxygen radicals generated by xanthine oxidase inactivated Na^+,K^+-ATPase from rat brain[257,258] and rat lung[259] presumably via the interaction with sulfhydryl groups. The inactivation was prevented by SOD. Similarly, the xanthine-xanthine oxidase system promoted the inactivation of Mn^{2+}- and Na^+,K^+-sensitive forms of ATPase and adenylate cyclase from gerbil cerebral cortex.[260] SOD, but not catalase, protected enzymes, indicating an important role of the superoxide ion in the inactivating process. Na^+,K^+-ATPase activity in the fraction from the outer medulla of porcine kidney was inhibited by a combination of Fe^{3+}(ADP) plus H_2O_2.[261] A decrease in the SH content and the turnover rate of Na^+,K^+-ATPase and an increase in the formation of MDA and conjugated dienes were observed. It is of interest that BHT inhibited lipid peroxidation, but did not normalize enzyme activity.

Xanthine oxidase in the presence of Fe(III)EDTA, the Fenton reagent (Fe^{2+} + H_2O_2), and the cupric ion-ascorbate system inactivated mitochondrial adenosine triphosphate from *Trypanosoma cruzi*.[262] SOD and catalase inhibited xanthine oxidase-stimulated inactivation, while mannitol inhibited inactivation stimulated by the Fenton reagent, but not the cupric ion-ascorbate system. It was concluded that the most effective inactivation process initiated by the cupric ion-ascorbate system was mediated by site-specific hydroxyl radicals.

It has been found[263] that a radiolytically generated superoxide ion inactivated penicillinase in the presence of copper ions. Since formate, a scavenger of hydroxyl radicals, did not prevent inactivation, it was proposed that the inactivation process was mediated by hydroxyl radicals formed during the site-specific interaction of the enzyme-bound cupric ions with hydrogen peroxide:

$$\text{Enzyme—Cu}^{2+} + O_2^{\overline{\cdot}} \rightarrow \text{Enzyme—Cu}^{+} + O_2 \quad (67)$$

$$\text{Enzyme—Cu}^{+} + H_2O_2 \rightarrow \text{Enzyme—Cu}^{+} + HO\cdot + HO^{-} \quad (68)$$

Similarly, the site-specific hydroxyl radicals are thought to mediate the inactivation of free and membrane-bound acetylcholinesterase by ascorbic acid and the superoxide ion in the presence of copper ions.[264]

Armstrong and Buchanon[265] have studied the mechanism of inactivation of the sulfhydryl enzymes papain, GAPDH, lactate dehydrogenase, and yeast alcohol dehydrogenase by radiolytically generated oxygen radicals. It was concluded that the activity of $O_2^{\overline{\cdot}}$ was the same as or even exceeded that of HO·. Glutathione, penicillamine, cysteine, and dithiothreitol protected against the inactivating effect of hydroxyl radicals, whereas SOD was the most effective inhibitor of the inactivating action of the superoxide ion.

S-Thiolation, i.e., the formation of a mixed disulfide with glutathione, is apparently an origin of inactivation of ornithine decarboxylase, an enzyme, participating in polyamine biosynthesis. It has been shown[266] that oxygen radicals generated by xanthine oxidase decreased ornithine decarboxylase activity in the cardiac supernatants from saline- or isoproterenol-treated rats. A parallel decrease in the sulfhydryl group content in the supernatants was also observed. Since SOD prevented inactivation and diethyldithiocarbamate, an inhibitor of SOD, enhanced the suppression of ornithine decarboxylase activity in both the *in vitro* and *in vivo* experiments, the involvement of the superoxide ion in the inactivation process was suggested.

Dean et al.[267] have shown that monoamine oxidase of the outer mitochondrial membrane was damaged by radiolytically generated oxygen radicals. The most active radicals were hydroxyl radicals; perhydroxyl radicals (at pH 4) manifested damaging activity close to that of HO·, while the superoxide ion was completely inactive.

E. PROTEINS WITHOUT ENZYMATIC ACTIVITY

In contrast to enzymes which, as we saw, can be activated or inactivated by oxygen radicals, the action of oxygen radicals on proteins without enzymatic activity leads to their damage. An exclusion is the interaction of the superoxide ion with iron-transport proteins (Chapter 2), triggering the release and transport of iron. Indeed, it was recently shown[268] that the superoxide ion produced by xanthine oxidase stimulates the transfer of iron ions from ferritin to apotransferrin and apolactoferrin. This process is thought to be of physiological significance. The superoxide ion apparently also mediates ascorbate-dependent iron release from ferritin because this process is inhibited by SOD.[269]

However, the damage of proteins by oxygen radicals should not be considered only as an origin of their uncontrollable destruction, since it is one of the most important pathways of the lysis of cellular proteins. Dean[270] has shown that limited oxygen radical attack upon the proteins induces amino acid modifications which lead to the formation of free amino groups at the sites of peptide bond cleavage. It was concluded[271] that the damage to even a small number of molecules of membrane proteins may stimulate a catastrophic disturbance in ionic homeostasis. Thus oxygen radicals may initiate an important process of catabolism of proteins.

Proteins (for example, histone, BSA, and collagen), as well as all biologically occurring amino acids, readily react with hydroxyl radicals.[272] Schuessler and Schilling[273] proposed that the fragmentation of BSA induced by hydroxyl radicals proceeds via the formation of peroxy radicals formed at the peptide α-carbon atoms. These data were confirmed by Wolff and Dean[274] and Dean et al.,[275] who observed the BSA fragmentation by hydroxyl radicals in the presence of dioxygen. However, in the absence of dioxygen, hydroxyl radicals caused the formation of high-molecular aggregates of BSA. The superoxide ion and perhydroxyl radical were unable to degrade BSA. However, contrary to this, it was recently concluded[276] that the superoxide ion mediated the alloxan-induced diminution of sulfhydryl groups in BSA. The Cu^{2+}-H_2O_2 system promoted BSA degradation similar to radiolytically generated hydroxyl radicals.[275]

It was proposed that the interaction of hydroxyl radicals with BSA leads to conformational changes which makes BSA more susceptible to proteolysis. In accord with this proposal, exposure of cultured cells to oxygen radicals produced radiolytically (or as a result of the prooxidant action of menadione or phenylhydrazine) stimulated intracellular proteolysis.[277]

Girotti et al.[278] have shown that in the presence of iron ions xanthine oxidase induced intermolecular cross-linking of polypeptides in the erythrocyte ghosts. There were two types of cross-linking: the formation of thiol-cleavable disulfide bonds, which did not require oxygen radicals, and nondisulfide cross-linking, which was observed after long periods of incubation with xanthine oxidase and was inhibited by SOD and catalase. The last process is believed to be mediated by hydroxyl radicals.

Davies[279] has studied the interaction of radiolytically generated HO· and a combination of HO· and $O_2^{\overline{\cdot}}$ with 17 proteins (BSA, lactate dehydrogenase, transferrin, SOD, catalase, cytochrome c, etc.). As in previous works,[274,275] hydroxyl radicals mainly induced protein aggregation. In contrast, the $O_2^{\overline{\cdot}}$ + HO· combination, which was less effective than HO·, stimulated the degradation of proteins. The most sensitive sites of BSA to oxygen radical attack were apparently tryptophan, tyrosine, histidine, and cysteine.[280] HO· and HO· + $O_2^{\overline{\cdot}}$ $(+O_2)$ distorted the secondary and tertiary structures of proteins, increasing their proteolytic susceptibility.[281]

The use of a combination of HO· and $O_2^{\overline{\cdot}}$ in the last works should be commented upon. These radicals react with each other with a diffusion-controlled rate constant ($10^{10}\ M^{-1}\ s^{-1}$, Volume I, Chapter 7).

$$O_2^{\overline{\cdot}} + HO\cdot \rightarrow O_2 + HO^- \tag{69}$$

Therefore, their combination will really be equal to a combination of HO· + O_2 (with excess HO·) or $O_2^{\overline{\cdot}}$ + O_2 (with excess $O_2^{\overline{\cdot}}$). Apparently, only the HO· + O_2 combination will be damaging to proteins.

Uchida and Kawakishi[282] have found that oxygen radicals generated by Cu^{2+} plus ascorbate preferably damaged the histidine residues in BSA. Using the model compound *N*-benzoylhistidine, these authors showed that oxygen radicals formed in this system (apparently hydroxyl radicals) selectively oxidized the imidazole ring of histidine residues.

Studying the action of xanthine oxidase or ascorbate plus iron ions on erythrocytes, Davies and Goldberg[283] have shown that oxygen radicals generated by these systems stimulated extensive protein degradation and hemoglobin oxidation. The most striking result was the fact that increased proteolysis measured by the production of free alanine was observed 5 min after the incubation of erythrocytes with xanthine oxidase, when no malondialdehyde could be detected. Thus proteins are degraded by oxygen radicals much more rapidly than lipids. It is also interesting that lipid peroxidation was inhibited by free radical scavengers (urate, glucose, or BHT), whereas protein degradation was not. However, mannitol inhibited the degradation of proteins in cell-free extracts.[284] Therefore, hydroxyl radicals or similar oxygen species apparently mediate protein damage.

Under anaerobic conditions, a composition of ascorbate and ferric or cupric ions promoted the oxidation of bovine lens crystallins.[285] It was suggested that this reaction which was accompanied by the cleavage of crystallins and the formation of nondisulfide cross-linkings was mediated by hydroxyl radicals formed by the Fenton reaction.

Griffiths et al.[286] have shown that hydroxyl radicals induced the aggregation of immunoglobulin G (IgG), whereas the superoxide ion was inert in this process. In subsequent work[287] it was found that hydroxyl radicals degraded the carbohydrate moiety of IgG; it is thought that this reaction may be an origin of agalactosylation of IgG, taking place in chronic inflammation.

The Cu^{2+}-ascorbate system promotes the denaturation of insulin.[288] The reaction was inhibited by catalase and hydroxyl radical scavengers, but not by SOD. On these grounds, it was proposed that site-specific hydroxyl radicals cleaved the disulfide bonds of the protein.

Greenwald and May[289] have shown that the superoxide ion generated by xanthine oxidase inhibited collagen gelation. These results were confirmed by Monboisse et al.,[290] who found that the xanthine-xanthine oxidase system promoted the cleavage of microfibrillar acid-soluble collagen into peptides of relatively small size. SOD, but not catalase, inhibited the collagen cleavage. It was also shown[291] that the superoxide ion reacted with collagen fibrils, releasing 4-hydroxyproline-containing peptides. Modified collagen became more susceptible to enzymatic proteolysis with trypsin. A rate constant for the reaction of a radiolytically generated superoxide ion with collagen was equal to $4.8 \cdot 10^6\ M^{-1}$.

The interaction of oxygen radicals with low-density lipoproteins (LDL) is of great interest in connection with a hypothesis that the modification of LDL by oxygen radicals stimulates their rapid uptake by macrophages. The last process initiates the transformation of macrophages into the lipid-laden foam cells which are found in atherosclerotic lesions. In addition, the LDL modification by free radicals converts LDL to a cytotoxin, participating in various pathological processes including diabetic tissue damage. It has been shown[292-294] that LDL modification may be promoted by oxygen radicals produced by phagocytes, including monocytes and neutrophils.

Steinbrocher[295] has shown that the LDL modification by endothelial cells, smooth muscle cells, and fibroblasts required dioxygen and was inhibited by SOD, which indicates the participation of superoxide ion in this process. On the other hand, catalase and hydroxyl radical scavengers had only a slight inhibitory effect. It was proposed that the superoxide ion was an indirect mediator of LDL modification by endothelial cells and that LDL were probably damaged by hydroxyl radicals formed by a site-specific mechanism.

LDL modification by human monocytes was also inhibited by SOD.[293] However, other than the superoxide ion free radicals, lipid peroxy radicals probably took part in the propagation of LDL modification, as the scavenger of neutral free radicals, BHT, was an effective inhibitor.

Heinecke et al.[296] have shown that monkey and human arterial smooth muscle cells induced the modification of LDL in the presence of iron or copper ions. These data were confirmed in subsequent work.[297] SOD, BHT, and desferrioxamine inhibited LDL modification, while catalase and mannitol did not. On these grounds, the importance of superoxide-dependent lipid peroxidation in the LDL modification by arterial smooth muscle cells was proposed. Hydroxyl radicals and hydrogen peroxide, it is thought, did not participate in this process.

Thornalley and Vasak[298] have studied the interaction of the superoxide ion and hydroxyl radicals produced by xanthine oxidase with rabbit liver metallothioneins (MT, low molecular weight metal- and sulfur-containing proteins). Although the physiological role of these proteins is not well understood, they apparently serve as cellular stores of zinc and copper. It was found that zinc- and cadmium-containing MT were exclusively reactive with hydroxyl radicals (rate constants of these reactions were about $10^{12}\ M^{-1}\ s^{-1}$, that is, more than the diffusion limit of about $10^{11}\ M^{-1}\ s^{-1}$), while the rate constants for their reactions with the superoxide ion were about $5 \cdot 10^5\ M^{-1}\ s^{-1}$. It was proposed that MT may defend cells against the hydroxyl radical attack.

V. INTERACTION OF OXYGEN RADICALS WITH DNA AND OXYGEN RADICAL-MEDIATED MUTAGENESIS

DNA is probably the most important potential target which is met by oxygen radicals entering or generated in the cell. DNA damage by oxygen radicals may be a cause of cell death or mutations. Because of this, in addition to the interaction of oxygen radicals with DNA, their ability to promote mutagenesis was extensively studied. A relatively long-lived superoxide ion is able to traverse the cytosol and directly interact with DNA, although it is questionable that the superoxide ion is an efficient DNA-damaging agent. On the other hand, very reactive hydroxyl radicals may be really dangerous only if they are generated close to a target DNA molecule by a site-specific mechanism. Other possible damaging agents are the precursors of free hydroxyl radicals in the Fenton reaction, the "crypto-hydroxyl" radicals which are expected to be sufficiently long-lived to achieve DNA and sufficiently active to damage DNA directly or after decomposing to free hydroxyl radicals.

A. DNA DAMAGE BY ENZYMATIC AND NONENZYMATIC GENERATORS OF THE SUPEROXIDE ION

There are numerous works concerning the role and structure of oxygen radicals in DNA damage and mutagenesis. Recently, two reviews on this subject were published.[299,300] At the beginning, the production of oxygen radicals enzymatically (by xanthine oxidase) and radiolytically was widely used. In 1975, Van Hemmen and Meuling[301] assumed that the superoxide ion or perhydroxyl radical produced by gamma ray irradiation inactivated DNA from bacteriophage F X174 in the presence of hydrogen peroxide because the DNA inactivation was prevented by SOD. However, in 1978, it was shown[302] that a radiolytically generated superoxide ion was unable to degrade the DNA component thymine, while oxygen radicals produced by autoxidation of 6-hydroxydopamine (apparently including other oxygen species) destroyed this compound.

The nature of active oxygen species participating in DNA damage and mediating mutagenesis was widely discussed in subsequent works. It has been shown[303,304] that oxygen radicals generated by xanthine oxidase or during the dissolution of potassium superoxide induced strand scission in isolated DNA (Col E1 DNA and bacteriophage DNA). DNA damage was inhibited by SOD, catalase, hydroxyl radical scavengers (benzoate, mannitol, ethanol, DMSO, and histidine), and the chelating agent DTPA. These findings possibly indicate the mediation of DNA scission by hydroxyl radicals produced by the superoxide-driven Fenton reaction. Conflicting results were obtained with EDTA. Lesko et al.[303] found that EDTA completely inhibited the strand scission of T7 bacteriophage DNA by $O_2^{\bar{\cdot}}$ plus H_2O_2, whereas Brawn and Fridovich[304] showed the accelerating effect of EDTA on the scission of Col E1 DNA. Cunningham and Lokesh[305] and Cunningham et al.[306] have shown that the superoxide ion generated by the dissolution of KO_2 produced an increase in the amount of thioguanine-resistant mutants of CHO cells and induced DNA single-strand breaks in CHO and human teratocarcinoma cells. Catalase and DTPA (partially) reduced the number of breaks.

Recently, Ito et al.[307] analyzed the difference in the damaging effects of KO_2 and xanthine oxidase on the transforming activity of *Bacillus subtilis* DNA. They showed that, whereas SOD enhanced DNA inactivation by KO_2, it completely inhibited DNA inactivation induced by xanthine oxidase. Diverse effects on DNA inactivation by these oxygen radical-generating systems were also observed when mannitol (a hydroxyl radical scavenger), copper and iron ions, and chelating agents were applied. It was concluded that in both cases the DNA damage is mediated by hydrogen peroxide formed as a result of superoxide dismuting. Conflicting results obtained in studying the action of xanthine oxidase and KO_2 on DNA are possibly explained by the difference in H_2O_2 concentrations formed during KO_2 dissolution and the dismutation of $O_2^{\overline{\cdot}}$ generated by xanthine oxidase. In both cases, a site-specific attack of hydroxyl radicals is apparently a source of DNA inactivation, but the KO_2 system seems to give more complicating results, and therefore the use of xanthine oxidase as a generator of oxygen radicals is preferable.

Emerit et al.[308] have shown that the superoxide ion generated by xanthine oxidase induced an increase in chromosome breaks and sister-chromatid exchanges (SCE) in human lymphocytes. SOD almost entirely prevented the damaging action of xanthine oxidase. Since extracellularly produced $O_2^{\overline{\cdot}}$ apparently cannot reach the nucleous without scavenging by intracellular cytosolic SOD, it was proposed that the mutagenic effect was manifested by its decomposition products, first of all, hydrogen peroxide. The same conclusion was achieved by studying the promotion by xanthine oxidase of chromosomal aberrations in cultured mammalian cells[309] because SOD did not inhibit the chromosomal aberrations, but catalase did. It was also proposed[310] that the products of the reaction of oxygen radicals with arachidonic acid (hydroperoxides and their decomposition products) can be a clastogenic factor stimulating DNA damage.

The formation of hydroxyl radicals in the reaction of xanthine oxidase with DNA was confirmed in experiments with plasmid and chromosomal DNA.[311] It was found that xanthine oxidase was able to diminish the capacity of DNA to transform bacteria only in the presence of ferric citrate, which apparently catalyzed the Fenton reaction. SOD, catalase, mannitol, and thiourea protected DNA against the toxic action of xanthine oxidase in the presence of Fe(II)citrate.

In contrast to extracellular generation, the superoxide ion is thought to be able to damage DNA by itself when it is generated intracellularly. We already discussed in Chapter 2 the ability of paraquat to catalyze the one-electron reduction of dioxygen into the superoxide ion in cells. Moody and Hassan[312] have shown that paraquat is highly mutagenic to *Salmonella typhimurium*. They concluded that the mutagenic effect of paraquat is mediated by the superoxide ion, because it was absent under anaerobic conditions. In addition, the cells with the enhanced SOD level were less susceptible to paraquat mutagenicity. The superoxide-mediated mechanism of paraquat mutagenicity was confirmed in subsequent works.[313,314] It was also found that the superoxide ion induces the SOS response to paraquat (an inducible DNA repair system).

Recently, Sofuni and Ishidate[315] and Sawada et al.[316] showed that although SOD and catalase did not prevent DNA damage in cultured Chinese hamster cells promoted by paraquat, the SOD inhibitor sodium diethyldithiocarbamate sharply increased chromatid breaks and exchanges induced by paraquat. The glutathione scavenger diethyl maleate also enhanced the paraquat induction of chromosomal aberrations. Therefore, it was concluded that endogenous SOD and probably glutathione peroxidase protect DNA against the damaging effect of the superoxide ion. At the same time, 3-aminotriazole, a catalase inhibitor, did not affect the paraquat-induced chromosomal aberrations, i.e., hydrogen peroxide apparently does not participate in DNA damage under the conditions studied. It was concluded that there is a difference in the nature of DNA damage induced intracellularly by paraquat and extracellularly by xanthine oxidase. The superoxide ion produced by paraquat induced chromatid gaps and breaks, whereas oxygen species produced by xanthine oxidase induced predominantly exchange-type aberrations.

Contrary to this, Phillips et al.[317] concluded that oxygen radicals generated by the xanthine-xanthine oxidase system promoted extensive chromosome breakage, SCE, and a definite increase in cell mutations (shown as an elevation of the frequency of thioguanine-resistant cells in the population) in Chinese hamster ovary (CHO) cells. SOD and catalase inhibited chromosome breakage, but only catalase prevented SCE and cell mutations. An important role of hydrogen peroxide in the stimulation of DNA damage by xanthine oxidase was stressed. This conclusion was recently confirmed with the aid of a modified fluorometric technique[318] which permits detection of the destruction of DNA immediately after incubation with xanthine oxidase, and not 24 or more hours later as in previous studies. Measurement of DNA damage was based on the ability of ethidium bromide to bind selectively to double-stranded DNA. The fluorometric technique also makes possible detection of the recovery of double-stranded DNA after exposure of CHO cells to oxygen radicals. It was confirmed that hydrogen peroxide and evidently hydroxyl radicals mediated the DNA damage because mannitol, dimethylurea, and catalase protected the cells.

The hypoxanthine-xanthine oxidase system or hydrogen peroxide also induced SCEs in Chinese hamster fibroblasts.[319] The iron-chelating agent *o*-phenanthroline inhibited SCE production, indicating that hydroxyl radicals formed by the Fenton reaction mediated the DNA damage. When menadione was used for the intracellular generation of the superoxide ion, the formation of SCE was prevented by the SOD mimetic agent copper diisopropyl-salicylate, which points out the occurrence of the superoxide-driven Fenton reaction.

Deng and Fridovich[320] have shown that oxygen radicals generated by xanthine oxidase also stimulated the formation of the endonuclease III-sensitive sites in DNA which are probably the residues of thymine glycol. SOD, catalase, hydroxyl radical scavengers, and chelating agents prevented this type of DNA damage. It was proposed that hydroxyl radicals formed in the Fenton reaction oxidized thymine residues into thymine glycol.

The DNA damage by enzymatically produced oxygen radicals may be prevented by uric acid, since it was shown in the experiments with rat liver nuclei.[321] Here, uric acid inhibited the DNA single-strand breaks induced by xanthine oxidase in the presence of ferrous ions and by hematin plus hydroperoxides. Surprisingly, uric acid did not inhibit the DNA damage promoted by the Fenton Reaction.

Recently, DeFlora et al.[322] have shown that the hypoxanthine-xanthine oxidase system was mutagenic to *S. typhimurium* bacteria. It is interesting that catalase inhibited the mutagenic action of this system only in the presence of SOD (which transformed $O_2^{\overline{\cdot}}$ into H_2O_2). On these grounds, it was concluded that both the superoxide ion and hydrogen peroxide manifested a mutagenic effect. Glutathione, *N*-acetylcysteine, and α-mercaptopropionylglycine protected against the mutagenic effect of xanthine oxidase on *S. typhimurium* cells.

B. DNA DAMAGE BY OXYGEN RADICALS FORMED IN THE FENTON REACTION

There are abundant data on the DNA damage by the $H_2O_2 + Cu^+$ system (see Reference 323 and references in it), which is apparently mediated by hydroxyl radicals formed by Reaction 70:

$$H_2O_2 + Cu^+ \rightarrow HO\cdot + HO^- + Cu^{2+} \tag{70}$$

Various copper complexes were used as catalysts for this reaction. For example, it has been shown[324] that the copper-1,10-phenanthroline complex induced the degradation of DNA in the presence of reducing agents and dioxygen. The reducing agents were ascorbate, glutathione or other thiols, and the superoxide ion. Hydroxyl radical scavengers (benzoate and acetate), catalase, and chelators (EDTA) inhibited DNA degradation. It is of interest that 1,10-phenanthroline inhibited the formation of DNA single-strand breaks produced by hydrogen peroxide.[325] Therefore, it was proposed that in contrast to copper ions, iron ions form an inactive complex with 1,10-phenanthroline which is unable to catalyze the DNA damage by hydrogen peroxide.

The reduction of Cu^{2+} ions is probably an origin of mutagenic action of ascorbic acid.[326] The ascorbate-Cu^{2+} system is also able to release bases (cytosine, thymine, guanine, and adenine) from DNA.[327] This process proceeds only under aerobic conditions and is inhibited by hydroxyl radical scavengers and catalase, which confirms the participation of hydroxyl radicals. DNA degradation was promoted by a mixture of 4′-(9-acridinylamino)methanesulfon-*m*-anisidide (mAMSA, a reductant) and cupric ions.[328] Anaerobic conditions, catalase, and 4,5-dihydroxy-1,2-benzenedisulfonate (tiron) inhibited DNA degradation, indicating the participation of both the superoxide ion and hydrogen peroxide. Similarly, aminosugars can be used as reductants during DNA degradation. Thus it was found[329] that D-glucosamine induced the DNA strand scission in the presence of copper ions. Since catalase, SOD (in part), tiron, and DTPA inhibited DNA degradation, it was proposed that the superoxide ion, hydrogen peroxide, and hydroxyl radicals mediated this process.

The above findings suggest an important role of free hydroxyl radicals in DNA damage induced by the Fenton-like systems. Recently, Schneider et al.[330] also concluded that the DNA damage induced by the ascorbate-iron ions system in aerobic conditions was mediated by free hydroxyl radicals as it was prevented by mannitol, glycerol, and desferrioxamine. However, there are many data which show that hydroxyl radicals formed by the Fenton-like reactions are not "free" and that their interaction with DNA is realized via a site-specific mechanism.[323]

A similar conclusion was achieved, for example, by Gutteridge and Halliwell[331] who found that the Cu^{2+}-phenanthroline complex in the presence of reductants induced the formation of DNA degradation products originated from the deoxyribose moiety which may be detected and measured as TBA-reactive products. Cu^{2+} ions were reduced by an enzymatically generated superoxide ion, mercaptoethanol, or NADH. A weak inhibitory effect of hydroxyl radical scavengers was explained by a site-specific mechanism of the formation of hydroxyl radicals.

Sagripanti and Kraemer[332] also concluded that the stimulation of strand breaks and the inactivation of the transforming ability in DNA by the Cu^{2+}-H_2O_2 system were mediated by site-specific hydroxyl radicals because their damaging effect was inhibited by metal chelators, catalase, and only by high concentrations of hydroxyl radical scavengers. It was also found that free radical attack was preferentially directed on polyguanosine sequences. Aronovitch et al.[333] have shown that DNA degradation by the ternary complex Cu(II)(phenenthroline)$_2$DNA is a more efficient process than DNA degradation by Cu(II)(phenanthroline)$_2$. These results confirm the site-specific mechanism of DNA destruction.

C. DNA DAMAGE BY OXYGEN RADICALS RELEASED BY PHAGOCYTES

We saw in Chapter 3 that phagocytes are powerful generators of oxygen radicals. Therefore, one may expect that, upon stimulation, they will promote DNA destruction and manifest a mutagenic effect in cells. Indeed, in 1981 Weitzman and Stossel[334] have shown that human leukocytes were mutagenic to *S. typhimurium* bacteria. Since leukocytes (which are unable to produce oxygen radicals) from a patient with chronic granulomatous disease (CGD) were much less effective, it was concluded that oxygen radicals are responsible for the mutagenic effect of leukocytes. In subsequent work[335] these authors have shown that the mutagenic action of human leukocytes on bacteria was inhibited by SOD, catalase, α-tocopherol, cysteine, and acetylcysteine, but was enhanced by ascorbate. Analogous results were obtained by Weitberg et al.[336]

The mutagenic effect of stimulated phagocytes was later confirmed by many workers. Weitberg and Weitzman[337] have shown that the superoxide ion produced by PMA-stimulated human leukocytes enhanced the number of SCE in CHO cells. The mutagenic effect of oxygen radicals on CHO cells and dermal fibroblasts was also shown[338] when PMA-stimulated mouse and rat neutrophils were applied. The appearance of chromosome abnormalities was inhibited by catalase.

Birnboim and Kanabus-Kaminska[339] and Birnboim[340] have studied the production of

DNA strand breaks in human leukocytes by oxygen radicals formed as a result of the stimulation of these phagocytes with PMA and TPA. They attempted to discriminate among the DNA damage caused by the superoxide ion, the hydroxyl radicals, and hydrogen peroxide. It was found that strand breaks induced by leukocytes are partially inhibited by SOD and partially by catalase. On these grounds, it was concluded that both H_2O_2 and $O_2^{\cdot-}$ promote DNA damage. However, while SOD afforded substantial protection of cellular DNA in stimulated leukocytes, there was no protection when the superoxide ion was generated by xanthine oxidase. By contrast, in the last case, catalase afforded complete protection.

In addition to the different effects of catalase and SOD, two other criteria of the superoxide- and hydroxyl radical-mediated DNA strand breaks were developed. It is known that the radiation-induced strand breaks which are mostly mediated by hydroxyl radicals are efficiently repaired in leukocytes. However, the PMA- and TPA-induced strand breaks in leukocytes are repaired more slowly. Thus while 55 to 60% of the radiation-induced strand breaks in leukocytes were repaired during 30 to 40 min, the number of breaks induced by TPA reached a plateau level after 40 min and remained unchanged for 1 h.[340] The second criterion is the response of $O_2^{\cdot-}$ - and HO·-induced strand breaks to some metabolic inhibitors. For example, fluoride and 2-deoxyglucose blocked the DNA damage induced by the superoxide ion and had little effect on that induced by hydrogen peroxide. All these data seem to be good evidence that the superoxide ion and hydrogen peroxide (and hydroxyl radicals formed from H_2O_2) may separately induce DNA strand breaks.

It is of interest that the superoxide ion produced by leukocytes is probably a far more effective destructive agent than $O_2^{\cdot-}$ generated enzymatically. Birnboim and Kanabus-Kaminska[339] proposed that when xanthine oxidase is applied, the superoxide ion is formed at some distance of a cell and dismutes before reaching the target. Because of this, hydrogen peroxide becomes the only DNA-damaging agent. In the case of leukocytes, the superoxide ion may survive transportation to a target cell due to adherence between leukocytes.

Recently, Birnboim[341] found that ionophore A 23187, fluoride, and 2-deoxyglucose inhibited the strand-break process in DNA induced by PMA and particular stimuli in human leukocytes supposedly by depleting intracellular calcium. On these grounds, it was concluded that the superoxide ion promotes strand breaks not as a result of direct interaction with DNA, but rather by activating a specific metabolic DNA strand-break pathway.

In addition to the antioxidative enzymes SOD and catalase, low-molecular free radical scavengers such as ascorbic acid[337,342,343] and the flavonoid rutin[343] are able to suppress the mutagenic effect of oxygen radicals. Weitberg and Weitzman[337] have found that ascorbic acid showed a double effect on SCE in DNA of CHO cells promoted by enzymatically generated oxygen radicals: at low ascorbate concentrations (<100 μM), a significant protective effect was observed, whereas at higher concentrations ascorbic acid enhanced the SCE number. Surprisingly, ascorbate did not affect the SCE number induced by PMA-stimulated human phagocytes. In contrast, ascorbic acid did affect the DNA damage in human lymphocytes induced by zeolite-stimulated rat macrophages.[343] It is interesting that the strongest inhibitory effect was observed at 100 μM ascorbate, while both an increase and a decrease in ascorbate concentrations diminished its protective effect. If the inhibitory effect of ascorbate is easily explained by scavenging oxygen radicals, its mutagenic effect is apparently a consequence of the reduction of transition metal ions (see above).

D. OXYGEN RADICALS AS MEDIATORS OF THE MUTAGENIC EFFECT OF XENOBIOTICS

Comparing the interaction of DNA with the superoxide ion generated extra- and intracellularly (Section V.A), we already discussed the mutagenic effect of paraquat. However, there are many other xenobiotics whose mutagenic effect is supposedly mediated by oxygen radicals. In 1980, Yamaguchi[344] showed that many potential generators of the superoxide ion (ascorbic acid, epinephrine, menadione, paraquat, etc.) were mutagenic in the Ames test. Mutagenicity of these compounds, as a rule, increased in the presence of Fe(EDTA). SOD inhibited the mutagenic effect of epinephrine, menadione, and methylene blue, but hydroxyl radical scavengers (ethanol, mannitol, and formate) were ineffective. Sakurai et al.[345] have shown that hemin-thiolate complexes, including the hemin-glutathione complex, induced the strand scission of supercoiled DNA. The authors proposed that this process was mediated by oxygen radicals. Surprisingly, hydroxyl radical scavengers (DMSO, mannitol, formate, ethanol, and benzoate) did not prevent the destruction of DNA, while SOD and catalase even increased DNA damage. Therefore, the true role of oxygen radicals in this process remains unclear.

It is known that benzene is metabolized in the liver into phenol, hydroquinone, 1,2,4-trihydroxybenzene, and some other toxic compounds which, it is believed, may be a source of the carcinogenic action of benzene. Indeed, it was found[346] that 1,2,4-trihydroxybenzene and hydroquinone reduced dioxygen to the superoxide ion and induced degradation and single- and double-strand breaks in supercoiled DNA. As SOD, catalase, and benzoate were the inhibitors of DNA breakage, this process was apparently mediated by oxygen radicals.

The anion of 2-nitropropane is mutagenic to some *Salmonella typhimurium* strains.[347] An inhibitory effect of DMSO possibly indicates the participation of hydroxyl radicals. Lesko et al.[348] proposed that the DNA damage induced by benzo[a]pyrene-3,6-dione is due to the generation of a superoxide ion during the one-electron oxidation of its hydroquinone or semiquinone because the mutagenic and cytotoxic activity of benzo[a]pyrene-3,6-dione increased with a decrease in the capacity of Syrian hamster fibroblasts to scavenge the superoxide ion.

Cysteamine manifests a double effect on the DNA damage in leukocytes: it enhances the number of strand breaks at 1 to 2 m*M* and inhibits the DNA damage at higher concentrations (about 10 m*M*).[349] The cysteamine-mediated DNA damage was inhibited by catalase and DTPA. It may be suggested that the damaging effect of low concentrations of cysteamine is due to the production of the superoxide ion and hydrogen peroxide during its autoxidation, while at higher concentrations cysteamine largely reacts as a free radical scavenger.

A very strange mutagenic effect was found in the case of the flavonoid quercetin. Quercetin is an antioxidant and a scavenger of oxygen radicals, so its mutagenic effect demonstrated by the Ames test with *S. typhimurium*[350] is unexpected. Surprisingly, the quercetin mutagenicity was enhanced by SOD and diminished by xanthine oxidase. Therefore, further studies are needed to understand the role of oxygen radicals in the quercetin mutagenicity.

Lick et al.[351] described the destruction of RNA in *Saccharomyces cerevisiae* by oxygen radicals generated by the oxidation of photochemically reduced riboflavin and by the hydrogen peroxide-peroxidase reaction. SOD and DMSO inhibited RNA decomposition.

E. FORMATION OF 8-HYDROXYDEOXYGUANOSINE FROM DNA MEDIATED BY OXYGEN RADICALS

The above results were obtained using mainly the experimental technique based on the detection of single-strand breaks, double-strand breaks, and SCE after exposure of DNA to oxygen radical attack. However, another experimental technique for studying the interaction of oxygen radicals with DNA was recently developed by Kasai and Nishimura.[352] These authors showed that hydroxyl radicals formed by various oxygen radical-generating systems react with the C-8 position of deoxyguanosine residues in DNA to produce 8-hydroxydeoxyguanosine (8-HOdG), which can be isolated after the DNA-enzymatic cleavage to deoxynucleosides. For example, ascorbic acid under aerobic conditions and hydroquinones in the presence of hydrogen peroxide and iron ions promote the formation of 8-HOdG from native calf thymus DNA.[353]

A great improvement in the experimental technique was achieved by using high-performance liquid chromatography with electrochemical detection.[354] By this method, it was shown that the stimulation of human granulocytes with TPA leads to the appearance in DNA of high levels of 8-HOdG.[354] SOD and catalase inhibited this process. It is of interest that SOD decreased the amount of 8-HOdG by 63% which is practically equal to the 61% decrease in the TPA-induced DNA strand breaks in human leukocytes observed in the presence of this enzyme.[339]

It has been shown[355] that 8-HOdG is formed in DNA isolated from HeLa cells and the livers of mice after irradiation of cells and whole bodies of mice with X-rays and gamma rays, respectively. It was proposed that this process is mediated by oxygen (hydroxyl) radicals. Richter et al.[356] have studied the effect of prooxidants (alloxan, ferric ions, etc.) and irradiation on the formation of 8-HOdG from mitochondrial and nuclear rat liver DNA. It was concluded that the 8-HOdG formation corresponds only to 5% of the total DNA damage. It was also found that the 8-HOdG level formed as a result of the oxygen radical attack on mitochondrial DNA is about one order of magnitude greater than that in nuclear DNA. Enhanced oxidation level in mitochondrial DNA is probably a consequence of a continuing flux of oxygen radicals in mitochondria due to its respiratory activity. Recently, Floyd et al.[357] confirmed that hydroxyl radicals generated by irradiation of hydrogen peroxide with UV light stimulated the formation of 8-HOdG within calf thymus DNA.

VI. ACTION OF OXYGEN RADICALS ON CELLULAR COMPONENTS

Many questions relevant to this section have already been discussed in preceding sections because lipid peroxidation, the destruction of proteins and DNA, the activation and inactivation of enzymes, etc., are various modes of the action of oxygen radicals on mitochondria, microsomes, nuclei, and other cellular components. Nonetheless, important information can also be obtained by studying the interaction of oxygen radicals with these components as a whole.

One may expect that the primary attack of oxygen radicals must be directed on the lipid membranes. Such an attack may induce morphological changes and an increase in the permeability of membranes. Indeed, it was shown as early as 1977[358] that oxygen radicals generated by xanthine oxidase enhance the release of chromate from multilamellar liposomes. Both SOD and catalase reduced the permeability of liposomes promoted by oxygen radicals. The possible origin of perturbing lipid bilayers by oxygen radicals is lipid peroxidation, but other modes of radical attack on the lipid membrane cannot also be excluded. Later on, Grankvist and Marklund[359] showed that oxygen radicals generated by xanthine oxidase increased the permeability of the plasma membranes of isolated islet cells to trypan blue. A combination of SOD and catalase significantly decreased a plasma membrane leakage to trypan blue, while hydroxyl radical scavengers (mannitol and benzoate) were ineffective. Thus the superoxide ion must play an important role in the increase of the permeability of islet cells.

Oxygen radicals may affect various functions and structures of mitochondria. It was proposed[360] that the Fe^{2+}-ascorbate system is able to impair mitochondrial energy-linked functions inducing lipid peroxidation. Ninnemann et al.[361] concluded that the superoxide ion is able to mediate electron transfer from flavoprotein to cytochrome b, because SOD inhibited the flavin-stimulated photoreduction of cytochrome b in beef heart mitochondria. Flamigni et al.[362] have found that oxygen radicals produced by xanthine oxidase decreased oxidative phosphorylation in both tocopherol-deficient and normal rabbit cardiac mitochondria. The addition of α-tocopherol improved the respiratory and phosphorylating activities of mitochondria.

In contrast to an earlier proposal,[360] Wiswedel et al.[363] recently showed that a decrease in active respiration in isolated rat liver mitochondria induced by Fe^{2+} plus ascorbate takes place before the start of lipid peroxidation. It was also found that there was no membrane alteration such as a proton or other ion leakage nor was there a change in a normal-level membrane potential at times when respiration was already substantially decreased. The breakdown of membrane potential was probably an ultimate step which was followed by intensive lipid peroxidation.

Roh et al.[364] have found that mitochondrial damage may be induced by oxygen radicals generated by the mitochondria themselves. Thus these authors observed lipid peroxidation and lysis of isocitrate dehydrogenase after the stimulation of superoxide production in rat brain mitochondria by antimycin. The reaction required iron ions, indicating the formation of hydroxyl radicals.

Oxygen radicals produced by xanthine oxidase strongly affected the pyruvate-dependent respiration in rat heart mitochondria.[365] This effect was partially prevented by SOD, catalase, or dithiothreitol. It was proposed that the superoxide ion is able to react directly with the SH groups of pyruvate dehydrogenase or to modify this and other enzymes by changing the availability of mitochondrial calcium. Indeed, it was earlier shown[366] that oxygen radicals produced by xanthine oxidase sharply diminished Ca^{2+} uptake by heart and liver mitochondria. Reduced ubiquinone, butylated hydroxytoluene, and mannitol prevented this effect of oxygen radicals. These authors proposed that oxygen radicals reduced the ability of mitochondria to take up calcium and increased the efflux of previously accumulated calcium due to the interaction with the sulfhydryl groups.

Mirabelli et al.[367] have shown that the xanthine-xanthine oxidase system promoted the oxidation of pyridine nucleotides, the collapse of transmembrane potential, the release of calcium, and large amplitude swelling in rat liver mitochondria. Since SOD accelerated these processes, one may propose that hydroxyl radicals forming from hydrogen peroxide, a product of superoxide dismutation, mediated mitochondrial damage. It is thought that the oxidation of pyridine nucleotides stimulated an increase in the rate of calcium cycling. In accord with this, it has been shown[368] that oxygen radicals produced by xanthine oxidase stimulated a rapid oxidation of NADPH followed by Ca^{2+} release in isolated renal mitochondria preloaded with calcium. A consequence of calcium efflux was an increase in state 4 respiration. Catalase, but not SOD, prevented the calcium efflux.

It should be noted that a change in the calcium level can influence the enzymatic (NADPH- and NADH-dependent) and Fe^{2+}-ascorbate-dependent lipid peroxidation of mitochondrial and microsomal fractions, egg lecithin, and linolenic acid.[369-371] Because of this, the oxygen radical-dependent calcium efflux can enhance their damaging effect. Indeed, it was recently shown[372] that xanthine oxidase alone did not affect the respiratory function of heart mitochondria from guinea pigs, whereas a combined action of oxygen radicals and calcium significantly damaged states 3 and 4 of mitochondria. It was proposed that oxygen radicals and Ca^{2+} ions inactivated cytochrome c or its reductase. Verapamil, a calcium channel blocker, and SOD preserved state 4, whereas phospholipase A_2 inhibitors and membrane stabilizers had no effect on state 3 and state 4.

It has been shown[373,374] that oxygen radicals decreased the gama-aminobutyric acid (GABA) uptake by brain synaptosomes. Braughler et al.[373] concluded that calcium and oxygen radicals generated by xanthine oxidase damaged synaptosomes synergistically under conditions where Ca^{2+} or the superoxide ion alone had no effect. It was proposed that the mechanisms of the synergistic effect of calcium relates to activation of Ca^{2+}-dependent phospholipases. Debler et al.[374] found that xanthine oxidase promoted the disruption of both nerve terminal plasma membranes and synaptic vesicle membranes. It is interesting that only SOD and a combination of SOD plus catalase protected synaptosomes against the action of xanthine oxidase, whereas catalase alone and mannitol had no effect on GABA uptake. These findings stress the important role of the superoxide ion in the oxygen radical damage of synaptosomes.

Okabe et al.[375] have shown that oxygen radicals produced by xanthine oxidase depressed Ca^{2+} uptake by canine cardiac sarcoplasmic reticulum vesicles. Since this process was inhibited by catalase, DMSO, and a combination of SOD and catalase, it was concluded that the inhibition of calcium uptake was mediated by hydroxyl radicals formed in the superoxide-driven Fenton reaction. It was also found that oxygen radicals affected only a calmodulin-dependent mechanism of sarcoplasmic reticulum calcium uptake supposedly via the regulation of a calcium release channel.

Kukrega et al.[95] and Das et al.[376] recently concluded that the depression of calcium uptake by sarcoplasmic reticulum mediated by hydroxyl radicals is a consequence of the inhibition of $Ca^{2+}Mg^{2+}$-ATPase activity and of an increase in membrane fluidity. However, the conclusion about the inhibition of Ca^{2+},Mg^{2+}-ATPase activity in sarcoplasmic reticulum by hydroxyl radicals is apparently erroneous and probably explained by the use of xanthine oxidase (as a source of oxygen radicals) contaminated by phospholypase A_2.[377] It was recently confirmed[378] that oxygen radicals directly modify the calmodulin-dependent component of Ca^{2+} fluxes in cardiac sarcoplasmic reticulum.

It has been shown[82] that oxygen radicals produced by microsomal NADPH-cytochrome P-450 reductase or xanthine oxidase in the presence of Fe^{3+}-ADP induced peroxidation and lysis of rat liver lysosomes. Both the release of acid phosphatase and MDA formation were enhanced in the presence of SOD and inhibited by hydroxyl radical scavengers. Therefore, it was concluded that hydroxyl radicals formed in the superoxide-driven Fenton reaction are main oxygen species promoting the lysis and peroxidation of lysosomes. The damaging effect of oxygen radicals on lysosomes was confirmed in subsequent works, although the stimulatory effect of SOD on the enzyme leakage did not. For example, the lysis of guinea pig liver lysosomes by oxygen radicals produced by xanthine oxidase or stimulated guinea pig PMNs was inhibited by SOD.[379]

The xanthine-xanthine oxidase system also induced the release of β-glucuronidase from the lysosome subcellular fraction prepared from rat liver.[380] SOD inhibited this process, indicating the mediation of lysosome damage by the superoxide ion. Similarly, these authors showed[381] that xanthine oxidase stimulated the release of hydroxylases from a lysosomal-enriched fraction from the homogenate of dog heart, which was also prevented by SOD. Olsson et al.[382] have found that SOD and catalase did not prevent enzyme leakage from rat liver lysosomes stimulated by xanthine oxidase. These authors proposed that this is a consequence of the presence in the lysosomes of cyanide-sensitive SOD.

VII. ACTION OF OXYGEN RADICALS ON CELLS

A. STIMULATORY AND REGULATORY EFFECTS OF OXYGEN RADICALS ON CELLS

At the beginning, the main attention of workers was given to studying the disrupting effects of oxygen radicals on cells. As we will see, there are plenty of examples of damage and killing of various cells by oxygen radicals. The most impressive example is possibly the mediation of cell destruction by oxygen radicals released from activated phagocytes. However, oxygen radicals may also stimulate various cellular processes and regulate some important physiological cellular functions.

We have seen that oxygen radicals induce mutagenic aberrations in lymphocytes (Section V). It has also been shown[383] that oxygen radicals produced by xanthine oxidase suppressed the transformation of lymphocytes to phytohemagglutinin, which was quantitated as the stimulation of DNA synthesis indicated by ^3H-thymidine uptake. The damaging effect of xanthine oxidase was inhibited by catalase, but not by SOD or mannitol. T-lymphocytes were especially susceptible to the attack of hydroxyl radicals forming during UV irradiation of human peripheral blood mononuclear cells.[384]

At the same time, it was found[385] that potassium superoxide induced high levels of uptake of tritiated thymidine by human lymphocytes. Lymphocytes were also stimulated by oxygen radicals released by zymosan-activated monocytes. SOD, but not catalase and mannitol, inhibited these processes. On these grounds, it was concluded that the superoxide ion is able to mediate the lymphocyte transformation. It was also found that the superoxide ion increased the adhesiveness of human peripheral blood mononuclear cells. Both phenomena, the lymphocyte transformation and the adhesiveness, are thought to be linked.

These contradictory findings are apparently explained by the ability of oxygen radicals to stimulate lymphocyte transformation or to damage lymphocytes, depending on the concentration level of radicals. Similar conclusions were achieved by studying the effects of oxygen radicals on other cells. Thus Murrell et al.[386] have shown that oxygen radicals produced by xanthine oxidase stimulated the proliferation of human fibroblasts at low concentrations and damaged them at high concentrations. Both the stimulatory and inhibitory effects of oxygen radicals were prevented by SOD and catalase.

Oxygen radicals generated by xanthine oxidase or released upon the activation of phagocytes with phorbol-12,13-dibutyrate or insulin in combination with hydrogen peroxide stimulated sugar transport (measured as the 2-deoxyglucose uptake) in mouse fibroblast cells.[387] The stimulation of sugar transport was inhibited by SOD + catalase and vitamin E. It was proposed that the superoxide ion stimulates sugar transport, increasing intracellular pH. On the other hand, the hypoxanthine-xanthine oxidase system promoted the damage in human fibroblasts.[388] The flavonoid silybin and butyl hydroxytoluene prevented the damage.

Oxygen radicals manifest a double effect on hepatocytes. Hussain et al.[389] have shown that the superoxide ion stimulated the prolyl hydroxylase and the collagen synthesis in rat hepatocytes. Therefore, it was proposed that oxygen radicals may enhance collagen synthesis during induced liver fibrosis. On the other hand, it has been found[390] that oxygen radicals produced by glucose oxidase or by redox cycling of menadione induced the killing of cultured hepatocytes. Pretreatment of the cells with desferrioxamine sharply diminished the killing; SOD, catalase, mannitol, thiourea, and benzoate were protective to the toxic effect of hydrogen peroxide produced by glucose oxidase, whereas iron and copper ions stimulated the killing. It was concluded that both the superoxide ion and iron ions mediate the killing of hepatocytes.

Oxygen radicals produced by xanthine oxidase stimulated the growth of a promotable 6clone of mouse epidermal cells, but were strongly toxic to a nonpromotable clone.[391] This difference is thought to be a consequence of increased DNA damage in the nonpromotable clone.

Larsson and Cerutti[392] recently showed that oxygen radicals generated extracellularly by xanthine oxidase or intracellulary by menadione stimulated the phosphorylation of the ribosomal S6 protein in mouse epidermal cells. This stimulatory effect was inhibited by catalase and not SOD, which indicates the mediation by hydrogen peroxide or (and) hydroxyl radicals. It was proposed that oxygen radicals enhanced Ca^{2+}/calmodulin-dependent protein kinase activity due to an increase in intracellular calcium level. Thus oxygen radicals are able to stimulate cellular mitogenesis via the phosphorylation of ribosomal protein S6.

Histamine release from mast cells is now believed to be an important step in developing many pathologies such as asthma, ischemia, etc. Among various histamine liberators, oxygen radicals may play an important role. As early as 1979, Ohmori et al.[393] showed that the hypoxanthine-xanthine oxidase system promoted the release of histamine from isolated rat mast cells. Mannaioni et al.[394] have studied the histamine release from mast cells induced by three drugs: paracetamol, cocaine, and mitomycin C. It was proposed that these drugs produced oxygen radicals as a result of redox cycling. Indeed, histamine release was observed only when the drugs were reduced by liver microsomes isolated from phenobarbital-treated rats or by prostaglandin synthase. Free radical scavengers (mannitol, glutathione, and α-tocopherol) inhibited the drug-dependent histamine release from mast cells, but SOD and catalase had no effect.

Oxygen radicals generated by xanthine oxidase or glucose oxidase transformed exogenous arachidonic acid into prostacyclin (prostaglandin I_2; PGI_2) in perinatal rat lung cells.[395] Catalase inhibited this process but SOD and dimethylthiourea did not. In contrast, pretreatment of the cells with xanthine oxidase inhibited the transformation of arachidonic acid into PGI_2; catalase again blocked it. It seems that hydrogen peroxide formed during the dismutation of the superoxide ion is mainly responsible both for the conversion of arachidonate to PGI_2 and the inhibition of this process in the case of pretreatment with xanthine oxidase. Maridonneau-Parini et al.[396] have shown that phenazine methosulfate (PMS) enhances the formation of 15-HETE (15-hydroxyeicosatetraenoic acid), prostaglandin E_2 (PGE_2), and lyso-PAF (a precursor of platelet activating factor, PAF) in rat renomedullary cells. This effect of PMS is thought to be mediated by oxygen radicals formed during the reduction of dioxygen by PMS inside the cells. SOD enhanced and catalase decreased the PMS action; therefore, it was concluded that hydrogen peroxide is a major stimulatory agent of the formation of these inflammatory lipid derivatives.

As early as 1977, Handin et al.[397] found that oxygen radicals generated by the xanthine-xanthine oxidase system induced platelet aggregation and the release of ^{14}C-serotonin. These processes were inhibited by SOD, but not by catalase and mannitol. Therefore, it was concluded that the superoxide ion can directly affect platelet function. This conclusion was recently confirmed by Salvemini et al.,[398] who showed that SOD and cytochrome c (but not catalase or mannitol) inhibited the thrombon-stimulated platelet adhesion to gelatin-coated plastic and the platelet aggregation. The mediation of these processes by the superoxide ion was supported by the accelerating effect of pyrogallol (a superoxide generator); the action of pyrogallol was inhibited by SOD.

Chaudhi and Clark[399] concluded that oxygen radicals mediated the *in vitro* and *in vivo* lipopolysaccharide-induced release of TNF from mouse peritoneal macrophages because free radical scavengers (BHA and BHT) and chelators such as desferrioxamine inhibited the enhanced secretion of TNF.

B. DAMAGING EFFECTS OF OXYGEN RADICALS ON CELLS

1. Erythrocytes

Erythrocytes are probably the most common target in studying the damaging action of oxygen radicals on cells. At the beginning, controversial data concerning the role of the superoxide ion in the disruption of erythrocytes were obtained. For example, Michelson and Durosay[400] concluded that the superoxide ion is inactive in the hemolysis of human erythrocytes induced by irradiation at 254 nm or by photoreduction of riboflavin. However, Kellogg and Fridovich[84] and Lynch and Fridovich[401] have shown that the acetaldehyde-xanthine oxidase system promoted hemolysis of human erythrocytes. Both SOD and catalase inhibited hemolysis; therefore, both the superoxide ion and hydrogen peroxide apparently mediated this process. It was also pointed out that earlier unsuccessful attempts to induce the hemolysis of erythrocytes by xanthine oxidase using xanthine as a substrate are probably explained by the inhibition of the hemolysis by xanthine and urate.

There are other examples of the damaging action of oxygen radicals on erythrocytes. Goldberg and Stern[402] have shown that the superoxide ion formed in the autoxidation of dihydroxyfumarate induced a rapid breakdown of hemoglobin in erythrocytes into methemoglobin and other green pigments. The reaction was inhibited by SOD and catalase. The hemoglobin destruction products reacted with the erythrocyte membrane and by this enhanced the osmotic fragility of erythrocytes. Jozwiak and Hellszer[403] found that a radiolytically produced superoxide ion inactivated ATPase in erythrocytes, while its contribution to hemolysis was negligible. Kong and Davison[404] have shown that a radiolytically produced superoxide ion and hydroxyl radicals damaged membrane-bound glyceraldehyde-3-phosphate dehydrogenase, lipid membranes, and erythrocyte ghosts.

Maridonneau et al.[405] have shown that the treatment of human erythrocytes with PMS promoted lipid peroxidation and methemoglobin binding to cell membranes as well as the loss of intracellular potassium. It was proposed that erythrocyte damage was mediated by the superoxide ion as it was enhanced by diethyldithiocarbamate, an SOD inhibitor. It was also shown[406] that the Fe^{2+}-ascorbate system decreased the filterability of human erythrocytes (supposedly due to the change in the fluidity of membrane lipids) and diminished the content of reduced glutathione in these cells.

Girotti and Thomas[407] have shown that xanthine oxidase damaged erythrocyte ghosts in the presence of iron ions, promoting both lipid peroxidation and the release of trapped markers, Na^+ and glucose-6-phosphate. The oxidative damage was inhibited by desferrioxamine, SOD, and catalase, whereas hydroxyl radical scavengers (ethanol and mannitol) had no effect. It was proposed that the oxidative damage of erythrocyte ghosts by xanthine oxidase is mediated by site-specific hydroxyl radicals formed in the superoxide-driven Fenton reaction.

The damaging effects of the superoxide ion and hydroxyl radicals apparently can be differentiated. Thus, Rosen et al.[408] have found that the superoxide ion and hydroxyl radicals manifest different actions on membrane organization in human erythrocyte ghosts. Superoxide ion (obtained by the dissolution of NMe_4O_2) elevated the disorder parameter (determined from an ESR spectrum of 4-maleimide-2,2,6,6-tetramethylpiperidinoxyl covalently bound to the erythrocyte ghost), indicating an increase in membrane fluidity, whereas exposure to hydroxyl radicals decreased this parameter and, respectively, the membrane fluidity. These effects were inhibited by SOD and mannitol. However, recently it was concluded[409] that the participation of the superoxide ion itself without other more active oxygen species in the erythrocyte damage is minimal.

In contrast to enzymatic and chemical sources of oxygen radicals, activated phagocytes are significantly more powerful promoters of erythrocyte hemolysis. In 1980, Weiss[410] showed that PMA- and zymosan-stimulated human neutrophils destroyed erythrocytes. The cytotoxic activity of stimulated neutrophils was inhibited by SOD, which indicates the importance of O_2^- participating in this process. In addition, the neutrophils isolated from patients with CGD which are unable to produce superoxide ion did not promote hemolysis. It was suggested[411] that the inability of xanthine oxidase to induce hemolysis of the erythrocytes under similar conditions is a consequence of the existence of a strong antioxidant defense system in erythrocytes which includes SOD, catalase, and glutathione peroxidase. (Origins of the strong oxidative activity of phagocytes are considered below.) This defense system can be suppressed by azide, which blocks catalase, or by excluding glucose, which blocks the regeneration of reduced glutathione.

The anion channel blockers, 4,4′-diisothiocyano-2,2′-disulfonic acid (DIDS) and 4-acetamido-4′-isothiocyano-2,2′-disulfonic acid stilbene (SITS) inhibited the destruction of erythrocytes by oxygen radicals.[401,411] Thus, although there are many doubts about the efficiency of superoxide traversing the cell membranes (see Chapter 4), nonetheless, the superoxide ion is apparently able to cross the erythrocyte membrane by traveling through the anion channels.

The mechanism of the strong damaging action of stimulated phagocytes on erythrocytes was discussed. Weiss[410] proposed that the attack of the superoxide ion on oxyhemoglobin after entering the cell is an initiating step of hemolysis. He also suggested that the formed methemoglobin and hydrogen peroxide form a peroxidative complex, which is a genuine hemolytic agent. In subsequent work, Weiss[411] showed that erythrocytes were not phagocytosed by neutrophils, i.e., the oxidation of hemoglobin occurred by a mechanism independent of phagocytosis. Since a combination of SOD and catalase strongly inhibited the oxidation of hemoglobin in the erythrocytes, both the superoxide ion and hydrogen peroxide apparently mediated the prooxidative action of neutrophils. The cell-cell contact between a neutrophil and an erythrocyte probably is also an important requirement for overcoming the defense antioxidant system of erythrocytes. It should be noted that the oxidation of hemoglobin in the erythrocytes by oxygen radials produced by neutrophils was recently studied with the aid of a new fluorescence microscopic technique.[412]

However, Vercellotti et al.[409] have shown that the formation of methemoglobin alone is insufficient for the initiation of erythrocyte hemolysis. For example, it was found that xanthine oxidase was able to oxidize hemoglobin in the erythrocytes, but was unable to promote hemolysis. This leads to the important conclusion that the superoxide ion may traverse the erythrocyte membrane without significant damage. If was further shown that the granule-depleted neutrophils which are efficient superoxide generators possessed a smaller hemolytic activity than neutrophils. In addition, the hemolytic activity of both neutrophils and neutroblasts was completely abolished by desferrioxamine. On these grounds, it was proposed that the iron-transporting protein lactoferrin contained in neutrophil-specific granules is an obligatory participant of the neutrophil-induced erythrocyte hemolysis. Its role is evidently to convert the superoxide ion to reactive hydroxyl or crypto-hydroxyl radicals by the Fenton-like reaction.

This proposal contradicts recent studies in which it was shown that lactoferrin is actually unable to stimulate the production of hydroxyl radicals by xanthine oxidase (Chapter 2). Nonetheless, it seems difficult to rule out the participation of lactoferrin in neutrophil-stimulated erythrocyte hemolysis. Possibly, the catalytic activity of lactoferrin in this process is due to its ability to bind to the erythrocyte membranes that may enhance the probability of hydroxyl radical generation.[409]

2. Endothelial Cells

Pulmonary endothelial cells are very sensitive to oxidative stress; thus they exhibit more injury than pulmonary fibroblasts upon exposure to dioxygen. It is believed that this injury induces the impairment of capillary integrity, the occlusion of capillaries, and the reduction of the effective capillary volume available for gas exchange. The elevated sensitivity of endothelial cells to an attack of oxygen radicals may be due to different reasons. For example, it may be due to their ability to generate oxygen radicals by themselves (Chapter 4). In addition, endothelial cells can have diminished activities of antioxidative enzymes in comparison with fibroblasts or be *in vivo* in the location which is the most vulnerable to the oxygen radical attack.

The effect of oxygen radicals on endothelial cells was studied in the experiments with cultured porcine aortic cells incubated with xanthine oxidase.[413,414] The damaging action of oxygen radicals was measured by the lactate dehydrogenase release and by their effect on the ESR spectra of 4-maleimido-2,2,6,6-tetramethylpiperidinooxyl (4-Mal-TEMPO) covalently bound to endothelial cells. It was found that the effects of oxygen radicals on cultured cells were strongly dependent on cell culture media constituents, cell types, and cell culture age, which may lead to wrong conclusions about the true magnitude of these effects. Although there is a real difference in the activities of antioxidative enzymes in endothelial cells and fibroblasts,[413] this difference seems not to explain the elevated sensitivity of endothelial cells to oxygen radical damage. It was found that the xanthine-xanthine oxidase system caused an increase in the mobility of 4-Mal-TEMPO bound to membrane proteins of endothelial cells. This effect of xanthine oxidase was inhibited by SOD.[414] It was concluded that the superoxide-mediated injury of endothelial cells is a consequence of the elevated disorder of cellular membranes.

At the same time, various other mechanisms of the interaction of oxygen radicals with endothelial cells apparently exist. Hirosumi et al.[415] have shown that xanthine oxidase stimulated an enhancement of cytosolic calcium in cultured endothelial cells. This effect was observed without significant cellular damage and was inhibited by SOD. The effect of catalase was insignificant; therefore, the superoxide ion and not hydrogen peroxide apparently plays a main role in this process. The lysis of endothelial cells promoted by xanthine oxidase is probably mediated by hydroxyl radicals formed from hydrogen peroxide. Thus, the damage of human endothelial cells (measured as the ^{3}H-2-deoxyribose release) by xanthine oxidase was inhibited by catalase, but not SOD.[416]

Endothelial cells can also be damaged by oxygen radicals released by other cells. Varani et al.[417] have shown that human blood neutrophils stimulated by immune complexes, opsonized zymosan, and TPA (but not FMLP or PAF) promoted the killing of bovine pulmonary artery endothelial cells. Again, hydroxyl radicals apparently mediated the killing because catalase, hydroxyl radical scavengers (dimethylthiourea and mannitol), and iron chelators (desferrioxamine mesitylate), but not SOD, were effective inhibitors. It was also proposed[416] that oxygen radicals mediated the damage of human endothelial cells promoted by FMLP-stimulated Walker carcinosarcoma cells.

In contrast to the cell lysis which occurs after prolonged incubation with the oxygen radical-generating systems, certain important damaging processes may proceed at significantly shorter time intervals. Thus the incubation of human umbilical vein endothelial cells with xanthine oxidase during 60 to 120 min induced an impairment of the receptor-dependent endocytosis of LDL in these cells.[418] The damage of cells was inhibited by catalase and not SOD, indicating the importance of HO· participating.

3. Other Animal Cells

Oxygen radicals are apparently able to damage and kill all animal cells, although the protective antioxidative abilities of various cells may differ to a considerable extent. For studying the damage of cultured murine myocardial cells by oxygen radicals, Scott et al.[419] applied fluorescence-activating sorting of antimyosin antibody-labeled cells. This method of measuring cell destruction is based on the capacity of antimyosin antibody to bind to intracellular myosin made available by the damage to the cell membrane. It was proposed that cell death upon exposure to dioxygen (which increased with increasing dioxygen tension) was due to the formation of a superoxide ion and hydroxyl radicals because SOD, mannitol, and DMSO reduced cellular injury.

Barrington et al.[420] have shown that oxygen radicals generated by two different systems (DHF and xanthine oxidase) altered the electrophysiological activity of isolated cardiocytes. The effect of oxygen radicals was suppressed by SOD and catalase. It was proposed that oxygen radicals may contribute to ischemia or reperfusion arrythmias.

The damaging effect of oxygen radicals on renal epithelial and neonatal rat myocardial cells was studied by measuring the membrane potential resulting from the permeabilities and electrochemical gradients of ions across the plasma membrane.[421] This method is apparently more sensitive than techniques based on the direct measurement of membrane permeability. It was found that ferrous chloride depolarized the membrane potential of renal epithelial and neonatal rat myocardial cells measured by fluorescence of cationic carbocyanine dye. Since the effect of ferrous chloride was inhibited by SOD, catalase, and DMSO, it was concluded that the superoxide ion and hydroxyl radicals together with hydrogen peroxide are active species which induce the membrane depolarization in renal and cardiac cells. This phenomenon is probably an early stage of the cellular destructive process promoted by oxygen radicals.

It has been shown[422] that oxygen radicals generated by xanthine oxidase suppressed the oxidation of exogenous substrates (glucose, lactate, and octanoate) by cardiac myocytes. SOD and catalase efficiently inhibited the action of oxygen radicals. It was proposed that a decrease in myocardial substrate oxidation induced by the superoxide ion and hydrogen peroxide may be an origin of functional and ultrastructural changes in the myocardium under reoxygenation conditions.

Oxygen radicals generated by the hypoxanthine-xanthine oxidase system killed pulmonary epithelial cells (rat alveolar type II pneumocytes).[423] They also impair the surfactant synthesis by these cells, apparently due to a decrease in the incorporation of precursors into phosphatidylcholine.

SOD, catalase, and DTPA gave significant protection against radiation-induced damage and death of rat alveolar macrophages.[424] It was suggested that the principal damaging agents were hydroxyl radicals forming by the superoxide-driven Fenton reaction. The photochemically generated superoxide ion inactivated white blood cells, principally granulocytes, monocytes, and macrophages developed from human bone marrow progenitor cells.[425] SOD, but not catalase, was protective.

Oxygen radicals produced by the Fenton reaction induced depolarization and lysis of human immature macrophage-like cells (U 937).[271] Both depolarization and lysis were inhibited by catalase and desferrioxamine. Therefore, it was concluded that depolarization is an early stage of oxygen radical damage which is next followed by lysis. Although significant lipid peroxidation occurred during the depolarization stage, it apparently was not a critical membrane event because BHT completely inhibited lipid peroxidation, but had no effect on depolarization and lysis. It was supposed that the origin of both phenomena is the destruction of membrane proteins, for example, Na^+,K^+-stimulated ATPase.

Olson et al.[426] have found that nontoxic concentrations of oxygen species produced by xanthine oxidase stimulate the production of PGE_2 by isolated gastric mucosal cells which predominantly consist of macrophages; this phenomenon can be considered an intercellular signal, promoting prostaglandin synthesis. Hydrogen peroxide is probably a main stimulatory agent of PGE_2 release, although the slight inhibition of activation by SOD and the enhancement of activation by ferric ions possibly indicate the participation of oxygen radicals.

Ascorbic acid induced lipid peroxidation and the leakage of lactate dehydrogenase after incubation with rat hepatocytes.[427] These processes were apparently mediated by iron ions because EDTA inhibited and ferric ions accelerated them. α-Tocopherol and phenolic antioxidants were protective.

Oxygen radials produced by xanthine oxidase induced minimal cytolysis of cultured canine gastric chief cells supposedly due to the strong endogenous antioxidant defense of these cells.[428] Indeed, the cell lysis measured by lactate dehydrogenase release was sharply enhanced by glutathione (GSH) depletion. Since catalase, thiourea, and desferrioxamine (but not SOD or mannitol) inhibited the cytolysis, it was concluded that it was mediated by hydrogen peroxide and hydroxyl radicals. Oxygen species formed during the dissolution of KO_2 were highly toxic to CHO cells.[305] Myers and Abney[429] proposed that oxygen radical damage may be a cause of the decreased steroidogenic capacity of cultured Leydig cells.

Recently, Ruch and Klanning[430] have shown that oxygen radicals may disrupt the intercellular communications through gap junctions (plasma membrane structures consisting of clusters of proteinaceous particles between adjacent cells). These structures mediate intercellular communications and therefore their degradation may lead to numerous pathological events. These authors found that oxygen radicals generated intracellularly (by paraquat) or extracellularly (by xanthine oxidase) inhibited hepatocyte intercellular communications determined as dye coupling. The mechanisms of the damaging effects of oxygen radicals on gap junctions are unknown, but they can involve lipid peroxidation, intracellular free calcium elevation, etc.

4. Bactericidal Activity of Oxygen Radicals

The study of the bactericidal activity of oxygen radicals was started at the very beginning of the oxygen radical era in biology because it seemed obvious to expect that active oxygen species may mediate the protective reactions of living organisms against microbe invasion. As we have seen (Chapter 4), such proposals were brilliantly confirmed by the discovery of oxygen radical production by activated phagocytes. In 1974, Klebanoff[431] showed that a combination of xanthine, xanthine oxidase, chloride, and myeloperoxidase exhibited a microbicidal effect on *E. coli, Staphylococous aureus,* and *Candida tropicalis*. This effect was inhibited by catalase, ethanol, and benzoate, but SOD reverted the inhibitory action of hydroxyl radical scavengers. It was proposed that the role of the superoxide ion in this bactericidal system is to produce hydrogen peroxide by dismutation.

Similar results were obtained by other authors. DeChatelet et al.[432] have shown that catalase, but not SOD, reversed the inhibition of the growths of *E. coli* and *S. aureus* by the superoxide ion formed during autoxidation of dialuric acid. Similarly, catalase was protective and SOD had no effect on the killing of *E. coli* bacteria by oxygen radicals produced by xanthine oxidase,[433] indicating that hydrogen peroxide was a main bactericidal agent for this microorganism. However, in the case of *S. epidermidis*, both SOD and catalase were protective. Therefore, it was proposed that in the last case, a bactericidal agent was formed in the reaction of $O_2^{\overline{\cdot}}$ with H_2O_2.

In contrast to the data obtained in Reference 432, Van Hemmen and Meuling[434] have found that SOD as well as catalase and EDTA protected *E. coli* cells against the oxygen radicals, forming in autoxidation of dialuric acid, whereas cupric ions enhanced the killing. It was concluded that cell inactivation was mediated by oxygen species formed in the Fenton-like reaction. DiGuiseppi and Fridovich[435] have shown that SOD suppressed the inhibition of the growth of *S. sanguis* cells induced by plumbagin. Since DMSO and catalase did not affect the plumbagin toxicity, hydroxyl radicals, it is thought, do not participate in the cell damage. Similarly to plumbagin, dioxathiadiaza-heteropentalenes inhibited the growth of *E. coli*, supposedly producing a superoxide ion as a result of redox cycling of these compounds.[436] In accord with this proposal, SOD and catalase suppressed the toxicity of heteropentalenes to *E. coli* cells.

There are different conclusions concerning the role of the Fenton reaction in the bactericidal action of oxygen radicals. Rosen and Klebanoff[437] have found that the bactericidal activity of xanthine oxidase depended completely on the presence of Fe(EDTA), which catalyzes the superoxide-driven Fenton reaction. The toxic effect of the acetaldehyde-xanthine oxidase-iron-EDTA system was inhibited by SOD, catalase, and hydroxyl radical scavengers. However, it was later shown[438] that the bactericidal activity of this system against *S. aureus, E. coli, Listeria monocytogenes,* and *Salmonella typhimurium* only slightly increased in the presence of Fe(EDTA). Catalase and mannitol almost completely inhibited bactericidal activity, whereas SOD was less inhibitory. The addition of hydrogen peroxide, as a rule, did not increase the killing. It was concluded that bactericidal activity was mediated by hydroxyl radicals.

Surprising results were obtained by Ewing and Jones,[439] who found that the damage of the *E. coli* cells from ionizing radiation within a range of low dioxygen concentrations (10^{-6} to 10^{-4} *M*) was inhibited by SOD and consequently was mediated by the superoxide ion. However, at higher dioxygen concentrations, the superoxide-mediated damage was insignificant. The authors explained these findings by assuming that the interaction of the superoxide ion with the cells competes with noncatalytic dismutation of the superoxide ion. At higher dioxygen concentrations (and higher superoxide concentrations, respectively), the last reaction is though to become predominant.

Lin et al.[440] concluded that the radiation killing of *E. coli* cells is directly mediated by the superoxide ion because formate (which converts HO· into $O_2^{\overline{\cdot}}$ under aerobic conditions) enhanced the radiosensitivity of bacteria. SOD and catalase did not affect cell survival, indicating that the superoxide ion was formed intracellularly. Nitrous oxide (which converts e^- into HO·) even decreased cell radiosensitivity. It was therefore assumed that hydroxyl radicals do not participate in the *E. coli* killing.

It was proposed[441] that oxygen radicals mediate the reductive repression in *E. coli*, presumably via the inhibition of adenylate cyclase with a concomitant repression of cyclic AMP-dependent phenotypes.

A radiolytically generated superoxide ion had no damaging effect on T4 and T7 bacteriophages.[442,443] However, there was a large damaging effect when the superoxide ion was generated in the presence of Cu^{2+} ions. The process was not inhibited by hydroxyl radical scavengers, SOD, and catalase. It was supposed that $O_2^{\overline{\cdot}}$ reduces Cu^{2+} bound to a protein molecule and by this initiates the site-specific formation of hydroxyl radicals. Murata et al.[444] proposed that hydroxyl radicals are responsible for inactivation of phages by ascorbic acid.

VIII. SOME EXTRACELLULAR FUNCTIONS OF OXYGEN RADICALS

It is obvious that oxygen radicals released by phagocytes or formed during some redox processes may interact with native substances outside the cells (in serum, sinovial liquid, etc.) Most of such interactions are apparently the same as those occurring in cells (see Chapter 4). However, oxygen radicals may promote specific extracellular interactions, participating in definite physiological functions or inducing some pathological effects. We will consider here two important extracellular processes proceeding with the participation of oxygen radicals: the activation by the superoxide ion of a chemotactic factor for neutrophils and the interaction of the superoxide ion with endothelium-derived vascular relaxing factor (EDRF).

McCord and Petrone[445] proposed that extracellular liquids contain a superoxide-activated chemotactic factor which allows resting neutrophils to recognize and respond to neutrophils activated by an immune complex. It has been shown that the superoxide ion produced by xanthine oxidase reacted with an albumin-bound lipid presented in plasma or released from activated neutrophils to form a powerful chemotactic factor for neutrophils. SOD, but not catalase, prevented the activation process. It was proposed that the superoxide-dependent chemotactic factor mediates the development of inflammation.

EDRF is an unstable humoral agent of unknown structure which is released by vascular endothelium. It was earlier proposed that EDRF may be the hydroperoxide of arachidonic acid, unstable ketone, aldehyde, or lactone, or even a free radical. However, recently, EDRF was identified as nitric oxide.[446] It has been shown[447,448] that SOD sharply increased the stability of EDRF, whereas ferrous ions decreased it. In addition, the protective role of SOD was augmented by lowering the oxygen tension.[448] Thus the superoxide ion may apparently promote the breakdown of EDRF, and SOD can probably protect EDRF *in vivo*. As the reaction of the superoxide ion with nitric oxide had been studied (Volume I, Chapter 6), their interaction must result in the formation of peroxonitrite:

$$O_2^{\overline{\cdot}} + NO \rightarrow ONOO^- \tag{71}$$

In accord with the disrupting effect of the superoxide ion on EDRF, it was recently found[449] that the superoxide ion is an endothelium-derived contracting factor because the xanthine-xanthine oxidase system induced in the presence of catalase the SOD-inhibitable contractions of the vascular smooth muscle. In addition, SOD inhibited the A 23187-dependent endothelium-mediated contractions in canine basilar arteries.[450]

IX. ADDITIONS

Keller et al.[451,452] concluded that the NADH oxidation catalyzed by vanadium compounds in the presence of hydrogen peroxide, the superoxide-generating system (xanthine oxidase), or thiols is mediated by hydroxyl radicals detected by spin trapping with DMPO. Using a pulse-radiolysis technique, Sabourault et al.[453] recalculated a rate constant for the reaction of the superoxide ion with desferrioxamine as $(1.3 \pm 0.1) \cdot 10^6 M^{-1} s^{-1}$ at pH 7.4. A wrong low value for this rate constant determined earlier by competition with cytochrome c[62] is thought to be explained by the interaction of reaction products with cytochrome c. Samuni et al.[454] have studied the reaction of the superoxide ion produced by xanthine oxidase with DMPO-OH and other nitroxide spin adducts.

Noda et al.[455] proposed that the oxidation of hydrazine into hydrazine free radical catalyzed by microsomal NADPH-cytochrome P-450 reductase is mediated by oxygen radicals because this reaction was inhibited by SOD. PMA stimulation of blood monocyte-derived macrophages sharply enhanced the conversion of ethanol to acetate.[456] SOD completely inhibited ethanol oxidation, indicating the formation of a superoxide ion. Oxygen radicals apparently mediate the oxidation of 2,3-mercaptopropane-1-sulfonate (a drug for the treatment of copper poisoning or Wilson's disease) by cupric ions because SOD and catalase inhibited the luminol-dependent CL accompanied the reaction.[457] Cavalieri et al.[458] have found that HRP and prostaglandin H synthase catalyzed the oxidation of the carcinogens benzo[a]pyrene and 6-fluorobenzo[a]pyrene into free radicals.

It has been shown[459] that acidosis strongly enhances lipid peroxidation in brain homogenates possibly due to an increase in iron dissociation from transferrin. Kettle and Winterbourn[460] have shown that the superoxide ion produced by xanthine oxidase enhanced the activity of myeloperoxidase apparently by preventing the accumulation of inactive compound II. Cabelli et al.[461] determined a rate constant for the reaction of HO_2^- with Cu(I)SOD as $2.6 \cdot 10^3 M^{-1} s^{-1}$. Bedwell et al.[462] compared the effects of HO·, HO_2·, and $O_2^{\overline{\cdot}}$ on human LDL structure and receptor activity. It was found that the superoxide ion (applied as KO_2) drastically decreased the binding of cytochrome c to cardiolipin.[463] Kobayashi et al.[464] have studied the effects of tryptophan and pH on the kinetics of binding of the superoxide ion to indoleamine-2,3-dioxygenase. The inactivating effects of oxygen radicals on the enzymes have been shown for mitochondrial carbomoyl phosphate synthase (in the reaction with the Fe^{3+}-O_2-ascorbate system),[465] glutamine synthase in mammalian brain,[466] and xanthine oxidase.[467] Nowak[468] has shown that oxygen radicals produced by xanthine oxidase or PMNs suppressed α_1-proteinase inhibitor activity.

Epe et al.[469] applied a new method for the identification of reactive oxygen species participating in DNA damage (hydroxyl radicals, singlet oxygen, and photoexcitation), based on measuring the repair enzyme activity in cellular extracts. It was found[470] that the Cu^{2+}-ascorbate system promotes two types of damage in bacteriophage DNA: a direct strand cleavage and base modification liable to alkali treatment. Both types were inhibited by catalase and metal chelators. Loeb et al.[471] proposed that the DNA damage and mutagenesis induced by ferrous ions is mediated by hydroxyl radicals forming in the Fenton reaction. A similar conclusion was achieved by studying the DNA damage by PMA-stimulated human neutrophils.[472] On the other hand, the DNA damage by the Cu^{2+}-H_2O_2 system is thought to be mediated by copper-peroxide complexes and not by free hydroxyl radicals.[473]

Shacter et al.[474] concluded that both the superoxide ion and hydrogen peroxide mediate the DNA damage in plasmacytoma cells by stimulated neutrophils. Resident (nonstimulated) and inflammatory peritoneal macrophages induced DNA strand breaks in tumor cells.[475] Catalase was protective in both cases, but SOD protected only against resident macrophages. It was found[476] that hydroxyl radicals, forming by the site-specific mechanism, enhanced the binding of anti-DNA antibodies to the denaturated DNA. Moseley[477] concluded that the destructive action of bleomycin and Fe^{2+}-bleomycin on DNA in viral minichromosomes is mediated by oxygen radicals. However, Pratviel et al.[478] proposed that high-valent iron-oxo species play an important role in bleomycin-DNA-damaging activity.

DNA single-strand breaks mediated by oxygen radicals were induced upon the illumination of anthrapyrazole antitumor agents with blue light.[479] The reduction of 3-amino-1,2,4-benzotriazine-1,4-dioxide (a hypoxic cell toxin) by xanthine oxidase to a free radical stimulated the single- and double-strand breaks in plasmid DNA.[480] Oxidative DNA damage was also promoted by β-lactam antibiotics in the presence of iron and copper ions[481] and nitroimidazole-aziridines (including misonidazole).[482] High concentrations of sodium selenite and ascorbate enhanced the number of single-strand breaks in DNA induced by oxygen radicals released by human mononuclear cells.[483] However, low concentrations of these compounds protected DNA against oxygen radical damage.

Hart et al.[484] have shown that the superoxide ion and hydroxyl radicals produced by PMA-activated human PMNs promoted the lysis of sheep and rabbit erythrocytes. Similarly, oxygen radicals generated by xanthine oxidase promoted CL, the formation of TBA-reactive compounds, and hemolysis of erythrocytes from vitamin E-deficient rats.[485] von Ruecker et al.[486] have shown that oxygen radicals generated by the Cu^{2+} -H_2O_2 system enhanced the membrane association of protein kinase C in cultured hepatocytes. It was proposed that membrane translocation and activation of this enzyme are responsible for the oxygen radical-dependent hepatocyte damage. Oxygen radicals produced by xanthine oxidase killed isolated pulmonary epithelial cells.[487] On the other hand, noncytosolic concentrations of oxygen radicals produced by xanthine oxidase stimulated the release of high-molecular weight glycoconjugates from epithelial cells supposedly through the cyclooxygenase-catalyzed metabolism of arachidonic acid.[488]

Oxygen radicals, in particular hydroxyl radicals, apparently mediate the injury of human endothelial cells promoted by thrombin-stimulated human platelets as it follows from the protective action of mannitol.[489] Jornot and Junod[490] have shown that oxygen radicals produced by xanthine oxidase markedly reduced the rate of [^{3}H] phenylalanine incorporation into total proteins in cultured endothelial cells. Kasama et al.[491] have shown that the superoxide ion stimulates the production of interleukin 1-like factor from human peripheral blood monocytes and PMNs. Krutmann et al.[492] proposed that the superoxide ion produced in porphyrin photosensitization inhibits the high-affinity Fc receptor on human monocytes.

Oxygen radicals promote the histamine release and MDA formation by mast cells.[493,494] Ueno et al.[495] have found that oxygen radicals stimulated the production of erythropoietin in renal carcinoma cells. Pretreatment of CHO and rat hepatoma cells with low concentrations of oxygen radicals produced by xanthine oxidase increased the resistance of the cells to the toxic effects of hydrogen peroxide and gamma rays.[496] Vincent et al.[497] have shown that oxygen radicals produced by xanthine oxidase inhibited the growth of rabbit and articular chondrocytes. Yamamoto et al.[498] suggested that the inhibitory effect of a combination of carbon tetrachloride and paraquat on the growth of *E. coli* cells is mediated by the $CCl_3OO\cdot$ radical, forming in the reaction of $O_2^{\bar{}}$ with CCl_4.

Burkitt and Gilbert[499] concluded that iron-induced damage in rat liver mitochondria is greater in state 4 than in state 3 due to the enhanced level of hydrogen peroxide in state 4. Kaneko et al.[500] have shown that oxygen radicals generated by xanthine oxidase or in the Fenton reaction inhibited calcium-pump activity in isolated rat heart sarcolemmal membranes. Kukreja et al.[501] concluded that oxygen radicals produced by PMA-stimulated human neutrophils diminish the calcium uptake rate and Ca^{2+}-ATPase activity in cardiac sarcoplasmic reticulum.

REFERENCES

1. **Land, E. J. and Swallow, A. J.,** One-electron reactions in biochemical systems as studied by pulse radiolysis. IV. Oxidation of dihydronicotinamide-adenine dinucleotide, *Biochim. Biophys. Acta,* 234, 34, 1971.
2. **Nadezhdin, A. and Dunford, H. B.,** Oxidation of nicotinamide dinucleotide by hydroperoxyl radicals. A flash photolysis study, *J. Phys. Chem.,* 83, 1957, 1979.
3. **Bielski, B. H. J. and Chan, P. C.,** Enzyme-catalyzed free radical reactions with nicotinamide-adenine nucleotides. I. Lactate dehydrogenase-catalyzed chain oxidation of bound NADH by superoxide radicals, *Arch. Biochem. Biophys.,* 159, 873, 1973.
4. **Chan, P. C. and Bielski, B. H. J.,** Enzyme-catalyzed free radical reactions with nicotinamide-adenine nucleotides. II. Lactate dehydrogenase-catalyzed oxidation of reduced nicotinamide adenine dinucleotide by superoxide radicals generated by xanthine oxidase, *J. Biol. Chem.,* 249, 1317, 1974.
5. **Bielski, B. H. J. and Chan, P. C.,** Kinetic study by pulse radiolysis of the lactate dehydrogenase-catalzyed chain oxidation of nicotinamide adenine dinucleotide by HO_2 and O_2^- radicals, *J. Biol. Chem.,* 250, 318, 1975.
6. **Chan, P. C. and Bielski, B. H. J.,** Glyceraldehyde-3-phosphate dehydrogenase-catalyzed chain oxidation of reduced nicotinamide adenine dinucleotide by perhydroxyl radicals, *J. Biol. Chem.,* 255, 874, 1980.
7. **Darr, D. and Fridovich, I.,** Vanadate and molybdate stimulate the oxidation of NADH by superoxide radical, *Arch. Biochem. Biophys.,* 232, 562, 1984.
8. **Darr, D. and Fridovich, I.,** Vanadate enhancement of the oxidation of NADH by O_2^-: effects of phosphate and chelating agents, *Arch. Biochem. Biophys.,* 243, 220, 1985.
9. **Liochev, S. and Fridovich, I.,** The vanadate-stimulated oxidation of NAD(P)H by biomembranes is a superoxide-initiated free radical chain reaction, *Arch. Biochem. Biophys.,* 250, 139, 1986.
10. **Halliwell, B. and DeRycker, J.,** Superoxide and peroxide-catalysed reactions. Oxidation of dihydroxyfumarate, NADH and dithiothreitol by horseradish peroxidase, *Photochem. Photobiol.,* 28, 757, 1978.
11. **Misra, H. P. and Fridovich, I.,** The role of superoxide anion in the autoxidation of epinephrine and a simple assay for superoxide dismustase, *J. Biol. Chem.,* 247, 3170, 1972.
12. **Powis, G.,** Hepatic microsomal metabolism of epinephrine and adrenochrome by superoxide-dependent and independent pathways, *Biochem. Pharmacol.,* 28, 83, 1979.
13. **Schenkman, J. B., Jansson, I., Powis, G., and Kappus, H.,** Active oxygen in liver microsomes: mechanism of epinephrine oxidation, *Mol. Pharmacol.,* 15, 428, 1979.
14. **Dybing, E., Nelson, S. D., Mitchell, J. R., Sasame, H. A., and Gillette, J. R.,** Oxidation of α-methyldopa and other catechols by cytochrome P-450-generated superoxide anion: possible mechanism of methyldopa hepatitis, *Mol. Pharmacol.,* 12, 911, 1976.
15. **Felix, C. C., Hyde, J. S., Sarna, T., and Sealy, R. C.,** Melamin photoreactions in aerated media: electron spin resonance evidence for production of superoxide and hydrogen peroxide, *Biochem. Biophys. Res. Commun.,* 84, 335, 1978.
16. **Korytowski, W., Kalyanaraman, B., Menon, I. A., Sarna, T., and Sealy, R. C.,** Reaction of superoxide anions with melamines: electron spin resonance and spin trapping studies, *Biochim. Biophys. Acta,* 882, 145, 1986.
17. **Sarna, T., Pilas, B., Land, E. J., and Truscott, T. G.,** Interaction of radicals from water radiolysis with melamins, *Biochim. Biophys. Acta,* 883, 162, 1986.
18. **Asada, K. and Kanematsu, S.,** Reactivity of thiols with superoxide radicals, *Agric. Biol. Chem.,* 40, 1891, 1976.
19. **Wefers, H. and Sies, H.,** Oxidation of glutathione by the superoxide radical to the disulfide and the sulfonate yielding singlet oxygen, *Eur. J. Biochem.,* 137, 29, 1983.
20. **Ross, D., Cotgreave, I., and Moldeus, P.,** The interaction of reduced glutathione with active oxygen species generated by xanthine-oxidase-catalyzed metabolism of xanthine, *Biochim. Biophys. Acta,* 841, 278, 1985.
21. **Park, E.-M. and Thomas, J. A.,** S-Thiolation of creatine kinase and glycogen phosphorylase b initiated by partially reduced oxygen species, *Biochim. Biophys. Acta,* 964, 151, 1988.
22. **Thomas, E. L., Learn, D. B., Jefferson, M. M., and Weatherred, W.,** Superoxide-dependent oxidation of extracellular reducing agents by isolated neutrophils, *J. Biol. Chem.,* 263, 2178, 1988.
23. **Shibata, H., Ochiai, H., Akiyama, M., Ishii, H., and Katoh, T.,** The reactivity of 4-thiouridine with peroxidase and superoxide radicals, *Agric. Biol. Chem.,* 44, 1427, 1980.
24. **Ito, S. and Fujita, K.,** Conjugation of dopa and 5-S-cystenyldopa with cysteine mediated by superoxide radical, *Biochem. Pharmacol.,* 31, 2887, 1982.
25. **Mashino, T. and Fridovich, I.,** Superoxide radical initiates the autoxidation of dihydroxyacetone, *Arch. Biochem. Biophys.,* 254, 547, 1987.
26. **Halliwell, B. and Gutteridge, J. M. C.,** Formation of a thiobarbituric-acid-reactive substance from deoxyribose in the presence of iron ions, *FEBS Lett.,* 128, 347, 1981.

27. **Aruoma, O. I., Grootveld, M., and Halliwell, B.,** The role of iron in ascorbate-dependent deoxyribose degradation. Evidence consistent with a site-specific hydroxyl radical generation caused by iron ions bound to the deoxyribose molecule, *J. Inorg. Biochem.*, 29, 289, 1987.
28. **Tadolini, B. and Cabrini, L.,** On the mechanism of OH· scavenger action, *Biochem. J.*, 253, 931, 1988.
29. **Gutteridge, J. M. C. and Halliwell, B.,** The deoxyribose assay: an assay both for "free" hydroxyl radical and for site-specific hydroxyl radical production, *Biochem. J.*, 253, 932, 1988.
30. **Gutteridge, J. M. C.,** Reactivity of hydroxyl and hydroxyl-like radicals discriminated by release of thiobarbituric acid-reactive material from deoxy sugars, nucleosides and benzoate, *Biochem. J.*, 224, 761, 1984.
31. **Uchida, K. and Kawakishi, S.,** Reactivity of amylose and dextran in metal-catalyzed hydroxyl radical generating systems, *Agric. Biol. Chem.*, 51, 605, 1987.
32. **Yamane, H., Yada, N., Katori, E., Mashino, T., Nagano, T., and Hirobe, M.,** Base liberation from nucleotides by superoxide and intramolecular enhancement effect of phosphate group, *Biochem. Biophys. Res. Commun.*, 142, 1104, 1987.
33. **McCord, J. M.,** Free radicals and inflammation: protection of synovial fluid by superoxide dismutase, *Science*, 185, 529, 1974.
34. **Greenwald, R. A. and Mai, W. W.,** Effect of oxygen-derived free radicals on hyaluronic acid, *Arthritis Rheum.*, 23, 455, 1980.
35. **McNeil, J. D., Wielkin, O. W., Betis, W. H., and Cleland, L. G.,** Depolymerisation products of hyaluronic acid after exposure to oxygen-derived free radicals, *Ann. Rheum. Dis.*, 44, 780, 1985.
36. **Halliwell, B.,** Superoxide-dependent formation of hydroxyl radicals in the presence of iron salts. Its role in degradation of hyaluronic acid by a superoxide-generating system, *FEBS Lett.*, 96, 238, 1978.
37. **Carlin, G. and Djursaeter, R.,** Xanthine oxidase induced depolymerization of hyaluronic acid in the presence of ferritin, *FEBS Lett.*, 177, 27, 1984.
38. **Wetz, Z., Moak, S. A., and Greenwals, R. A.,** Degradation of hyaluronic acid by neutrophil derived oxygen radicals is stimulus dependent, *J. Rheumatol.*, 15, 1250, 1988.
39. **Littarru, G. P., Lippa, S., DeSole, P., and Oradei, A.,** In vitro effect of different ubiquinones on the scavenging of biologically generated O_2^- *Drug Exp. Clin. Res.*, 11, 529, 1985.
40. **Nakamura, M., Murakami, M., Umei, T., and Minakami, S.,** Ubiquinone-5 is reduced by superoxide in the aerobic state by NADPH oxidase of guinea pig macrophages, *FEBS Lett.*, 186, 215, 1985.
41. **Nagase, S., Aoyagi, K., Narita, M., and Tojo, S.,** Active oxygen in methylguanidine synthesis, *Nephron*, 44, 299, 1986.
42. **Aoyagi, K., Nagase, S., Narita, M., and Tojo, S.,** Role of active oxygen in methylguanidine synthesis in isolated rate hepatocytes, *Kidney Int.*, 32 (Suppl. 22), S229, 1987.
43. **Uemura, T., Kanashiro, M., Yamano, T., Hirai, K., and Miyazaki, N.,** Isolation, structure, and properties of the β-carboline formed from 5-hydroxytryptamine by the superoxide anion-generating system, *J. Neurochem.*, 51, 710, 1988.
44. **Winterbourn, C. C., Benatti, U., and DeFlora, A.,** Contributions of superoxide, hydrogen peroxide, and transition metal ions to autoxidation of the favism-inducing pyrimidine aglycone, divicine, and its reactions with hemoglobin, *Biochem. Pharmacol.*, 35, 2009, 1986.
45. **Winterbourn, C. C., Cowden, W. B., and Sutton, H. C.,** Auto-oxidation of dialuric acid, divicine and isouramil. Superoxide dependent and independent mechanisms, *Biochem. Pharmacol.*, 38, 611, 1989.
46. **Maynard, M. S. and Cho, A. K.,** Oxidation of N-hydroxyphentermine to 2-methyl-2-nitro-1-phenylpropane by liver microsomes, *Biochem. Pharmacol.*, 30, 1115, 1981.
47. **Duncan, J. D., DiStefano, E. W., Miwa, G. T., and Cho, A. K.,** Role of superoxide in the N-oxidation of N-(2-methyl-1-phenyl-2-propyl)hydroxyamine by the rat liver cytochrome P-450 system, *Biochemistry*, 24, 4155, 1985.
48. **Fukuto, J. M., DiStefano, E. W., Burstyn, J. N., Valentine, J. S., and Cho, A. K.,** Mechanism of oxidation of N-hydroxyphentermine by superoxide, *Biochemistry*, 24, 4161, 1985.
49. **Bernhardt, F.-H. and Kuthan, H.,** Kinetics of reduction of putidamonooxin by NADH-putidamonooxin oxidoreductase, sodium dithionite and superoxide dismutase, *Eur. J. Biochem.*, 130, 99, 1983.
50. **Scislowski, P. W. D. and Davis, E. J.** Sulfur oxidation of free methionine by oxygen free radicals, *FEBS Lett.*, 224, 177, 1987.
51. **Younes, M.,** Involvement of reactive oxygen species in the microsomal S-oxidation of thiobenzamide, *Experientia*, 41, 479, 1985.
52. **Sagone, A. L., Jr. and Husney, R. M.,** Oxidation of salicylates by stimulated granulocytes: evidence that these drugs act as free radical scavengers in biological systems, *J. Immunol.* 138, 2177, 1987.
53. **Winston, G. W. and Cederbaum, A. I.,** Oxidative decarboxylation of benzoate to carbon dioxide by rat liver microsomes: a probe for oxygen radical production during microsomal electron transfer, *Biochemistry*, 21, 4265, 1982.
54. **Dull, B. J., Salata, K., VanLangenhove, A., and Goldman, P.,** 5-Aminosalicylate: oxidation by activated leucocytes and protection of cultured cells from oxidative damage, *Biochem. Pharmacol.*, 36, 2467, 1987.

55. **Cederbaum, A. I., Dicker, E., and Cohen, G.,** Effect of hydroxyl radical scavengers on microsomal oxidation of alcohols and on associated microsomal reaction, *Biochemistry,* 17, 3058, 1978.
56. **Ohnishi, K. and Lieber, C. S.,** Respective role of superoxide and hydroxyl radical in the activity of the reconstituted microsomal ethanol-oxidizing system, *Arch. Biochem. Biophys.,* 191, 798, 1978.
57. **Puntarulo, S. and Cederbaum, A. I.** Temperature dependence of the microsomal oxidation of ethanol by cytochrome P-450 and hydroxyl radical-dependent reactions, *Arch. Biochem. Biophys.,* 269, 569, 1989.
58. **Cederbaum, A. I., Dicker, E., and Cohen, G.,** Role of hydroxyl radicals in the iron-ethylenediamine-tetraacetic acid mediated stimulation of microsomal oxidation of ethanol, *Biochemistry,* 19, 3698, 1980.
59. **Albano, E., Tomasi, A., Goria-Gatti, L., Poli, G., Vannini, V., and Dianzani, M. U.,** Free radical metabolism of alcohols by rat liver microsomes, *Free Rad. Res. Commun.,* 2, 243, 1987.
60. **Shaw, S., Jayatilleka, E., Herbert, V., and Colman, N.,** Cleavage of folates during ethanol metabolism. Role of acetaldehyde/xanthine oxidase-generated superoxide, *Biochem. J.,* 257, 277, 1989.
61. **Sinaceur, J., Ribiere, C., Nordmann, I., and Nordman, R.,** Superoxide degrading activity of desferrioxamine, in *Oxygen Radicals in Chemistry and Biology,* Bors, W., Saran, M., and Tait, D., Eds., Walter de Gruyter, Berlin, 1984, 211.
62. **Halliwell, B.,** Use of desferrioxamine as a "probe" for iron-dependent formation of hydroxyl radicals. Evidence for a direct reaction between desferal and the superoxide radical, *Biochem. Pharmacol.,* 34, 229, 1985.
63. **Ribiere, C., Sabourault, D., Sinaceur, J., Nordmann, R., Houee-Levin, C., and Ferradinia, C.,** Radiolysis study of the reaction of desferrioxamine with O_2^-, in *Superoxide and Superoxide Dismutase in Chemistry, Biology and Medicine,* Rotilio, G., Ed., Elsevier, Amsterdam, 1986, 47.
64. **Davies, M. J., Donkor, R., Dunster, C. A., Gee, C. A., Jonas, S., and Willson, R. L.,** Desferrioxamine (desferal) and superoxide free radicals. Formation of an enzyme-damaging nitroxide, *Biochem. J.,* 246, 725, 1987.
65. **Sevanian, A. and Hochstein, P.,** Mechanisms and consequences of lipid peroxidation in biological systems, *Annu. Rev. Nutrition,* 5, 365, 1985.
66. **Vladimirov, Yu. A.,** Free radial lipid peroxidation in biomembranes: mechanism, regulation, and biological consequences, in *Free Radicals, Aging, and Degenerative Diseases,* Johnson, J. E., Jr., Walford, R., Harman, D., and Miquel, J., Eds., Alan R. Liss, New York, 1985, 141.
67. **Gutteridge, J. M. C.,** Aspects to consider when detecting and measuring lipid peroxidation, *Free Rad. Res. Commun.,* 1, 173, 1986.
68. **Cheeseman, K. H., Beavis, A., and Esterbauer, H.,** Hydroxyl-radical-induced iron-catalyzed degradation of 2-deoxyribose. Quantitative determination of malondialdehyde, *Biochem. J.,* 252, 649, 1988.
69. **Cohen, G.,** Production of ethane and pentane during lipid peroxidation: biphasic effect of oxygen, in *Lipid Peroxides in Biology and Medicine,* Yagi, K., Ed., Academic Press, New York, 1982, 23.
70. **Kostrucha, J. and Kappus, H.,** Inverse relationship of ethane or n-pentane and malondialdehyde formed during lipid peroxidation in rat liver microsomes with different oxygen concentrations, *Biochim. Biophys. Acta,* 879, 120, 1986.
71. **Rosen, G. M. and Rauckman, E. J.,** Spin trapping of free radicals during hepatic microsomal lipid peroxidation, *Proc. Natl. Acad. Sci. U.S.A.,* 78, 7346, 1981.
72. **Osipov, A. N., Savov, V. M., Zubarev, V. E., Azizova, O. A., and Vladimirov, Yu. A.,** Participation of Fe in the formation of OH radicals in superoxide-generating systems, *Biofizika,* 26, 193, 1981.
73. **Osipov, A. N., Savov, V. M., Yakhyaev, A. V., Zubarev, V. E., Azizova, O. A., Kagan, V. E., Kozlov, Yu. P., and Vladimirov, Yu. A.,** Study of the role of free radicals formed in the Fe-ascorbate system under lipid peroxidation, *Biofizika,* 28, 204, 1983.
74. **Azizova, O. A., Osipov, A. N., Savov, V. M., Zubarev, V. E., Kagan, V. E., and Vladimirov, Yu. A.,** Study of the mechanism of linoleic acid free radical formation under lipoperoxidation initiation in Fenton's reagent using spin traps, *Biofizika,* 29, 766, 1984.
75. **Azizova, O. A., Osipov, A. N., Savov, V. M., Yakhyaev, A. V., Zubarev, V. E., Kagan, V. E., and Vladimirov, Yu. A.,** Induction of non-enzyme lipid peroxidation in Fe^{2+}-ascorbate-linolenate system, *Biofizika,* 30, 36, 1985.
76. **Weimann, A., Hildebrandt, A. G., and Kahl, R.,** Different efficiency of various synthetic antioxidants towards NADPH induced chemiluminescence in rat liver microsomes, *Biochem. Biophys. Res. Commun.,* 125, 1033, 1984.
77. **Goddard, J. G. and Sweeney, G. D.,** Delayed ferrous ion-dependent peroxidation of rat liver microsomes, *Arch. Biochem. Biophys.,* 259, 372, 1987.
78. **Teebor, G. W., Boorstein, R. J., and Cadet, J.,** The repairability of oxidative free radical mediated damage to DNA: a review, *Int. J. Radiat. Biol.,* 54, 131, 1988.
79. **Noll, T., de Groot, H., and Sies, H.,** Distinct temporal relation among oxygen uptake, malondialdehyde formation, and low-level chemiluminescence during microsomal lipid peroxidation, *Arch. Biochem. Biophys.,* 252, 284, 1987.

80. **Pederson, T. C. and Aust, S. D.,** The role of superoxide and singlet oxygen in lipid peroxidation promoted by xanthine oxidase, *Biochem. Biophys. Res. Commun.*, 52, 107, 1973.
81. **Noguchi, T. and Nakano, M.,** Effect of ferrous ions on microsomal phospholipid peroxidation and related light emission, *Biochim. Biophys. Acta*, 368, 446, 1974.
82. **Fong, K. L., McCay, P. B., Poyer, J. L., Keele, B. B., and Misra, H.,** Evidence that peroxidation of lysosomal membranes is initiated by hydroxyl free radicals produced during flavin enzyme activity, *J. Biol. Chem.*, 248, 7792, 1973.
83. **Tyler, D. D.,** Role of superoxide radicals in the lipid peroxidation of intracellular membranes, *FEBS Lett.*, 51, 180, 1975.
84. **Kellogg, E. W. and Fridovich, I.,** Liposome oxidation and erythrocyte lysis by enzymically generated superoxide and hydrogen peroxide, *J. Biol. Chem.*, 252, 6721, 1977.
85. **Fridovich, S. E. and Porter, N. A.,** Oxidation of arachidonic acid in micelles by superoxide and hydrogen peroxide, *J. Biol. Chem.*, 256, 260, 1981.
86. **Thomas, M. J., Mehl, K. S., and Pryor, W. A.,** The role of superoxide in xanthine oxidase-induced autoxidation of linoleic acid, *J. Biol. Chem.*, 257, 8343, 1982.
87. **Girotti, A. W. and Thomas, J. P.,** Superoxide and hydrogen peroxide-dependent lipid peroxidation in intact and Triton-dispersed erythrocyte membranes, *Biochem. Biophys. Res. Commun.*, 118, 474, 1984.
88. **Tien, M., Svingen, B. A., and Aust, S. D.,** An investigation of the role of hydroxyl radical in xanthine oxidase-dependent lipid peroxidation, *Arch. Biochem. Biophys.*, 216, 142, 1982.
89. **Gutteridge, J. M. C., Beard, A. P. C., and Quinlau, G. J.,** Superoxide-dependent lipid peroxidation. Problems with the use of catalase as a specific probe for Fenton-derived hydroxyl radicals, *Biochem. Biophys. Res. Commun.*, 117, 901, 1983.
90. **Miura, T. and Ogiso, T.,** Lipid peroxidation of erythrocyte membrane induced by xanthine oxidase system: modification of superoxide dismutase effect by hemoglobin, *Chem. Pharm. Bull.*, 30, 3662, 1982.
91. **Miura, T. and Ogiso, T.,** Luminol chemiluminescence and peroxidation of unsaturated fatty acid induced by the xanthine oxidase system: effect of oxygen radical scavengers, *Chem. Pharm. Bull.*, 33, 3402, 1985.
92. **Janero, D. R. and Burghardt, B.,** Thiobarbituric acid-reactive malondialdehyde formation during superoxide-dependent, iron-catalyzed lipid peroxidation: influence of peroxidation conditions, *Lipids*, 24, 125, 1989.
93. **Shaw, S. and Jayatilleke, E.,** Acetaldehyde-mediated hepatic lipid peroxidation: role of superoxide and ferritin, *Biochem. Biophys. Res. Commun.*, 143, 984, 1987.
94. **Girroti, A. W., Thomas, J. P., and Jordan, J. E.,** Lipid photooxidation in erythrocyte ghosts: sensitization of the membranes toward ascorbate- and superoxide-induced peroxidation and lysis, *Arch. Biochem. Biophys.*, 236, 238, 1985.
95. **Kukreja, R. C., Okabe, E., Schrier, G. M., and Hess, M. L.,** Oxygen radical- mediated lipid peroxidation and inhibition of Ca^{2+}-ATPase activity of cardiac sarcoplasmic reticulum, *Arch. Biochem. Biophys.*, 261, 447, 1988.
96. **Arroyo, C. M., Mak, I. T., and Weglicki, W. B.,** Spin trapping of free radicals formed during peroxidation of sarcolemmal membranes (SLM), *Free Rad. Res. Commun.*, 5, 369, 1989.
97. **Stossel, T. P., Mason, R. J., and Smith, A. L.,** Lipid peroxidation by human blood phagocytes, *J. Clin. Invest.*, 54, 638, 1974.
98. **Carlin, G.** Peroxidation of linolenic acid promoted by human polymorphonuclear leukocytes, *J. Free Rad. Biol. Med.*, 1, 255, 1985.
99. **Carlin, G. and Arfors, K. E.,** Peroxidation of liposomes promoted by human polymorphonuclear leukocytes, *J. Free Rad. Biol. Med.*, 1, 437, 1984.
100. **Hochstein, P. and Ernster, L.,** ADP-activated lipid peroxidation coupled to the TPNH oxidase system of microsomes, *Biochem. Biophys. Res. Commun.*, 12, 388, 1963.
101. **Svingen, B. A., Buege, J. A., O'Neal, F. O., and Aust, S. D.,** The mechanism of NADPH-dependent lipid peroxidation, *J. Biol. Chem.*, 254, 5892, 1979.
102. **Ursini, F., Maiorino, M. Hochstein, P., and Ernster, L.,** Microsomal lipid peroxidation: mechanisms of initiation. The role of iron and iron chelators, *Free Rad. Biol. Med.*, 6, 31, 1989.
103. **Lai, C. and Piette, L. H.,** Hydroxyl radical production involved in lipid peroxidation of rat liver microsomes, *Biochem. Biophys. Res. Commun.*, 78, 51, 1977.
104. **Rosen, G. M. and Rauckman, E. J.,** Spin trapping of free radicals during hepatic microsomal lipid peroxidation, *Proc. Natl. Acad. Sci. U.S.A.*, 78, 7346, 1981.
105. **Koster, J. F. and Slee, R. G.,** Lipid peroxidation of rat liver microsomes, *Biochim. Biophys. Acta*, 620, 489, 1980.
106. **Kameda, K., Ono, T., and Imai, Y.,** Participation of superoxide, hydrogen peroxide and hydroxyl radicals in NADPH-cytochrome P-450 reductase-catalyzed peroxidation of methyl linoleate, *Biochim. Biophys. Acta*, 572, 77, 1979.
107. **Kornbrust, D. J. and Mavis, R. D.,** Microsomal lipid peroxidation. I. Characterization of the role of iron and NADPH, *Mol. Pharmacol.*, 17, 400, 1980.

108. **Klimek, J.,** The involvement of superoxide and iron ions in the NADPH-dependent lipid peroxidation in human placental mitochondria, *Biochim. Biophys. Acta,* 958, 31, 1988.
109. **Morehouse, L. A., Thomas, C. E., and Aust, S. D.,** Superoxide generation by NADPH-cytochrome P-450 reductase: the effect of iron chelators and the role of superoxide in microsomal lipid peroxidation, *Arch. Biochem. Biophys.,* 232, 366, 1984.
110. **Takayanagi, R., Takeshige, K., and Minakami, S.,** NADH- and NADPH-dependent lipid peroxidation in bovine heart submitochondrial particles. Dependence of the rate of electron flow in the respiratory chain and an antioxidant role of ubiquinone, *Biochem. J.,* 192, 853, 1980.
111. **Takeshige, K., Takayanagi, R., and Minakami, S.,** Lipid peroxidation and the reduction of ADP-Fe^{3+} chelate by NADH-ubiquinone reductase preparation from bovine heart mitochondria, *Biochem. J.,* 192, 861, 1980.
112. **Cavallini, L., Valente, M., and Bindoli, A.,** Comparison of cumene hydroperoxide- and NADPH/Fe^{3+}/ADP-induced lipid peroxidation in heart and liver submitochondrial particles, *Biochim. Biophys. Acta,* 795, 466, 1984.
113. **Fukuzawa, K., Tadakoro, T., Kishikawa, K., Mukai, K., and Gebicki, J. M.,** Site-specific induction of lipid peroxidation by iron in charged micelles, *Arch. Biochem. Biophys.,* 260, 146, 1988.
114. **Gutteridge, J. M. C.,** The role of superoxide and hydroxyl radicals in phospholipid peroxidation catalysed by iron salts, *FEBS Lett.,* 150, 454, 1982.
115. **Afanas'ev, I. B., Dorozhko, A. I., Brodskii, A. V., Kostyuk, V. A., and Potapovitch, A. I.,** Chelating and free radical scavenging mechanisms of inhibitory action of rutin and quercetin in lipid peroxidation, *Biochem. Pharmacol.,* 38, 1763, 1989.
116. **Miller, D. M. and Aust, S. D.,** Studies of ascorbate-dependent, iron-catalyzed lipid peroxidation, *Arch. Biochem. Biophys.,* 271, 113, 1989.
117. **Searle, A. J. F. and Willson, R. L.,** Stimulation of microsomal lipid peroxidation by iron and cysteine. Characterization and the role of free radicals, *Biochem. J.,* 212, 549, 1983.
118. **Kanner, J., Harel, S., and Hazan, B.,** Muscle membranal lipid peroxidation by an "iron redox cycle" system: initiation by oxy radicals and site-specific mechanism, *J. Agric. Food Chem.,* 34, 506, 1986.
119. **Slater, T. F. and Sawyer, B. C.,** The stimulatory effects of carbon tetrachloride and other halogenalkanes on peroxidative reactions in rat liver fractions *in vitro, Biochem. J.,* 123, 805, 1971.
120. **Kornbrust, D. J. and Mavis, R. D.,** Microsomal lipid peroxidation. II. Stimulation by carbon tetrachloride, *Mol. Pharmacol.,* 17, 408, 1980.
121. **Keller, R. J., Sharma, R. P., Grover, T. A., and Piette, L. H.,** Vanadium and lipid peroxidation; evidence for involvement of vanadyl and hydroxyl radical, *Arch. Biochem. Biophys.,* 265, 524, 1988.
122. **Mason, R. P., Kalyanaraman, B., Tainer, B. E., and Eling, T. E.,** A carbon-centered free radical intermediate in the prostaglandin synthetase oxidation of arachidonic acid. Spin trapping and oxygen uptake studies, *J. Biol. Chem.,* 255, 5019, 1980.
123. **Connor, H. D., Fischer, V., and Mason, R. P.,** A search for oxygen-centered free radicals in the lipoxygenase/linoleic acid system, *Biochem. Biophys. Res. Commun.,* 141, 614, 1986.
124. **Vries, J. D. and Verboom, C. N.,** Effect of scavengers of superoxide radicals, hydrogen peroxide, singlet oxygen and hydroxyl radicals on malondialdehyde generation from arachidonic acid by bovine seminal vesicle microsomes, *Experientia,* 36, 1339, 1980.
125. **Svingen, B. A., O'Neal, F. O., and Aust, S. D.,** The role of superoxide and singlet oxygen in lipid peroxidation, *Photochem. Photobiol.,* 28, 803, 1978.
126. **Tien, M., Svingen, B. A., and Aust, S. D.,** Superoxide dependent lipid peroxidation, *Fed. Proc. Fed. Am. Soc. Exp. Biol.,* 40, 179, 1981.
127. **Bast, A. and Steeghs, M. H. M.,** Hydroxyl radicals are not involved in NADPH dependent microsomal lipid peroxidation, *Experientia,* 42, 555, 1986.
128. **Gutteridge, J. M. C.,** Ferrous ion-EDTA-stimulated phospholipid peroxidation. A reaction changing from alkoxyl-radical- to hydroxyl-radical-dependent initiation, *Biochem. J.,* 224, 697, 1984.
129. **Vile, G. F. and Winterbourn, C. C.,** Iron binding to microsomes and liposomes in relation to lipid peroxidation, *FEBS Lett.,* 215, 151, 1987.
130. **Marton, A., Sukosd-Roazlosnik, N., Vertes, A., and Horvath, I.,** The effect of EDTA-Fe(III) complexes with different chemical structure on the lipid peroxidation in brain microsomes, *Biochem. Biophys. Res. Commun.,* 145, 211, 1987.
131. **Puntarulo, S. and Cederbaum, A. I.,** Comparison of the ability of ferric complexes to catalyze microsomal chemiluminescence, lipid peroxidation, and hydroxyl radical generation, *Arch. Biochem. Biophys.,* 264, 482, 1988.
132. **Bucher, J. R., Tien, M., and Aust, S. D.,** The requirement for ferric in the initiation of lipid peroxidation by chelated ferrous iron, *Biochem. Biophys. Res. Commun.,* 111, 777, 1983.
133. **Braughler, J. M., Duncan, L. A., and Chase, R. L.,** The involvement of iron in lipid peroxidation. Importance of ferric to ferrous ratios in initiation, *J. Biol. Chem.,* 261, 10282, 1986.

134. **Minotti, G. and Aust, S. D.,** The requirement for iron(III) in the initiation of lipid peroxidation by iron(II) and hydrogen peroxide, *J. Biol. Chem.*, 262, 1098, 1987.
135. **Goddard, J. G. and Sweeney, G. D.,** Delayed, ferrous ion-dependent peroxidation of rat liver microsomes, *Arch. Biochem. Biophys.*, 259, 372, 1987.
136. **Minotti, G. and Aust, S. D.,** An investigation into the mechanism of citrate-Fe^{2+}-dependent lipid peroxidation, *Free Rad. Biol. Med.*, 3, 379, 1987.
137. **Aust, S. D., Morehouse, L. A., and Thomas, C. E.,** Role metals in oxygen radical reactions, *J. Free Rad. Biol. Med.*, 1, 3, 1985.
138. **Gutteridge, J. M. C., Halliwell, B., Treffry, A., Harrison, P. M., and Blake, D.,** Effect of ferritin-containing fractions with different iron loading on lipid peroxidation, *Biochem. J.*, 209, 557, 1983.
139. **Thomas, C. E., Morehouse, L. A., and Aust, S. D.,** Ferritin and superoxide-dependent lipid peroxidation, *J. Biol. Chem.*, 260, 3275, 1985.
140. **Koster, J. F. and Slee, R. G.,** Ferritin, a physiological donor for microsomal lipid peroxidation, *FEBS Lett.*, 199, 85, 1986.
141. **Reif, D. W., Samokyszyn, V. M., Miller, D. M., and Aust, S. D.,** Alloxan- and glutathione-dependent ferritin iron release and lipid peroxidation, *Arch. Biochem. Biophys.*, 269, 407, 1989.
142. **Gutteridge, J. M. C.,** Iron promotors of the Fenton reaction and lipid peroxidation can be released from haemoglobin by peroxides, *FEBS Lett.*, 201, 291, 1986.
143. **Kanner, J. and Harel, S.,** Initiation of membranal lipid peroxidation by activated metmyoglobin and methemoglobin, *Arch. Biochem. Biophys.*, 237, 314, 1985.
144. **McCord, J. M. and Fridovich, I.,** The reduction of cytochrome c by milk xanthine oxidase, *J. Biol. Chem.*, 243, 5753, 1968.
145. **Vandewalle, P. L. and Peterson, N. O.,** Oxidation of reduced cytochrome c by hydrogen peroxide: implications for superoxide assays, *FEBS Lett.*, 210, 195, 1987.
146. **Turrens, J. F. and McCord, J. M.,** How relevant is the reoxidation of ferrocytochrome c by hydrogen peroxide when determining superoxide anion production?, *FEBS Letts.*, 277, 43, 1988.
147. **Ballou, D., Palmer, G., and Massey, V.,** Direct demonstration of superoxide anion production during the oxidation of reduced flavin and its catalytic decomposition by erythrocuprein, *Biochem. Biophys. Res. Commun.*, 36, 898, 1969.
148. **Land, E. J. and Swallow, A. J.,** One-electron reactions in biochemical systems as studied by pulse radiolysis. V. Cytochrome c, *Arch. Biochem. Biophys.*, 145, 365, 1971.
149. **Simic, M., Taub, I. A., Tocci, J., and Hurwitz, P. A.,** Free radical reduction of ferricytochrome-c, *Biochem. Biophys. Res. Commun.*, 62, 161, 1975.
150. **Seki, H., Ilan, Y. A., Ilan, Y., and Stein, G.,** Reactions of the ferri-ferrocytochrome-c system with superoxide/oxygen and CO_2^-/CO_2 studied by fast pulse radiolysis, *Biochim. Biophys. Acta*, 440, 573, 1976.
151. **Butler, J., Jayson, G. G., and Swallow, A. J.,** The reaction between the superoxide anion radical and cytochrome c, *Biochim. Biophys. Acta*, 408, 215, 1975.
152. **Koppenol, W. H., van Buren, K. J. H., Butler, J., and Braams, R.,** The kinetics of the reduction of cytochrome c by the superoxide anion radical, *Biochim. Biophys. Acta*, 449, 157, 1976.
153. **Butler, J., Koppenol, W. H., and Margohash, E.,** Kinetics and mechanism of the reduction of ferricytochrome c by the superoxide anion, *J. Biol. Chem.*, 257, 10789, 1982.
154. **Heijman, M. G. J., Nauta, H., and Levine, Y. K.,** The influence of a detergent on the reactivity of cytochrome c toward the superoxide radical as measured by pulse radiolysis, *Biochim. Biophys. Acta*, 704, 560, 1982.
155. **Azzi, A., Montecucco, C., and Richter, C.,** The use of acetylated ferri-cytochrome c for the detection of superoxide radicals produced in biological membranes, *Biochem. Biophys. Res. Commun.*, 65, 597, 1975.
156. **Ilan, Y., Shafferman, A., Feinberg, B. A., and Lau, Y.,** Partitioning of electrostatic and conformational contributions in the redox reactions of modified cytochromes c, *Biochim. Biophys. Acta*, 548, 565, 1979.
157. **Kuthan, H., Ullrich, V., and Eastabrook, R. A.,** A quantitative test for superoxide radicals produced in biological systems, *Biochem. J.*, 203, 551, 1982.
158. **Finkelstein, E., Rosen, G. M., Patton, S. E., Cohen, M. S., and Rauckman, E. J.,** Effect of modification of cytochrome c on its reactions with superoxide and NADPH cytochrome P-450 reductase, *Biochem. Biophys. Res. Commun.*, 102, 1008, 1981.
159. **Morel, F., Dianoux, A. C., and Vignais, P. V.,** Superoxide anion measurement by sulfonated phenyl isothiocyanate cytochrome c, *Biochem. Biophys. Res. Commun.*, 156, 1175, 1988.
160. **Debey, P., Land, E. J., Santus, R., and Swallow, A. J.,** Electron transfer from pyridinyl radicals, hydrated electrons, $CO_2^{\cdot}$, $O_2^{\cdot}$ to bacterial cytochrome P_{450}, *Biochem. Biophys. Res. Commun.*, 86, 953, 1979.
161. **Ferradini, C., Foos, J., Gilles, L., Haristoy, D., and Pucheault, J.,** Gamma and pulse radiolysis studies of the reactions between superoxide ions and oxyhemoglobin-methemoglobin system, *Photochem. Photobiol.*, 28, 851, 1978.

162. **Lynch, R. E., Thomas J. E., and Lee, G. R.,** Inhibition of methemoglobin formation from purified oxyhemoglobin by superoxide dismutase, *Biochemistry,* 16, 4563, 1977.
163. **Winterbourn, C. C., McGrath, B. M., and Carrell, R. W.,** Reactions involving superoxide and normal and unstable haemoglobins, *Biochem. J.,* 155, 493, 1976.
164. **Sutton, H. C., Roberts, P. B., and Winterbourn, C. C.,** The rate of reaction of superoxide radical ion with oxyhaemoglobin and methaemoglobin, *Biochem. J.,* 155, 503, 1976.
165. **Thillet, J. and Michelson, A. M.,** Oxidation and cross-linking of human hemoglobins by activated oxygen species, *Free Rad. Res. Commun.,* 1, 89, 1986.
166. **Rister, M. and Baehner, R. L.,** The alteration of superoxide dismutase, catalase, glutathione peroxidase, and NADPH cytochrome c reductase in guinea pig polymorphonuclear leukocytes and alveolar macrophages during hyperoxia, *J. Clin. Invest.,* 58, 1174, 1986.
167. **Kono, Y. and Fridovich, I.,** Superoxide radical inhibits catalase, *J. Biol. Chem.,* 257, 5751, 1982.
168. **Shimizu, N., Kobayashi, K., and Hayashi, K.,** The mechanism of superoxide radical reaction with catalase. Mechanism of the inhibition of catalase by superoxide radical, *J. Biol. Chem.,* 259, 4414, 1984.
169. **Kobayashi, K., Shimizu, N., and Hayashi, K.,** The reactions of superoxide radicals with catalase, in *Superoxide and Superoxide Dismutase in Chemistry, Biology and Medicine,* Rotilio, G., Ed., Elsevier, Amsterdam, 1986, 228.
170. **Gebicka, L., Metodiewa, D., and Gebicki, J. L.,** Pulse radiolysis of catalase in solution. I. Reactions of superoxide anion with catalase and its compound I, *Int. J. Radiat. Biol.,* 55, 45, 1989.
171. **Davison, A. J., Kettle, A. J., and Fatur, D. J.,** Mechanism of the inhibition of catalase by ascorbate. Roles of active oxygen species, copper semidehydroascorbate, *J. Biol. Chem.,* 261, 1193, 1986.
172. **Odajima, T. and Yamazaki, I.,** Myeloperoxidase of the leukocyte of normal blood. III. The reaction of ferric myeloperoxidase with superoxide ion, *Biochim. Biophys. Acta,* 284, 355, 1972.
173. **Rabani, J., Klug, D., and Fridovich, I.,** Decay of the HO_2 and O_2^- radicals catalyzed by superoxide dismutase. A pulse radiolytic investigation, *Isr. J. Chem.,* 10, 1095, 1972.
174. **Cuperus, R. A., Muijsers, A. O., and Wever, R.,** The superoxide dismutase activity of myeloperoxidase; formation of Compound III, *Biochim. Biophys. Acta,* 871, 78, 1986.
175. **Metodiewa, D. and Dunford, H. B.,** The reactions of horseradish peroxidase, lactoperoxidase, and myeloperoxidase with enzymatically generated superoxide, *Arch. Biochem. Biophys.,* 272, 245, 1989.
176. **Kettle, A. J., Sangster, D. F., Gebicki, J. M., and Winterbourn, C. C.,** A pulse radiolysis investigation of the reactions of myeloperoxidase with superoxide and hydrogen peroxide, *Biochim. Biophys. Acta,* 956, 58, 1988.
177. **Kettle, A. J. and Winterbourn, C. C.,** Superoxide modulates the activity of myeloperoxidase and optimizes the production of hypochlorous acid, *Biochem. J.,* 252, 529, 1988.
178. **Shimizu, N., Kobayashi, K., and Hayashi, K.,** Kinetics of the reaction of superoxide anion with ferric horseradish peroxidase, *Biochim. Biophys. Acta,* 995, 133, 1989.
179. **Bielski, B. H. J. and Gebicki, J. M.,** Study of peroxidase mechanisms by pulse radiolysis. III. The rate of reaction of O_2^- and HO_2 radicals with horseradish peroxidase compound I, *Biochim. Biophys. Acta,* 364, 233, 1974.
180. **Jenzer, H., Kohler, H., and Broger, C.,** The role of hydroxyl radicals in irreversible inactivation of lactoperoxidase by excess H_2O_2. A spin-trapping/ESR and absorption spectroscopy study, *Arch. Biochem. Biophys.,* 258, 381, 1987.
181. **Hirata, F. and Hayaishi, O.,** Possible participation of superoxide anion in the intestinal tryptophan 2,3-dioxygenase reaction, *J. Biol. Chem.,* 246, 7825, 1971.
182. **Hayaishi, O., Hirato, F., Ohnishi, T., Henry, J.-P., Rosenthal, I., and Katoh, A.,** Indoleamine 2, 3-dioxygenase. Incorporation of $^{18}O_2^-$ and $^{18}O_2$ into reaction products, *J. Biol. Chem.,* 252, 3548, 1977.
183. **Hirata, F., Ohnishi, T., and Hayaishi, O.,** Indoleamine 2,3-dioxygenase. Characterization and properties of enzyme·O_2^- complex, *J. Biol. Chem.,* 252, 4637, 1977.
184. **Ohnishi, T., Hirata, F., and Hayaishi, O.,** Indoleamine 2,3-dioxygenase. Potassium superoxide as substrate, *J. Biol. Chem.,* 252, 4643, 1977.
185. **Taniguchi, T., Sono, M., Hirata, F., Hataishi, O., Tamuro, M., Hayashi, K., Iizuka, T., and Ishimura, Y.,** Indoleamine 2,3-dioxygenase. Kinetic studies on the binding of superoxide anion and molecular oxygen to enzyme, *J. Biol. Chem.,* 254, 3288, 1979.
186. **Taniguchi, T., Hirata, F., and Hayaishi, O.,** Intracellular utilization of superoxide anion by indoleamine 2,3-dioxygenase of rabbit enterocytes, *J. Biol. Chem.,* 252, 2777, 1977.
187. **Sono, M.,** The roles of superoxide anion and methylene blue in the reductive activation of indoleamine 2,3-dioxygenase by ascorbic acid or by xanthine oxidase-hypoxanthine, *J. Biol. Chem.,* 264, 1616, 1989.
188. **Ozaki, Y., Nichol, C. A., and Duch, D. S.,** Utilization of dihydroflavin mononucleotide and superoxide anion for the decyclization of L-tryptophan by murine epididymal indoleamine 2,3-dioxygenase, *Arch. Biochem. Biophys.,* 257, 100, 1987.
189. **Brady, F. O., Forman, H. J., and Feigelson, P.,** The role of superoxide and hydroperoxide in the reductive activation of tryptophan-2,3-dioxygenase, *J. Biol. Chem.,* 246, 7119, 1971.

190. **Oberley, L. W., Ed.,** *Superoxide Dismutase,* Vol. 1, 2, and 3, CRC Press, Boca Raton, FL, 1981, 1982, and 1985.
191. **McCord, J. M. and Fridovich, I.,** Superoxide dismutase. An enzymic function for erythrocuprein (hemocuprein), *J. Biol. Chem.*, 244, 6049, 1969.
192. **Fridovich, I.,** Superoxide radical and superoxide dismutase, *Acc. Chem. Res.*, 5, 321, 1972.
193. **Forman, H. J. and Fridovich, I.,** Superoxide dismutase: a comparison of rate constants, *Arch. Biochem. Biophys.*, 158, 396, 1973.
194. **Klug-Roth, D., Fridovich, I., and Rabani, J.,** Pulse-radiolytic investigations of superoxide catalyzed disproportionation. Mechanism for bovine superoxide dismutase, *J. Am. Chem. Soc.*, 95, 2786, 1973.
195. **Bannister, J. V., Bannister, W. H., Bray, R. C., Fielden, E. M., Roberts, P. B., and Rotilio, G.,** The superoxide dismutase activity of human erythrocuprein, *FEBS Lett.*, 32, 303, 1973.
196. **Brigelius, R., Spoettl, R., Bors, W., Lengfelder, E., Saran, M., and Weser, U.,** Superoxide dismutase activity of low molecular weight Cu^{2+}-chelates studied by pulse radiolysis, *FEBS Lett.*, 47, 72, 1974.
197. **Takahashi, M. and Asada, K.,** A flash-photometric method for determination of reactivity of superoxide: application to superoxide dismutase assay, *J. Biochem.*, 91, 889, 1982.
198. **Rigo, A., Tomat, R., and Rotilio, G.,** Determination of superoxide dismutase activity and rate constants by the polarographic catalytic currents method, *J. Electroanal. Chem.*, 57, 291, 1974.
199. **Fee, J. A. and Bull, C.,** Steady-state kinetic studies of superoxide dismutases. Sturative behavior of the copper- and zinc-containing protein, *J. Biol. Chem.*, 261, 13000, 1986.
200. **Viglino, P., Rigo, A., Argese, E., Calabrese, L., Cocco, D., and Rotilio, G.,** ^{19}F Relaxation as a probe of the oxidation state of Cu,Zn superoxide dismutase. Studies of the enzyme in steady-state turnover, *Biochem. Biophys. Res. Commun.*, 100, 125, 1981.
201. **Fielden, E. M., Roberts, P. B., Bray, R. C., Lowe, D. J., Mautner, G. N., Rotilio, G., and Calabrese, L.,** The mechanism of action of superoxide dismutase from pulse radiolysis and electron paramagnetic resonance. Evidence that only half the active sites function in catalysis, *Biochem. J.*, 139, 49, 1974.
202. **Symons, M. C. R. and Stephenson, J. M.,** Mechanism of the reaction of bovine copper-zinc superoxide dismutase. An electron spin resonance study, *J. Chem. Soc., Faraday Trans.* 1, 2983, 1983.
203. **Rigo, A., Viglino, P., and Rotilio, G.,** Kinetic study of $O_2^{\cdot}$ dismutation by bovine superoxide dismutase. Evidence for saturation of the catalytic site by O_2^-, *Biochem. Biophys. Res. Commun.*, 63, 1013, 1975.
204. **Viglino, P., Scarpa, M., Coin, F., Rotilio, G., and Rigo, A.,** Oxidation of reduced copper-zinc superoxide dismutase by molecular oxygen. A kinetic study, *Biochem. J.*, 237, 305, 1986.
205. **Pick, M., Rabani, J., Yost, F., and Fridovich, I.,** The catalytic mechanism of the manganese-containing superoxide dismutase of *Escherichia coli* studied by pulse radiolysis, *J. Am. Chem. Soc.*, 96, 7329, 1974.
206. **McAdam, M. E., Lavelle, F., Fox, R. A., and Fielden, E. M.,** A pulse radiolysis study of the manganese-containing superoxide dismutase from *Bacillus stearothermophibis.* A kinetic model for the enzyme action, *Biochem. J.*, 165, 71, 1977.
207. **Lavelle, F., McAdam, M. E., Fielden, E. M., Roberts, P. B., Puget, K., and Michelson, A. M.,** A pulse-radiolysis study of the catalytic mechanism of the iron-containing superoxide dismutase from *Photobacterium leignathi, Biochem. J.*, 161, 3, 1977.
208. **Fee, J. A., McClune, G. J., O'Neill, P., and Fielden, E. M.,** Saturated behavior of superoxide dismutation catalyzed by the iron containing superoxide dismutase from *E. coli* B, *Biochem. Biophys. Res. Commun.*, 100, 377, 1981.
209. **Terech, A., Pucheault, J., and Ferradini, C.,** Saturation behavior of the manganese-containing superoxide dismutase from *Paracoccus denitrificans, Biochem. Biophys. Res. Commun.*, 113, 114, 1983.
210. **Rigo, A., Viglino, P., Rotilio, G., and Tomat, R.,** Effect of ionic strength on the activity of bovine superoxide dismutase, *FEBS Lett.*, 50, 86, 1975.
211. **McAdam, M. E.,** A consideration of the effects of added solutes on the activity of bovine superoxide dismutase, *Biochem. J.*, 161, 697, 1977.
212. **Cudd, A. and Fridovich, I.,** Electrostatic interactions in the reaction mechanism of bovine erythrocyte superoxide dismutase, *J. Biol. Chem.*, 257, 11443, 1982.
213. **Benovic, J., Tillman, T., Cudd, A., and Fridovich, I.,** Electrostatic facilitation of the reaction catalyzed by the manganese-containing and iron-containing superoxide dismutases, *Arch. Biochem. Biophys.*, 221, 329, 1983.
214. **Getzoff, E. D., Tainer, J. A., Weiner, P. K., Kollman, P. A., Richardson, J. S., and Richardson, D. C.,** Electrostatic interaction between superoxide and copper, zinc superoxide dismutase, *Nature,* 306, 287, 1983.
215. **Semsei, I. and Zs.-Nagy, I.,** Effect of ionic strength on the activity of superoxide dismutase in vitro, *Arch. Gerontol. Geriatr.*, 3, 287, 1984.
216. **O'Neill, P., Davies, S., Fielden, E. M., Calabrese, L., Capo, C., Marmocchi, F., Natoli, G., and Rotilio, G.,** The effects of pH and various salts upon the activity of a series of superoxide dismutases, *Biochem. J.*, 251, 41, 1988.

217. **Hodgson, E. K. and Fridovich, I.,** The interaction of bovine erythrocyte superoxide dismutase with hydrogen peroxide: inactivation of the enzyme, *Biochemistry,* 14, 5294, 1975.
218. **Richardson, J. S., Thomas, K. A., Rubin, B. H., and Richardson, D. C.,** Crystal structure of bovine Cu,Zn superoxide dismutase at 3A resolution: chain tracing and metal ligands, *Proc. Natl. Acad. Sci. U.S.A.,* 72, 1349, 1975.
219. **McAdam, M. E., Fielden, E. M., Lavelle, F., Calabrese, L., Cocco, D., and Rotilio, G.,** The involvement of the bridging imidazolate in the catalytic mechanism of action of bovine superoxide dismutase, *Biochem. J.,* 167, 271, 1977.
220. **Fee, J. A. and Ward, R. L.,** Evidence for a coordination position available to solute molecules on one of the metals at the active center of reduced superoxide dismutase, *Biochem. Biophys. Res. Commun.,* 71, 427, 1976.
221. **Bull, C. and Fee, J. A.,** Steady-state kinetic studies of superoxide dismutases: properties of the iron containing protein from *Escherichia coli, J. Am. Chem. Soc.,* 107, 3295, 1985.
222. **Argese, E., Viglino, P., Rotilio, G., Scarpa, M., and Rigo, A.,** Electrostatic control of the rate-determining step of the copper, zinc superoxide dismutase catalytic reaction, *Biochemistry,* 26, 3224, 1987.
223. **Fielden, E. M., Roberts, P. B., Bray, R. C., and Rotilio, G.,** Mechanism and inactivation of superoxide dismutase activity, *Biochem. Soc. Trans.,* 1, 52, 1973.
224. **Bray, R. C., Cockle, S. A., Fielden, E. M., Roberts, P. B., Rotilio, G., and Calabrese, L.,** Reduction and inactivation of superoxide dismutase by hydrogen peroxide, *Biochem. J.,* 139, 43, 1974.
225. **Beyer, W. F. and Fridovich, I.,** Effect of hydrogen peroxide on the iron-containing superoxide dismutase of *Escherichia coli, Biochemistry,* 26, 1251, 1987.
226. **Blech, D. M. and Borders, C. L.,** Hydroperoxide anion, HO_2^- is an affinity reagent for the inactivation of yeast Cu,Zn superoxide dismutase: modification of one histidine per subunit, *Arch. Biochem. Biophys.,* 224, 579, 1983.
227. **Viglino, P., Scarpa, M., Rotilio, G., and Rigo, A.,** A kinetic study of the reactions between hydrogen peroxide and copper-zinc superoxide dismutase; evidence for an electrostatic control of the reaction rate, *Biochim. Biophys. Acta,* 952, 77, 1988.
228. **Cleveland, L. and Davis, L.,** Superoxide dismutase activity of galactose oxidase, *Biochim. Biophys. Acta,* 341, 517, 1974.
229. **Markossian, K. A., Poghossian, A. A., Paitian, N. A., and Nalbandian, R. M.,** Superoxide dismutase activity of cytochrome oxidase, *Biochem. Biophys. Res. Commun.,* 81, 1336, 1978.
230. **Naqui, A. and Chance, B.,** Enhanced superoxide dismutase activity of pulsed cytochrome oxidase, *Biochem. Biophys. Res. Commun.,* 136, 433, 1986.
231. **Anderson, R. F., Massey, V., and Schopfer, L. M.,** Rate constants for the reaction of superoxide with glucose oxidase and its flavosemiquinone, in *Superoxide and Superoxide Dismutase in Chemistry, Biology and Medicine,* Rotilio, G., Ed., Elsevier, Amsterdam, 1986, 79.
232. **Henry, J.-P., Hirata, F., and Hayaishi, O.,** Superoxide anion as a cofactor of dopamine-β-hydroxylase, *Biochem. Biophys. Res. Commun.,* 81, 1091, 1978.
233. **Myllyla, R., Schubotz, L. M., Weser, U., and Kiveirikko, K. I.,** Involvement of superoxide in the prolyl and lysyl hydroxylase reactions, *Biochem. Biophys. Res. Commun.,* 89, 98, 1979.
234. **Aniya, Y. and Anders, M. W.,** Activation of rat liver microsomal glutathione S-transferase by reduced oxygen species, *J. Biol. Chem.,* 264, 1998, 1989.
235. **Mittal, C. K. and Murad, F.,** Activation of guanylate cyclase by superoxide dismutase and hydroxyl radical: a physiological regulator of guanosine 3′,5′-monophosphate formation, *Proc. Natl. Acad. Sci. U.S.A.,* 74, 4360, 1977.
236. **Schallreuter, K. U. and Wood, J. M.,** The role of thioreductase in the reduction of free radicals at the surface of the epidermus, *Biochem. Biophys. Res. Commun.,* 136, 630, 1986.
237. **Schallreuter, K. U., Pittelkow, M. R., and Wood, J. M.,** Free radical reduction by thioredoxin reductase at the surface of normal and vitiliginious human keratinocytes, *J. Invest. Dermatol.,* 87, 728, 1986.
238. **Schallreuter, K. U. and Wood, J. M.,** Free radical reduction in the human epidermis, *Free Rad. Biol. Med.,* 6, 519, 1989.
239. **Mayer, R., Widom, J., and Que, L.,** Involvement of superoxide in the reactions of the catechol dioxygenases, *Biochem. Biophys. Res. Commun.,* 92, 285, 1980.
240. **Bateman, R. C., Jr., Youngblood, W. W., Busby, W. H., Jr., and Kizer, J. S.,** Nonenzymic peptide α-amidation. Implications for a novel enzyme mechanism, *J. Biol. Chem.,* 260, 9088, 1985.
241. **Asami, S. and Akazawa, T.,** Enzymic formation of glycolate in *Chromatiun.* Role of superoxide radical in a transketolase-type mechanism, *Biochemistry,* 16, 2202, 1977.
242. **Takabe, T., Asami, S., and Akazawa, T.,** Glycolate formation catalyzed by spinach leaf transketolase utilizing the superoxide radical, *Biochemistry,* 19, 3985, 1980.
243. **Guissani, A., Henry, Y., and Gilles, L.,** Radical scavenging and electron-transfer reactions in *Polyporus versicolor* laccase. A pulse radiolysis study, *Biophys. Chem.,* 15, 177, 1982.
244. **Searle, A. J. and Willson, R. L.,** Glutathione peroxidase: effect of superoxide, hydroxyl and bromine free radicals on enzyme activity, *Int. J. Rad. Biol.,* 37, 213, 1980.

245. **Blum, J. and Fridovich, I.**, Inactivation of glutathione peroxidase by superoxide radical, *Arch. Biochem. Biophys.*, 240, 500, 1985.
246. **Henry, L. E. A., Adams, M. W. W., Rao, K. K., and Hall, D. O.**, The effect of oxygen species on the enzymatic activity of hydrogenase, *FEBS Lett.*, 122, 211, 1980.
247. **Huang, T. S., Boado, R. J., Chopra, I. J., Solomon, D. H., and Teco, G. N. C.**, The effect of free radicals on hepatic 5'-monodeiodination of thyroxine and 3,3',5'-triiodothyronine, *Endocrinology*, 121, 498, 1987.
248. **Kim, H.-S., Minard, P., Legoy, M.-D., and Tomas, D.**, Inactivation of 3α-hydroxysteroid dehydrogenase by superoxide radicals. Modification of histidine and cysteine residues causes the conformational change, *Biochem. J.*, 233, 493, 1986.
249. **Higson, F. K., Kohen, R., and Chevion, M.**, Iron enhancement of ascorbate toxicity, *Free Rad. Res. Commun.*, 5, 107, 1988.
250. **Deshpande, V. V. and Joshi, J. G.**, Vit.C-Fe(III) induced loss of the covalently bound phosphate and enzyme activity of phosphoglucomutase, *J. Biol. Chem.*, 260, 757, 1985.
251. **Carp, H. and Janoff, A.**, In vitro suppression of serum elastase-inhibitory capacity of reactive oxygen species generated by phagocytosing polymorphonuclear leukocytes, *J. Clin. Invest.*, 63, 793, 1979.
252. **Aruoma, O. I. and Halliwell, B.**, Inactivation of α-antiproteinase by hydroxyl radicals: the effect of uric acid, *FEBS Lett.*, 244, 76, 1989.
253. **Kuo, C. F., Mashino, T., and Fridovich, I.**, α,β-Dihydroxyisovalerate dehydratase. A superoxide-sensitive enzyme, *J. Biol. Chem.*, 262, 4724, 1987.
254. **Elliot, A. J., Munk, P. L., Stevenson, K. J., and Armstrong, D. A.**, Reactions of oxidizing and reducing radical probes with lipoamide dehydrogenase, *Biochemistry*, 19, 4945, 1980.
255. **Kim, K., Rhee, S. G. and Stadtman, E. R.**, Nonenzymatic cleavage of proteins by reactive oxygen species generated by dithiothreitol and iron, *J. Biol. Chem.*, 260, 15394, 1985.
256. **Kang, J. O., Chan, P. C., and Kesner, L.**, Peroxidation of lysozyme treated with Cu(II) and hydrogen peroxide, *Inorg. Chim. Acta*, 107, 253, 1985.
257. **Hexum, T. D. and Fried, R.**, Effects of superoxide radicals on transport (sodium-potassium)-dependent adenosine triphosphatase and protection by superoxide dismutase, *Neurochem. Res.*, 4, 73, 1979.
258. **Chang, M. H., Kim, J. S., Son, Y. S., Choi, K. S., and Park, C. W.**, Studies on the mechanism of reactive oxygen species inactivation of brain microsomal sodium(+)-potassium(+)-ATPase, *Seoul J. Med.*, 26, 22, 1985.
259. **Das, D. K. and Neogi, A.**, Effects of superoxide anions on the (sodium, potassium)ATPase system in rat lung, *Clin. Physiol. Biochem.*, 2, 32, 1984.
260. **Palmer, G. C.**, Free radicals generated by xanthine oxidase-hypoxanthine damage adenylate cyclase and ATPase in gerbil cerebral cortex, *Metab. Brain Dis.*, 2, 243, 1987.
261. **Kako, K., Kato, M., Matsuoka, T., and Mustapha, A.**, Depression of membrane-bound Na^+-K^+-ATPase activity induced by free radicals and by ischemia of kidney, *Am. J. Physiol.*, 254, C330, 1988.
262. **de Flombaum, M. A. C. and Stoppani, A. O. M.**, Inactivation of mitochondrial adenosine triphosphatase from *Trypanosoma cruzi* by oxygen radicals, *Biochem. Int.*, 12, 785, 1986.
263. **Samuni, A., Chevion, M., and Czapski, G.**, Unusual copper-induced sensitization of the biological damage due to superoxide radicals, *J. Biol. Chem.*, 256, 12632, 1981.
264. **Shinar, E., Navok, T., and Chevion, M.**, The analogous mechanisms of enzymatic inactivation induced by ascorbate and superoxide in the presence of copper, *J. Biol. Chem.*, 258, 14778, 1983.
265. **Armstrong, D. A. and Buchanon, J. D.**, Reactions of O_2^-, H_2O_2 and other oxidants with sulfhydryl enzymes, *Photochem. Photobiol.*, 28, 743, 1978.
266. **Guarnieri, G., Lugaresi, A., Flamigni, F., Muscari, C., and Calderera, C. M.**, Effect of oxygen radicals and hyperoxia on rat heart ornithine decarboxylase activity, *Biochim. Biophys. Acta*, 718, 157, 1982.
267. **Dean, R. T., Thomas, S. M., and Garner, A.**, Free-radical-mediated fragmentation of monoamine oxidase in the mitochondrial membrane. Role for lipid radicals, *Biochem. J.*, 240, 489, 1986.
268. **Monteiro, H. P. and Winterbourn, C. C.**, The superoxide-dependent transfer of iron from ferritin to transferrin and lactoferrin, *Biochem. J.*, 256, 923, 1988.
269. **Boyer, R. F. and McCleary, C. J.**, Superoxide ion as a primary reductant in ascorbate-mediated ferritin iron release, *Free Rad. Biol. Med.*, 3, 389, 1987.
270. **Dean, R. T.**, A mechanism for accelerated degradation of intracellular proteins after limited damage by free radicals, *FEBS Lett.*, 220, 278, 1987.
271. **Richards, D. M. C., Dean, R. T., and Jessup, W.**, Membrane proteins are critical targets in free radical mediated cytolysis, *Biochim. Biophys. Acta*, 946, 281, 1988.
272. **Zs-Nagy, I. and Floyd, R. A.**, Hydroxyl free radical reactions with amino acids and proteins studied by electron spin resonance spectroscopy and spin trapping, *Biochim. Biophys. Acta*, 790, 238, 1984.
273. **Schuessler, N. and Schilling, K.**, Oxygen effect in the radiolysis of proteins. Part 2. Bovine serum albumin, *Int. J. Radiat. Biol.*, 45, 267, 1984.

274. **Wolff, S. P. and Dean, R. T.,** Fragmentation of proteins by free radicals and its effect on their susceptibility to enzyme hydrolysis, *Biochem. J.*, 234, 399, 1986.
275. **Dean, R. T., Thomas, S. M., Vince, G., and Wolff, S. P.,** Oxidation induced proteolysis and its possible restriction by some secondary protein modifications, *Biomed. Biochim. Acta*, 45, 1563, 1986.
276. **Sakurai, K. and Miura, T.,** Generation of free radicals by alloxan in the presence of bovine serum albumin: a role of protein sulfhydryl groups in alloxan cytotoxicity, *Biochem. Int.*, 19, 402, 1989.
277. **Leevstad, R. A.,** Copper catalyzed oxidation of ascorbate (vitamin C). Inhibitory effect of catalase, superoxide dismutase, serum proteins (ceruloplasmin, albumin, apotransferrin) and amino acids, *Int. J. Biochem.*, 19, 309, 1987.
278. **Girotti, A. W., Thomas, J. P., and Jordan, J. E.,** Xanthine oxidase-catalyzed crosslinking of cell membrane proteins, *Arch. Biochem. Biophys.*, 251, 639, 1986.
279. **Davies, K. J. A.,** Protein damage and degradation by oxygen radicals. I. General aspects, *J. Biol. Chem.*, 262, 9895, 1987.
280. **Davies, K. J. A., Delsignore, M. E., and Line, S. W.,** Protein damage and degradation by oxygen radicals. II. Modification of amino acids, *J. Biol. Chem.*, 262, 9902, 1987.
281. **Davies, K. J. A. and Delsignore, M. E.,** Protein damage and degradation by oxygen radicals. III. Modification of secondary and tertiary structure, *J. Biol. Chem.*, 262, 9802, 1987.
282. **Uchida, K. and Kawakishi, S.,** Selective oxidation of imidazole ring in histidine residues by the ascorbic acid copper ion system, *Biochem. Biophys. Res. Commun.*, 138, 659, 1986.
283. **Davies, K. J. A. and Goldberg, A. L.,** Oxygen radicals stimulate intracellular proteolysis and lipid peroxidation by independent mechanisms in erythrocytes, *J. Biol. Chem.*, 262, 8220, 1987.
284. **Davies, K. J. A. and Golding, A. L.,** Proteins damaged by oxygen radials are rapidly degraded in extracts of red blood cells, *J. Biol. Chem.*, 262, 8227, 1987.
285. **Garland, D., Zigler, J. S., and Kinoshita, J.,** Structural changes in bovine lens crystallins induced by ascorbate, metal, and oxygen, *Arch. Biochem. Biophys.*, 251, 771, 1986.
286. **Griffiths, H. R., Lunec, J., Gee, C. A., and Willson, R. L.,** Oxygen radical induced alterations in polyclonal IgG, *FEBS Lett.*, 230, 155, 1988.
287. **Griffiths, H. R. and Lunec, J.,** The effect of oxygen free radicals on the carbohydrate moiety of IgG, *FEBS Lett.*, 245, 95, 1989.
288. **Inoue, H. and Hirobe, M.,** Disulphide clearage and insulin denaturation by active oxygen in the copper(II)/ascorbic acid system, *Chem. Pharm. Bull.*, 34, 1075, 1986.
289. **Greenwald, R. A. and Moy, W. W.,** Inhibition of collagen gelation by action of superoxide radical, *Arthritis Rheum.*, 22, 251, 1979.
290. **Monboisse, J. C., Braquet, P., Randoux, A., and Borel, J. P.,** Non-enzymatic degradation of acid-soluble calf skin collagen by superoxide ion: protective effect of flavonoids, *Biochem. Pharmacol.*, 32, 53, 1982.
291. **Monboisse, J. C., Cardes-Albert, M., Randoux, A., Bord, J. P., and Ferradini, C.,** Collagen degradation by superoxide anion in pulse and gamma radiolysis, *Biochim. Biophys. Acta*, 965, 29, 1988.
292. **Hiramatsu, K., Rosen, H., Heinecke, J. W., Wolfbauer, G., and Chait, A.,** Superoxide initiates oxidation of low density lipoprotein by human monocytes, *Arteriosclerosis-J. Vasc. Biol.*, 7, 55, 1987.
293. **Cathcart, M. K., McNally, A. M., Morel, D. W., Chisolm, G. M., III,** Superoxide anion participation in human monocyte-mediated oxidation of low-density lipoprotein and conversion of low-density lipoprotein to a cytotoxin, *J. Immunol.*, 142, 1963, 1989.
294. **Cathcart, M. K., Morel, D. W., and Chisolm, G. M.,** Monocytes and neutrophils oxidize low density lipoprotein making it cytotoxic, *J. Leukocyte Biol.*, 38, 341, 1985.
295. **Steinbrocher, U. P.,** Role of superoxide in endothelial-cell modification of low-density lipoproteins, *Biochim. Biophys. Acta*, 959, 20, 1988.
296. **Heinecke, J. W., Baker, L., Rosen, H., and Chait, A.,** Superoxide-mediated modification of low density lipoprotein by arterial smooth muscle cells, *J. Clin. Invest.*, 77, 757, 1986.
297. **Parthasarathy, S., Wieland, E., and Steinberg, D.,** A role for endothelial cell lipoxygenase in the oxidative modification of low density lipoprotein, *Proc. Natl. Acad. Sci. U.S.A.*, 86, 1046, 1989.
298. **Thornally, P. J. and Vasak, M.,** Possible role for metallothionein in protection against radiation-induced oxidative stress. Kinetics and mechanism of its reaction with superoxide and hydroxyl radicals, *Biochim. Biophys. Acta*, 827, 36, 1985.
299. **Meneghini, R.,** Genotoxicity of active oxygen species in mammalian cells, *Mutat. Res.*, 195, 215, 1988.
300. **Imlay, J. A. and Linn, S.,** DNA damage and oxygen radical toxicity, *Science* , 240, 1302, 1988.
301. **van Hemmen, J. J. and Meuling, W. J. A.,** Inactivation of biologically active DNA by γ-ray-induced superoxide radicals and their dismutation products singlet molecular oxygen and hydrogen peroxide, *Biochim. Biophys. Acta*, 402, 133, 1975.
302. **Cadet, J. and Teoule, R.,** Comparative study of oxidation of nucleic acid components by hydroxyl radicals, singlet oxygen and superoxide anion radicals, *Photochem. Photobiol.*, 28, 661, 1978.

303. **Lesko, S. A., Lorentzen, R. J., and Ts'o, P. O. P.,** Role of superoxide in deoxyribonucleic acid strand scission, *Biochemistry,* 19, 3023, 1980.
304. **Brawn, K. and Fridovisch, I.,** DNA strand scission by enzymically generated oxygen radicals, *Arch. Biochem. Biophys.,* 206, 414, 1981.
305. **Cunningham, M. L. and Lokesh, B. R.,** Superoxide anion generated by potassium superoxide is cytotoxic and mutagenic to Chinese hamster ovary cells, *Mutat. Res.,* 121, 299, 1983.
306. **Cunningham, M. L., Peak, J. G., and Peak, M. J.,** Single-strand DNA breaks in rodent and human cells produced by superoxide anion or its reduction products, *Mutat. Res.,* 184, 217, 1987.
307. **Ito, A., Krinsky, N. I., Cunningham, M. L., and Peak, M. J.,** Comparison of the inactivation of Bacillus subtilis transforming DNA by the potassium superoxide and xanthine-xanthine oxidase systems for generating superoxide, *Free Rad. Biol. Med.,* 3, 111, 1987.
308. **Emerit, I., Keck, M., Levy, A., Feingold, J., and Michelson, A. M.,** Activated oxygen species at the origin of chromosome breakage and sister-chromatide exchanges, *Mutat. Res.,* 103, 165, 1982.
309. **Sofuni, T. and Ishidate, M., Jr.,** Induction of chromosomal abberations in cultured Chinese hamster cells in a superoxide-generating system, *Mutat. Res.,* 140, 27, 1984.
310. **Emerit, I., Khan, S. H., and Cerutti, P. A.,** Treatment of lymphocyte cultures with a hypoxanthine-xanthine oxidase system induces the formation of transferable clastogenic material, *J. Free Rad. Biol. Med.,* 1, 51, 1985.
311. **Rosenberg-Arska, M., Van Asbeck, B. S., Martens, T. E. J., and Verhoef, J.,** Damage to chromosomal and plasmid DNA by toxic oxygen species, *J. Gen. Microbiol.,* 131, 3325, 1985.
312. **Moody, C. S. and Hassan, H. M.,** Mutagenicity of oxygen free radicals, *Proc. Natl. Acad. Sci. U.S.A.,* 79, 2855, 1982.
313. **Brawn, M. K. and Fridovich, I.,** Increased superoxide radical production evokes inducible DNA repair in *E. coli, J. Biol. Chem.,* 260, 922, 1985.
314. **Yonei, S., Noda, A., Tachibana, A., and Akasaka, S.,** Mutagenic and cytotoxic effects of oxygen free radicals generated by methylviologen (paraquat) on Escherichia coli with different DNA-repair capacities, *Mutat. Res.,* 163, 15, 1986.
315. **Sofuni, T. and Ishidate, M., Jr.,** Induction of chromosomal aberrations in active oxygen-generating systems. I. Effects of paraquat in Chinese hamster cells in culture, *Mutat. Res.,* 197, 127, 1988.
316. **Sawada, M., Sofuni, T., and Ishidate, M., Jr.,** Induction of chromosomal aberrations in active oxygen-generating systems. II. A study with hydrogen peroxide-resistant cells in culture, *Mutat. Res.,* 197, 133, 1988.
317. **Phillips, B. J., James, T. E. B., and Anderson, D.,** Genetic damage in CHO cells exposed to enzymically generated active oxygen species, *Mutat. Res.,* 126, 265, 1984.
318. **Hall, A. H., Jr., Eanes, R. Z., Waymack, P. P., Jr., and Patterson, R. M.,** Acute effects of a superoxide radical-generating system on DNA double-strand stability in Chinese hamster ovary cells. Determination by a modified fluorometric procedure, *Mutat. Res.,* 198, 161, 1988.
319. **Larramendy, M., Mello-Filho, A. C., Martins, E. A. L., and Meneghini, R.,** Iron-mediated induction of sister-chromatid exchanges by hydrogen peroxide and superoxide anion, *Mutat. Res.,* 178, 57, 1987.
320. **Deng, R.-Y., and Fridovich, I.,** Formation of endonuclease III-sensitive sites as a consequence of oxygen radical attack on DNA, *Free Rad. Biol. Med.,* 6, 123, 1989.
321. **Cohen, A. M., Aberdroth, R. E., and Hochstein, P.,** Inhibition of free radical-induced DNA damage by uric acid, *FEBS Lett.,* 174, 147, 1984.
322. **DeFlora, S., Bennicelli, C., Zanacchi, P., D'Agostini, F., and Camoirano, A.,** Mutagenicity of active oxygen species in bacteria and its enzymatic or chemical inhibition, *Mutat. Res.,* 214, 153, 1989.
323. **Stoewe, R. and Pruetz, W. A.,** Copper-catalyzed DNA damage by ascorbate and hydrogen peroxide: kinetics and yield, *Free. Rad. Biol. Med.,* 3, 97, 1987.
324. **Que, B. G., Downey, K. M., and So, A. G.,** Degradation of deoxyribonucleic acid by a 1,10-phenanthroline-copper complex: the role of hydroxyl radicals, *Biochemistry,* 19, 5987, 1980.
325. **Filho, A. C. M., Hoffmann, M. E., and Meneghini, R.,** Cell killing and DNA damage by hydrogen peroxide are mediated by intracellular iron, *Biochem. J.,* 218, 273, 1984.
326. **Stich, H. F., Karim, J., Koropatnick, J., and Lo, L.,** Mutagenic action of ascorbic acid, *Nature,* 260, 722, 1983.
327. **Fujimoto, S., Adachi, Y., Ishimitsu, S., and Ohara, A.,** Release of bases from deoxyribonucleic acid by ascorbic acid in the presence of copper, *Chem. Pharm. Bull.,* 34, 4848, 1986.
328. **Wong, A., Huang, C.-H., and Crooke, S. T.,** Mechanism of deoxyribonucleic acid breakage induced by 4'-(9-acridinylamino)methanesulfon-m-anisidide and copper: role for cuprous ion and oxygen free radicals, *Biochemistry,* 23, 2946, 1984.
329. **Watanabe, K., Kashige, N., Nakashima, Y., Hayashida, M., and Sumoto, K.,** DNA-Strand scission by D-glucosamine and its phosphate in plasmid pBR322, *Agric. Biol. Chem.,* 50, 1459, 1986.
330. **Schneider, J. E., Browning, M. M., Zhu, X., Eneff, K. L., and Floyd, R. A.,** Characterization of hydroxyl free radical mediated damage to plasmid pBR322 DNA, *Mutat. Res.,* 214, 23, 1989.

331. **Gutteridge, J. M. C. and Halliwell, B.,** The role of superoxide and hydroxyl radicals in the degradation of DNA and deoxyribose induced by a copper-phenanthroline complex, *Biochem. Pharmacol.*, 31, 2801, 1982.
332. **Sagripanti, J.-L. and Kraemer, K. H.,** Site-specific oxidative DNA damage of polyguanosines produced by copper plus hydrogen peroxide, *J. Biol. Chem.*, 264, 1729, 1989.
333. **Aronovitch, J., Godinger, D., Samuni, A., and Czapski, G.,** Ascorbic acid oxidation and DNA scission catalyzed by iron and copper chelates, *Free Rad. Res. Commun.*, 2, 241, 1987.
334. **Weitzman, S. A. and Stossel, T. P.,** Mutation caused by human phagocytes, *Science*, 212, 546, 1981.
335. **Weitzman, S. A. and Stossel, T. P.,** Effects of oxygen radical scavengers and antioxidants on phagocyte-induced mutagenesis, *J. Immunol.*, 128, 2770, 1982.
336. **Weitberg, A. B., Weitzman, S. A., Destrempes, M., Latt, S. A., and Stossel, T. P.,** Stimulated human phagocytes produce cytogenetic changes in cultured mammalian cells, *N. Engl. J. Med.*, 308, 26, 1983.
337. **Weitberg, A. B. and Weitzman, S. A.,** The effect of vitamin C on oxygen radical-induced sister-chromatid exchanges, *Mutat. Res.*, 144, 23, 1985.
338. **Phillips, B. J., Anderson, D., and Gandoli, S. D.,** The respiratory burst of phagocytic cells as a potential source of mutagenic species, in *Free Radicals in Liver Injury*, Poli, G., Cheeseman, K. H., Dianzani, M. U., and Slater, T. F., Eds., IRL Press, Oxford, 1985, 21.
339. **Birnboim, H. C. and Kanabus-Kaminska, M.,** The production of DNA strand breaks in human leukocytes by superoxide anion may involve a metabolic process, *Proc. Natl. Acad. Sci. U.S.A.*, 82, 6820, 1985.
340. **Birnboim, H. C.,** DNA Strand breaks in human leukocytes induced by superoxide anion, hydrogen peroxide and tumor promoters are repaired slowly compared to breaks induced by ionizing radiation, *Carcinogenesis*, 7, 1511, 1986.
341. **Birnboim, H. C.,** A superoxide anion-induced DNA strand-break metabolic pathway in human leukocytes: effect of vanadate, *Biochem. Cell Biol.*, 66, 374, 1988.
342. **Weitberg, A. B.,** Antioxidants inhibit the effect of vitamin C on oxygen radical-induced sister chromatid exchanges, *Mutat. Res.*, 191, 53, 1987.
343. **Korkina, L. G., Durnev, A. D., Daudel-Dauge, N. O., Suslova, T. B., Cheremisina, Z. P., and Afanas'ev, I. B.,** Oxygen radical-mediated mutagenic effect of dust on human lymphocytes: suppression by antioxidants, *Mutat. Res.*, in press.
344. **Yamaguchi, T.,** Mutagenicity of low molecular substances in various superoxide generating systems, *Agric. Biol. Chem.*, 45, 327, 1981.
345. **Sakurai, H., Shibuya, M., Shimizu, C., Akimoto, S., Maeda, M., and Kawasaki, K.,** DNA-strand scission by hemin-thiolate complexes as models of cytochrome P-450, *Biochem. Biophys. Res. Commun.*, 136, 645, 1986.
346. **Lewis, L. G., Stewart, W., and Adams, D. O.,** Role of oxygen radicals in induction of DNA damage by metabolites of benzene, *Cancer Res.*, 48, 4762, 1988.
347. **Fiala, E. S., Conaway, C. C., Biley, W. T., and Johnson, B.,** Enhanced mutagenicity of 2-nitropropane nitronate with respect to 2-nitropropane-possible involvement of free radical species, *Mutat. Res.*, 179, 15, 1987.
348. **Lesko, S. A., Trpis, L., and Zheng, R.,** Somatic mutation, DNA damage and cytotoxicity induced by benzo a pyrene dione/benzo a pyrene diol redox couples in cultivated mammalian cells, *Mutat. Res.*, 161, 173, 1986.
349. **Kanabus-Kominska, J. M., Feeley, M., and Birnboim, H. C.,** Simultaneous protective and damaging effects of cysteamine on intracellular DNA of leukocytes, *Free Rad. Biol. Med.*, 4, 141, 1988.
350. **Ueno, I., Kohno, M., Haraikawa, K., and Hirono, I.,** Interaction between quercetin and superoxide radicals. Reduction of the quercetin mutagenity, *J. Pharm. Din.*, 7, 798, 1984.
351. **Lickl, E., Alth, G., Ebermann, R., and Tuma, K.,** Fundamental model assays on chemical damage to ribonucleic acids by oxygen radicals in aqueous solution, *Strahlentherapie Onkol.*, 162, 775, 1986.
352. **Kasai, H. and Nishimura, S.,** Hydroxylation of deoxyguanosine at the C-8 position by ascorbic acid and other reducing agents, *Nucleic Acids Res.*, 12, 2137, 1984.
353. **Kasai, H. and Nishimura, S.,** Hydroxylation of guanine in nucleosides and DNA at the C-8 position by heated glucose and oxygen radical-forming agents, *Environ. Health Perspect.*, 67, 111, 1986.
354. **Floyd, R. A., Watson, J. J., Harris, J., West, M., and Wong, P. K.,** Formation of 8-hydroxydeoxyguanosine, hydroxyl free radical adduct of DNA in granulocytes exposed to the tumor promoter, tetradeconylphorbol acetate, *Biochem. Biophys. Res. Commun.*, 137, 841, 1986.
355. **Kasai, H., Crain, P. F., Kuchino, Y., Nishimura, S., Ootsuyama, A., and Tanooka, H.,** Formation of 8-hydroxyguanine moiety in cellular DNA by agents producing oxygen radicals and evidence for its repair, *Carcinogenesis*, 7, 1849, 1986.
356. **Richter, C., Park, J. W., and Ames, B. N.,** Normal oxidative damage to mitochondrial and nuclear DNA is extensive, *Proc. Natl. Acad. Sci. U.S.A.*, 85, 6465, 1988.
357. **Floyd, R. A., West, M. S., Eneff, K. L., Hogsett, W. E., and Tingey, D. T.,** Hydroxyl free radical mediated formation of 8-hydroxyguanine in isolated DNA, *Arch. Biochem. Biophys.*, 262, 266, 1988.

358. **Goldstein, I. M. and Weissman, G.,** Effects of the generation of superoxide anion on permeability of liposomes, *Biochem. Biophys. Res. Commun.*, 75, 604, 1977.
359. **Grankvist, K. and Marklund, S. L.,** Effect of extracellularly generated free radicals on the plasma membrane permeability of isolated pancreatic B-cells, *Int. J. Biochem.*, 18, 109, 1986.
360. **Vladimirov, Yu. A. and Cheremisina, Z. P.,** The effect of the lipid peroxidation on the respiratory control in rat liver mitochondria, *Stud. Biophys.*, 49, 161, 1975.
361. **Ninnemann, H., Strasser, R. J., and Butler, W. L.,** The superoxide anion as electron donor to mitochondrial electron transport chain, *Photochem. Photobiol.*, 26, 41, 1977.
362. **Flamigni, F., Guarnieri, C., Toni, R., and Caldarera, C. M.,** Effect of oxygen radicals on heart mitochondrial function in α-tocopherol deficient rabbits, *Int. J. Vitam. Nutr. Res.*, 52, 401, 1982.
363. **Wiswedel, I., Trümper, L., Schild, L., and Augustin, W.,** Injury of mitochondrial respiration and membrane potential during iron/ascorbate-induced peroxidation, *Biochim. Biophys. Acta*, 934, 80, 1988.
364. **Roh, J. K., Pyo, J. G., Chung, M. H., Lim, J. K., and Myung, H. J.,** Generation of superoxide radical from rat brain mitochondria and mechanism of its toxic action on mitochondrial and extramitochondrial components, *Taehan Yakrihak Chapchi*, 21, 12, 1985; *Chem. Abstr.*, 104, 1842, 1986.
365. **Guarnieri, C., Muscari, C., Ceconi, C., Flamigni, F., and Caldarera, C. M.** Effect of superoxide generation on rat heart mitochondrial pyruvate utilization, *J. Mol. Cell. Cardiol.*, 15, 859, 1983.
366. **Harris, E. J., Booth, R., and Cooper, M. B.,** The effect of superoxide generation on the ability of mitochondria to take up and retain Ca^{2+}, *FEBS Lett.*, 146, 267, 1982.
367. **Mirabelli, F., Richelmi, P., Salis, A., Marinoni, V., Bianchi, A., Berte, F., and Bellomo, G.,** Superoxide anion generation causes oxidation of pyridine nucleotides, collapse of transmembrane potential, calcium release and membrane damage in liver mitochondria, *Med. Biol. Environ.*, 15, 313, 1987.
368. **Vlessis, A. A. and Mela-Riker, L.,** Potential role of mitochondrial calcium metabolism during reperfusion injury, *Am. J. Physiol.*, 256, c1196, 1989.
369. **Kagan, V. E., Savov, V. M., Didenko, V. V., Arkhipenko, Yu.V., and Meerson, F. Z.,** Calcium and lipid peroxidation in heart mitochondrial and microsomal membranes, *Bull. Exp. Biol. Med.*, N 4, 46, 1983.
370. **Savov, V. M., Babizhae, M. A., and Kagan, V. E.,** Mechanisms of the effects of Ca^{2+} ions on lipid peroxidation processes, *Bull. Exp. Biol. Med.*, N 6, 693, 1986.
371. **Babizhaev, M. A.,** The biphasic effect of calcium on lipid peroxidation, *Arch. Biochem. Biophys.*, 266, 446, 1988.
372. **Toyooka, T., Arisaka, H., Sanma, H., Shin, W. S., Dau, Y., and Sugimoto, T.,** Synergistic deleterious effect of micromolar calcium ions and free radicals on respiratory function of heart mitochondria at cytochrome c and its salvage trail, *Biochem. Biophys. Res. Commun.*, 163, 1397, 1989.
373. **Braughler, J. M., Duncan, L. A., and Goodman, T.,** Calcium enhances in vitro free radical-induced damage to brain synaptosomes, mitochondria, and cultured spinal cord neurons, *J. Neurochem.*, 45, 1288, 1985.
374. **Debler, E. A., Sershen, H., Lajtha, A., and Gennaro, J. E., Jr.,** Superoxide radical-mediated alteration of synaptosome membrane structure and high-affinity $\gamma[^{14}C]$ aminobutyric acid uptake, *J. Neurochem.*, 47, 1804, 1986.
375. **Okabe, E., Kato, Y., Sasaki, H., Saito, G., Hess, M. L., and Ito, H.,** Calmodulin participation in oxygen radical-induced cardiac sarcoplasmic reticulum calcium uptake reduction, *Arch. Biochem. Biophys.*, 255, 464, 1987.
376. **Das, D. K., Bagchi, M., Prasad, M. R., Engleman, R. M.,** Calcium uptake by sarcoplasmic reticulum and changes in membrane fluidity during free radical exposure, *Biochem. Soc. Trans.*, 17, 698, 1989.
377. **Okabe, E., Odajima, C., Taga, R., Kukreja, R. C., Hess, M. L., and Ito, H.,** The effect of oxygen free radicals on calcium permeability and calcium loading at steady state in cardiac sarcoplasmic reticulum, *Mol. Pharmacol.*, 34, 388, 1988.
378. **Okabe, E., Sugihava, M., Tanaka, K., Sasaki, H., and Ito, H.,** Calmodulin and free oxygen radicals interaction with steady-state calcium accumulation and passive calcium permeability of cardiac sarcoplasmic reticulum, *J. Pharmacol. Exp. Ther.*, 250, 286, 1989.
379. **Kafy, A. M. L. and Lewis, D. A.,** Antioxidant effects of exogenous polyamines in damage of lysosomes inflicted by xanthine oxidase or stimulated polymorphonuclear leukocytes, *Agents Actions*, 24, 145, 1988.
380. **Kalra, J., Lautner, D., Massey, K. L., and Prasad, K.,** Oxygen free radicals induced release of lysosomal enzymes in vitro, *Mol. Cell. Biochem.*, 84, 233, 1988.
381. **Kalra, J., Chaudhary, A. K., and Prasad, K.,** Role of oxygen free radicals and pH on the release of cardiac lysosomal enzymes, *J. Mol. Cell. Cardiol.*, 21, 1125, 1989.
382. **Olsson, M., Svensson, I., Zdolsek, J. M., and Brunk, U. T.,** Lysosomal enzyme leakage during the hypoxanthine/xanthine oxidase reaction, *Virchows Arch.*, B56, 385, 1989.
383. **Sagone, A. L., Jr., Kamps, S., and Campbell, R.,** The effect of oxidant injury on the lymphoblastic transformation of human lymphocytes, *Photochem. Photobiol.*, 28, 909, 1978.

384. **Allan, I. M., Lunec, J., Salmon, M., and Bacon, P. A.,** Selective lymphocyte killing by reactive oxygen species, *Agents Actions,* 19, 351, 1986.
385. **Gallagher, R. B. and Curtis, A. S.,** The superoxide ion in lymphocyte transformation, *Immunol. Lett.,* 8, 329, 1984.
386. **Murrell, G. A. C., Francis, M. J. O., and Bromley, L.,** Oxygen free radicals stimulate fibroblast proliferation, *Biochem. Soc. Trans.,* 17, 484, 1989.
387. **Kitagawa, K., Nishino, H., Ogiso, Y., and Iwashima, A.,** Mechanism of superoxide anion- and hydrogen peroxide-induced stimulation of sugar transport in mouse fibroblast BALB/3T3 cells, *Biochim. Biophys. Acta,* 972, 293, 1988.
388. **Noel-Hudson, M. S., DeBelilovsky, C., Petit, N., Lindenbaum, A., and Wepierre, J.,** In vitro cytotoxic effects of enzymatically induced oxygen radicals in human fibroblasts: experimental procedures and protection by radical scavengers, *Toxicol. In Vitro,* 3, 103, 1989.
389. **Hussain, M. Z., Watson, J. A., and Bhatnager, R. S.,** Increased prolyl hydroxylase activity and collagen synthesis in hepatocyte cultures exposed to superoxide, *Hepatology,* 7, 502, 1987.
390. **Starke, P. E. and Farber, J. L.,** Ferric iron and superoxide ions are required for the killing of cultured hepatocytes by hydrogen peroxide. Evidence for the participation of hydroxyl radicals formed by an iron-catalzyed Haber-Weiss reaction, *J. Biol. Chem.,* 260, 10099, 1985.
391. **Muehlematter, D., Larsson, R., and Cerutti, P.,** Active oxygen induced DNA strand breakage and poly ADP-ribosylation in promotable and non-promotable JB 6 mouse epidermal cells, *Carcinogenesis,* 9, 239, 1988.
392. **Larsson, R. and Cerutti, P.,** Oxidants induce phosphorylation of ribosomal protein S 6, *J. Biol. Chem.,* 263, 17452, 1988.
393. **Ohmori, H., Komoriya, K., Azuma, A., Kurozumi, S., and Hoshimoto, Y.,** Xanthine oxidase-induced histamine release from isolated rat peritoneal cells: involvement of hydrogen peroxide, *Biochem. Pharmacol.,* 28, 333, 1979.
394. **Mannaioni, P. F., Gainnella, E., Palmerani, B., Pistelli, A., Gambassi, F., Bani-Sacchi, T., Bianchi, S., and Masini, E.,** The release of histamine by free radicals, *Agents Actions,* 24, 129, 1988.
395. **Lee, D. S. C., McCallum, E. A., and Olson, D. M.,** Effect of reactive oxygen species on prostacyclin production in perinatal rat lung cells, *J. Appl. Physiol.,* 66, 1321, 1989.
396. **Maridonneau-Parini, I., Fradin, A., Tonqui, L., and Russo-Marie, F.,** Effect of intracellular oxygen-free radicals on the formation of lipid derived mediators in rat renomedullary interstital cells, *Biochem. Pharmacol.,* 34, 4137, 1985.
397. **Handin, R. I., Karabin, R., and Boxer, G. J.,** Enhancement of platelet function by superoxide anion, *J. Clin. Invest.,* 59, 959, 1977.
398. **Salvemini, D., DeNucci, G., Sueddon, J. M., and Vane, J. R.,** Superoxide anions enhance platelet adhesion and aggregation, *Br. J. Pharmacol.,* 97, 1145, 1989.
399. **Chaudhri, G. and Clark, I. A.,** Reactive oxygen species facilitate the in vitro and in vivo lipopolysaccharide-induced release of tumor necrosis factor, *J. Immunol.,* 143, 1290, 1989.
400. **Michelson, A. M. and Durosay, P.,** Hemolysis of human erythrocytes by active oxygen, *Photochem. Photobiol.,* 25, 55, 1977.
401. **Lynch, R. E. and Fridovich, I.,** Effects of superoxide on the erythrocyte membrane, *J. Biol. Chem.,* 253, 1838, 1978.
402. **Goldberg, B. and Stern, A.,** The role of the superoxide anion as a toxic species in the erythrocyte, *Arch. Biochem. Biophys.,* 178, 218, 1977.
403. **Jozwiak, Z. and Helszer, Z.,** Participation of free oxygen radicals in damage of porcine erythrocytes, *Radiat. Res.,* 88, 11, 1981.
404. **Kong, S. and Davison, A. J.,** The relative effectiveness of ·OH, H_2O_2, O_2^-, and reducing free radicals in causing damage to biomembranes. A study of radiation damage to erythrocyte ghosts using selective radical scavengers, *Biochim. Biophys. Acta,* 640, 313, 1981.
405. **Maridonneau, I., Braquet, P., and Garay, R. P.,** Na^+ and K^+ transport damage induced by oxygen free radicals in human red cell membranes, *J. Biol. Chem.,* 258, 3107, 1983.
406. **Jones, S. A., Dempsey, T., Jones, J. G., and Rice-Evans, C.,** Oxygen radicals, iron and erythrocyte filterability, *Biochem. Soc. Trans.,* 16, 292, 1988.
407. **Girotti, A. W. and Thomas, J. P.,** Damaging effects of oxygen radicals on released erythrocyte ghosts, *J. Biol. Chem.,* 259, 1744, 1984.
408. **Rosen, G. M., Barber, M. J., and Rauckman, E. J.,** Disruption of erythrocyte membranal organization by superoxide, *J. Biol. Chem.,* 258, 2225, 1983.
409. **Vercellotti, G. M., Vanasbeek, B. S., and Jacob, H. S.,** Oxygen radical-induced erythrocyte hemolysis by neutrophils—critical role of iron and lactoferrin, *J. Clin. Invest.,* 76, 956, 1985.
410. **Weiss, S. J.,** The role of superoxide in the destruction of erythrocyte targets by human neutrophils, *J. Biol. Chem.,* 255, 9912, 1980.

411. **Weiss, S. J.,** Neutrophil-mediated methemoglobin formation in the erythrocyte. The role of superoxide and hydrogen peroxide, *J. Biol. Chem.*, 257, 2947, 1982.
412. **Francis, J. W., Boxer, L. A., and Petty, H. R.,** Optical microscopy of antibody-dependent phagocytosis and lysis of erythrocytes by living normal and chronic granulomatous disease neutrophils: a role of superoxide anions in extra- and intra-cellular lysis, *J. Cell. Physiol.*, 135, 1, 1988.
413. **Bishop, C. T., Mirza, Z., Crapo, J. D., and Freeman, B. A.,** Free radical damage to cultured porcine aortic endothelial cells and lung fibroblasts: modulation by culture conditions, *In Vitro*, 21, 229, 1985.
414. **Freeman, B. A., Rosen, G. M., and Barber, M. J.,** Superoxide perturbation of the organization of vascular endothelial cell membranes, *J. Biol. Chem.*, 261, 6590, 1986.
415. **Hirosumi, J., Ouchi, Y., Watanabe, M., Kusunoki, J., Nakamura, T., and Orimo, H.,** Effect of superoxide and lipid peroxide on cytosolic free calcium concentration in cultured pig aortic endothelial cells, *Biochem. Biophys. Res. Commun.*, 152, 301, 1988.
416. **Shaughnessy, S. G., Buchanan, M. R., Turple, S., Richardson, M., and Orr, F. W.,** Walker carcinosarcoma cells damage endothelial cells by the generation of reactive oxygen species, *Am. J. Pathol.*, 134, 787, 1989.
417. **Varani, J., Fligel, S. E. G., Till, G. O., Kunkel, R. G., Ryan, U. S., and Ward, P. A.,** Pulmonary endothelial cell killing by human neutrophils. Possible involvement of hydroxyl radicals, *Lab. Invest.*, 53, 656, 1985.
418. **Poumay, Y. and Ronveaux-Dupal, M.-F.,** Incubation of endothelial cells in a superoxide-generating system: impaired low-density lipoprotein receptor-mediated endocytosis, *J. Cell. Physiol.*, 136, 289, 1988.
419. **Scott, J. A., Khaw, B. A., Locke, E., Haber, E., and Homcy, C.,** The role of free radical-mediated processes in oxygen-related damage in cultured murine myocardial cells, *Circ. Res.*, 56, 62, 1985.
420. **Barrington, P. L., Meier, C. F., Jr., and Weglicki, W. B.,** Abnormal electrical activity induced by free radical generating systems in isolated cardiocytes, *J. Mol. Cell. Cardiol.*, 20, 1163, 1988.
421. **Scott, J. A., Fischman, A. J., Khaw, B.-A., Homey, C. J., and Rabito, C. A.,** Free radical-mediated membrane depolarization in renal and cardiac cells, *Biochim. Biophys. Acta*, 899, 76, 1987.
422. **McDonough, K. H., Henry, J. J., and Spitzer, J. J.,** Effects of oxygen radicals on substrate oxidation by cardiac myocytes, *Biochim. Biophys. Acta*, 926, 127, 1987.
423. **Crim, C. and Simon, R. H.,** Effect of oxygen metabolites on rat alveolar type II cell viability and surfactant metabolism, *Lab. Invest.*, 58, 428, 1988.
424. **McLennan, G., Oberley, L. W., and Autor, A. P.,** The role of oxygen-derived free radicals in radiation-induced damage and death of nondividing eucaryotic cells, *Radiat. Res.*, 84, 122, 1980.
425. **Petkau, A., Chelack, W. S., Palamer, E., and Gerrard, J.,** In vitro response of human bone marrow progenitor cells to superoxide radicals, *Res. Commun. Chem. Pathol. Pharmacol.*, 57, 107, 1987.
426. **Olson, C. E., Chen, M. C., Amirian, D. A., and Soll, A. H.,** Oxygen metabolites modulate prostaglandin E_2 production by isolated gastric mucosal cells, *Am. J. Physiol.*, 256, G925, 1989.
427. **Chen, L. H.,** Ascorbic acid-stimulated peroxidation in hepatocytes and inhibition by antioxidants, *Biochem. Arch.*, 4, 373, 1988.
428. **Olson, C. E.,** Glutathione modulates toxic oxygen metabolite injury of canine chief cell monolayers in primary culture, *Am. J. Physiol.*, 254, G49, 1988.
429. **Myers, R. B. and Abney, T. O.,** The effects of reduced oxygen and antioxidants on steroidogenic capacity of cultured rat Leydig cells, *J. Steroid Biochem.*, 31, 305, 1988.
430. **Ruch, R. J. and Klauning, J. E.,** Inhibition of mouse hepatocyte intercellular communication by paraquat-generated oxygen free radicals, *Toxicol. Appl. Pharmacol.*, 94, 427, 1988.
431. **Klebanoff, S. J.,** The role of the superoxide anion in the myeloperoxidase-mediated antimicrobial system, *J. Biol. Chem.*, 249, 3724, 1974.
432. **DeChatelet, L. R., Shirley, P. S., Goodson, P. R., and McCall, C. E..,** Bactericidal activity of superoxide anion and of hydrogen peroxide: investigations employing dialuric acid, a superoxide generating drug, *Antimicrob. Agents Chemother.*, 8, 146, 1975.
433. **Babior, B., Curnutte, J. T., and Kipnes, R. S.,** Biological defence mechanisms. Evidence for the participation of superoxide in bacterial killing by xanthine oxidase, *J. Lab. Clin. Med.*, 85, 235, 1975.
434. **van Hemmen, J. J. and Meuling, W. J. A.,** Inactivation of *Escherichia coli* by superoxide radicals and their dismutation products, *Arch. Biochem. Biophys.*, 182, 743, 1977.
435. **DiGuiseppi, J. and Fridovich, I.,** Oxygen toxicity in *Streptococcus sanguis*. The relative importance of superoxide and hydroxyl radicals, *J. Biol. Chem.*, 257, 4046, 1982.
436. **Takahashi, M., Nagano, T., and Hirobe, M.,** Dioxathiadiaza-heteropentalenes mediate superoxide and hydrogen peroxide production in *Escherichia coli*, *Arch. Biochem. Biophys.*, 268, 137, 1989.
437. **Rosen, H. and Klebanoff, S. J.,** Role of iron and ethylenediaminetetraacetic acid in the bacterial activity of a superoxide anion-generating system, *Arch. Biochem. Biophys.*, 208, 512, 1982.
438. **Yamada, Y.,** Susceptibility of micro-organisms to active oxygen species: sensitivity to the xanthine-oxidase-mediated antimicrobial system, *J. Gen. Microbiol.*, 133, 2007, 1987.
439. **Ewing, D. and Jones, S. R.,** Superoxide removal and radiation protection in bacteria, *Arch. Biochem. Biophys.*, 254, 53, 1987.

440. **Lin, W. S., Wong, F., and Anderson, R.,** Role of superoxide in radiation-killing of *Escherichia coli* and in thymine release from thymidine, *Biochem. Biophys. Res. Commun.*, 147, 778, 1987.
441. **Hertz, R. and Bar-Tana, J.,** Reductive repression in *Escherichia coli* K-12 is mediated by oxygen radicals, *Arch. Biochem. Biophys.*, 250, 54, 1986.
442. **Samuni, A., Chevion, M., Halpern, Y. S., Ilan, Y. A., and Czapski, G.,** Radiation-induced damage in T4 bacteriophage: the effect of superoxide radicals and molecular oxygen, *Radiat. Res.*, 75, 489, 1978.
443. **Samuni, A., Chevion, M., and Czapski, G.,** Roles of copper and O_2^- in the radiation-induced inactivation of T7 bacteriophage, *Radiat. Res.*, 99, 562, 1984.
444. **Murata, A., Suenaga, H., Hideshima, S., Tanaka, Y., and Kato, F.,** Virus-inactivating effect of ascorbic acid. XI. Hydroxyl radical as the reactive species in the inactivation of phages by ascorbic acid, *Agric. Biol. Chem.*, 50, 1481, 1986.
445. **McCord, J. M. and Petrone, W. F.,** A superoxide-activated lipid-albumin chemotactic factor for neutrophils, in *Lipid Peroxides in Biology and Medicine*, Yagi, K., Ed., Academic Press, New York, 1982, 123.
446. **Halliwell, B.,** Invited commentary: superoxide, iron, vascular endothelium and reperfusion injury, *Free Rad. Res. Commun.*, 5, 315, 1989.
447. **Gryglewski, R. J., Palmer, R. M. J., and Moncada, S.,** Superoxide anion is involved in the breakdown of endothelium-derived vascular relaxing factor, *Nature*, 320, 454, 1986.
448. **Rubanyi, G. M. and Vahoutte, P. M.,** Oxygen-derived free radicals, endothelium, and responsiveness of vascular smooth muscle, *Am. J. Physiol.*, 250, H815, 1986.
449. **Katusic, Z. S. and Vanhoutte, P. M.,** Superoxide anion is an endothelium-derived contractive factor, *Am. J. Physiol.*, 257, H33, 1989.
450. **Rubanyi, G. M.,** Vascular effects of oxygen-derived free radicals, *Free Rad. Biol. Med.*, 4, 107, 1988.
451. **Keller, R. J., Coulombe, R. A., Jr., Sharma, R. P., Grover, T. A., and Piette, L. H.,** Importance of hydroxyl radical in the vanadium-stimulated oxidation of NADH, *Free Rad. Biol. Med.*, 6, 15, 1989.
452. **Keller, R. J., Coulombe, R. A., Sharma, R. P., Glover, T. A., and Piette, L. H.,** Oxidation of NADH by vanadium compounds in the presence of thiols, *Arch. Biochem. Biophys.*, 271, 40, 1989.
453. **Sabourault, D., Ribiere, C., Nordmann, R., Houce-Levin, C., and Ferradini, C.,** Gamma and pulse radiolysis investigation of the reaction of desferrioxamine with superoxide anions, *Int. J. Radiat. Biol.*, 56, 911, 1989.
454. **Samuni, A., Krishna, C. M., Riesz, P., Finkelstein, E., and Russo, A.,** Superoxide reaction with nitroxide spin-adducts, *Free Rad. Biol. Med.*, 6, 141, 1989.
455. **Noda, A., Noda, H., Misaka, A., Sumimoto, H., and Tatsumi, K.,** Hydrazine radical formation catalyzed by rat microsomal NADPH-cytochrome P-450 reductase, *Biochem. Biophys. Res. Commun.*, 153, 256, 1988.
456. **Wickramasinghe, S. N.,** Role of superoxide anion radicals in ethanol metabolism by blood monocyte-derived human macrophages, *J. Exp. Med.*, 169, 755, 1989.
457. **Aaseth, J., Ribarov, S., and Bochev, P.,** The interaction of copper ($Cu(2^+)$) with erythrocyte membrane and 2,3-dimercaptopropanesulfonate in vitro: a source of activated oxygen species, *Pharmacol. Toxicol.*, 61, 250, 1987.
458. **Cavalieri, E. L., Devanesan, P. D., and Rogan, E. G.,** Radical cations in the horseradish peroxidase and prostaglandin H synthase mediated metabolism and binding of benzo a pyrene to deoxyribonucleic acid, *Biochem. Pharmacol.*, 37, 2183, 1988.
459. **Rehncrona, S., Hauge, H. M., and Siesjoe, B. K.,** Enhancement of iron-catalyzed free radical formation by acidosis in brain homogenates: difference in effect by lactic acid and carbon dioxide, *J. Cereb. Blood Flow Metab.*, 9, 65, 1989.
460. **Kettle, A. J. and Winterbourn, C. C.,** Superoxide modulates the activity of myeloperoxidase and optimizes the production of hypochlorous acid, *Biochem. J.*, 252, 529, 1988.
461. **Cabelli, D. E., Allen, D., and Bielski, B. H. J.,** The interaction between copper(I) superoxide dismutase and hydrogen peroxide, *J. Biol. Chem.*, 264, 9967, 1989.
462. **Bedwell, S., Dean, R. T., and Jessup, W.,** The action of defined oxygen-centered free radicals on human low-density lipoproteins, *Biochem. J.*, 262, 707, 1989.
463. **Soussi, B., Bylund-Fellenius, A. C., Schersten, T., and Aangstroem, J.,** Proton-NMR evaluation of the ferricytochrome c-cardiolipin interaction. Effect of superoxide radicals, *Biochem. J.*, 265, 227, 1990.
464. **Kobayashi, K., Hayashi, K., and Sono, M.,** Effect of tryptophan and pH on the kinetics of superoxide radical binding to indoleamine 2,3-dioxygenase studied by pulse radiolysis, *J. Biol. Chem.*, 264, 15280, 1989.
465. **Alonso, E. and Rubio, V.,** Inactivation of mitochondrial carbamoyl phosphate synthetase induced by ascorbate, oxygen and Fe^{3+} in the presence of acetylglutamate: protection by ATP and HCO_3^- and lack of inactivation of ornithine transcarbamylase, *Arch. Biochem. Biophys.*, 258, 342, 1987.
466. **Schor, N. F.,** Inactivation of mammalian brain glutamine synthetase by oxygen radicals, *Brain Res.*, 456, 17, 1988.

467. **Terada, L. S., Beehler, C. J., Banerjee, A., Brown, J. M., Grosso, M. A., Harken, A. H., McCord, J. M., and Repine, J. E.,** Hyperoxia and self- or neutrophil-generated oxygen metabolites inactivate xanthine oxidase, *J. Appl. Physiol.*, 65, 2349, 1988.
468. **Nowak, D.,** The comparative study of reactive oxygen species generated by polymorphonuclear leukocytes as α1-proteinase inhibitor inactivators—possible application for antioxidant prevention of emphysema, *Arch. Immunol. Ther. Exp.*, 36, 723, 1988.
469. **Epe, B., Mützel, P., and Adam, W.,** DNA damage by oxygen radicals and excited state species: a comparative study using enzymatic probes in vitro, *Chem. Biol. Interact.*, 67, 149, 1988.
470. **Kobayashi, S., Ueda, K., Morita, J., Sakai, H., and Komano, T.,** DNA damage by ascorbate in the presence of copper (2^+), *Biochim. Biophys. Acta*, 949, 143, 1988.
471. **Loeb, L. A., James E. A., Waltersdorph, A. M., and Klebanoff, S. J.,** Mutagenesis by the autoxidation of iron with isolated DNA, *Proc. Natl. Acad. Sci. U.S.A.*, 85, 3918, 1988.
472. **Jackson, J. H., Gajewski, E., Schraufstatter, I. U., Hyslop, P. A., Fuciarelli, A. F., Cochrane, C. G., and Dizdaroglu, M.,** Damage to the bases in DNA induced by stimulated human neutrophils, *J. Clin. Invest.*, 84, 1644, 1989.
473. **Yamamoto, K. and Kawanishi, S.,** Hydroxyl free radical is not the main active species in site-specific DNA damage induced by copper(II) ion and hydrogen peroxide, *J. Biol. Chem.*, 264, 15435, 1989.
474. **Shacter, E., Beecham, E. J., Covey, J. M., Kohn, K. W., and Potter, M.,** Activated neutrophils induced prolonged DNA damage in neighboring cells, *Carcinogenesis*, 9, 2297, 1988.
475. **Chong, Y. C., Heppner, G. H., Paul, L. A., and Fulton, A. M.,** Macrophage-mediated induction of DNA strand breaks in target tumor cells, *Cancer Res.*, 49, 6652, 1989.
476. **Blount, S., Griffiths, H. R., and Lunec, J.,** Reactive oxygen species induce autogenic changes in DNA, *FEBS Lett.*, 245, 100, 1989.
477. **Moseley, P. L.,** Augmentation of bleomycin-induced DNA damage in intact cells, *Am. J. Physiol.*, 257, C882, 1989.
478. **Pratviel, G., Bernadou, J., and Meunier, B.,** Evidence for high-valent iron-oxo species active in the DNA breaks mediated by iron-bleomycin, *Biochem. Pharmacol.*, 38, 133, 1989.
479. **Hartley, J. A., Reszka, K., and Lown, J. W.,** Photosensitization of antitumor agents. 4. Anthrapyrazole-photosensitized formation of single strand-breaks in DNA, *Free Rad. Biol. Med.*, 4, 337, 1988.
480. **Laderoute, K., Wardman, P., and Rauth, A. M.,** Molecular mechanisms for the hypoxia-dependent activation of 3-amino-1,2,4-benzotriazine-1,4-dioxide (SR 4233), *Biochem. Pharmacol.*, 37, 1487, 1988.
481. **Quinlan, G. J. and Gutteridge, J. M. C.,** Oxidative damage to DNA and deoxyribose by β-lactam antibiotics in the presence of iron and copper salts, *Free. Rad. Res. Commun.*, 5, 149, 1988.
482. **Jenner, T. J., Sapora, O., O'Neill, P., and Fielden, E. M.,** Enhancement of DNA damage in mammalian cells upon bioreduction of the nitroimidazole-aziridines RSU-1069 and RSU-1131, *Biochem. Pharmacol.*, 37, 3837, 1988.
483. **Hung, S. J., Sousa, D. M., and Weitberg, A. B.,** The effects of selenium and vitamin C on oxygen radical-induced DNA strand breaks, *J. Exp. Clin. Cancer Res.*, 7, 267, 1988.
484. **Hart, L. A., Van Enckevort, P., Van Kessel, K. P. M., Van Dijk, H., and Labadie, R. P.,** Evidence that superoxide anion produced by PMA-activated human polymorphonuclear leukocytes is the cytolytic agent for rabbit, but not for sheep red blood cells, *Immunol. Lett.*, 18, 139, 1988.
485. **Yasuda, H., Miki, M., Takenaka, Y., Tamai, H., and Mino, M.,** Changes in membrane constituents and chemiluminescence in vitamin E-deficient red blood cells induced by the xanthine oxidase reaction, *Arch. Biochem. Biophys.*, 272, 81, 1989.
486. **Von Ruecker, A. A., Hau-Jeon, B. G., Wild, M., and Bidlingmaier, F.,** Protein kinase C involvement in lipid peroxidation and cell membrane damage induced by oxygen-based radicals in hepatocytes, *Biochem. Biophys. Res. Commun.*, 163, 836, 1989.
487. **Crim, C. and Simon, R. H.,** Effect of oxygen metabolites on rat alveolar type II cell viability and surfactant metabolism, *Lab. Invest.*, 58, 428, 1988.
488. **Adler, K. B., Holden-Stauffer, W. J., and Repine, J. E.,** Oxygen metabolites stimulate release of high-molecular-weight glycoconjugates by cell and organ cultures of rodent respiratory epithelium via an arachidonic acid-dependent mechanism, *J. Clin. Invest.*, 85, 75, 1990.
489. **Larsen, T., Soerensen, M. B., Olsen, R., and Joergensen, L.,** Effect of scavengers of active oxygen species and pretreatment with acetylsalicylic acid on the injury to cultured endothelial cells by thrombin-stimulated platelets, *In Vitro Cell. Dev. Biol.*, 25, 276, 1989.
490. **Jornot, L. and Junod, A. F.,** Hypoxanthine-xanthine oxidase-related defect in polypeptide chain initiation by endothelium, *J. Appl. Physiol.*, 66, 450, 1989.
491. **Kasama, T., Kobayashi, K., Fukushima, T., Tabata, M., Ohno, I., Negishi, M., Ide, H., Takahashi, T., and Niwa, Y.,** Production of interleukin 1-like factor from human peripheral blood monocytes and polymorphonuclear leukocytes by superoxide anion: the role of interleukin 1 and reactive oxygen species in inflamed sites, *Clin. Immunol. Immunopathol.*, 53, 439, 1989.

492. **Krutmann, J., Athar, M., Mendel, D. B., Khan, I. U., Guyre, P. M., Mukhtar, H., and Elmets, C. A.,** Inhibition of the high affinity Fc receptor (Fcγ RI) on human monocytes by porphyrin photosensitization is highly specific and mediated by the generation of superoxide radicals, *J. Biol. Chem.*, 264, 11407, 1989.
493. **Masini, E., Giannella, E., Pistelli, A., Palmerani, B., Cambassi, F., Occupati, B., and Mannaioni, P. F.,** Histamine release by free radicals: the relationship with the signal transduction systems, *Agents Actions*, 27, 72, 1989.
494. **Gushchin, I. S., Voitenko, V. G., Petyaev, I. M., and Tsynkalovskii, O. R.,** Activation of oxygen metabolism during histamine secretion from rat mast cell induced by a specific allergen, substance 48/80 and ionophore A 23187, *Immunologia*, N 5, 33, 1989.
495. **Ueno, M., Brookins, J., Beckman, B. S., and Fisher, J. W.,** Effects of reactive oxygen metabolites on erythropoietin production in renal carcinoma cells, *Biochem. Biophys. Res. Commun.*, 154, 773, 1988.
496. **Laval, F.,** Pretreatment with oxygen species increases the resistance of mammalian cells to hydrogen peroxide and γ-rays, *Mutat. Res.*, 201, 73, 1988.
497. **Vincent, F., Brun, H., Clain, E., Ronot, X., and Adolphe, M.,** Effect of oxygen-free radicals on proliferation kinetics of cultured rabbit articular chondrocytes, *J. Cell. Physiol.*, 141, 262, 1989.
498. **Yamamoto, H., Nagano, T., and Hirobe, T.,** Carbon tetrachloride toxicity on Escherichia coli exacerbated by superoxide, *J. Biol. Chem.*, 263, 12224, 1988.
499. **Burkitt, M. J. and Gilbert, B. C.,** The control of iron-induced oxidative damage in isolated rat liver mitochondria by respiratory state and ascorbate, *Free Rad. Res. Commun.*, 5, 333, 1989.
500. **Kaneko, M., Beamish, R. E., and Dhalla, N. S.,** Depression of heart sarcolemmal calcium-pump activity by oxygen free radicals, *Am. J. Physiol.*, 256, H 368, 1989.
501. **Kukreja, R. C., Weaver, A. B., and Hess, M. L.,** Stimulated human neutrophils damage cardiac sarcoplasmic reticulum, *Biochim. Biophys. Acta*, 990, 198, 1989.

APPENDIX I

EXPERIMENTAL TECHNIQUES USED IN OXYGEN RADICAL STUDIES

1. Detection of superoxide ion by SOD-inhibitable autoxidation of epinephrine, References 20,95—97 (Chapter 2).
2. Detection of superoxide ion by SOD-inhibitable reduction of cytochrome c, Reference 41 (Chapter 2); Reference 12 (Chapter 4).
3. Detection of superoxide ion by the reduction of modified cytochromes c, References 108,109,113,118 (Chapter 2); References 158,159 (Chapter 5).
4. Detection of superoxide ion by SOD-inhibitable reduction of NBT, References 95—97,104 (Chapter 2); Reference 265 (Chapter 3).
5. Detection of superoxide ion on the basis of its ESR spectrum, using the rapid freezing method, References 27,43,44 (Chapter 2).
6. Detection of superoxide ion via the SOD-inhibitable formation of lactoperoxidase compound III, References 38,98 (Chapter 2).
7. Detection of superoxide ion via the formation of compound III of diacetyldeuteroheme-substituted HRP, Reference 41 (Chapter 4).
8. Detection of superoxide ion by SOD-inhibitable, lucigenin-dependent CL, References 48,151,152 (Chapter 2); References 29,32 (Chapter 4).
9. Detection of superoxide ion by SOD-inhibitable, luminol-dependent CL, References 28,31 (Chapter 4).
10. Application of ultrasensitive video intensifier microscopy for measuring luminol-dependent CL, Reference 30 (Chapter 4).
11. Continuous and simultaneous detection of superoxide production and dioxygen consumption by phagocytes via SOD-inhibitable cytochrome c reduction and polarography with a Clark-type electrode, Reference 21 (Chapter 4).
12. Detection of superoxide ion by the manganese-dependent SOD-inhibitable oxidation of diaminobenzidine, Reference 23 (Chapter 4).
13. Detection of superoxide ion by oxidation of 2-ethyl-1-hydroxy-2,5,5-trimethyl-3-oxazolidine, Reference 40 (Chapter 4).
14. Spin trapping of oxygen radicals, References 2,46,47,79 (Chapter 2); Reference 28 (Chapter 3).
15. Pulse radiolysis study of oxygen radicals, Reference 51 (Chapter 2).
16. Detection of hydroxyl radicals by the oxidation of reduced cytochrome c, Reference 121 (Chapter 2).
17. Detection of hydroxyl radicals on the basis of hydroxylation of aromatic compounds, References 123,134 (Chapter 2); Reference 42 (Chapter 3).
18. Detection of active oxygen radicals by measuring the ethylene formation from methional, Reference 81 (Chapter 3).
19. Detection of active oxygen radicals by measuring the ethylene formation from 2-keto-4-thiomethylbutyric acid, Reference 135 (Chapter 2); Reference 39 (Chapter 3).
20. Detection of hydroxyl radicals via the formation of methane from DMSO, Reference 129 (Chapter 2); References 77—80 (Chapter 3).
21. Detection of hydroxyl radicals by measuring $^{14}CO_2$ formation during the decarboxylation of ^{14}C benzoate, Reference 71 (Chapter 3).
22. DNA-nickling as a test on oxygen radicals, Reference 119 (Chapter 2).
23. Studying of oxygen radical-mediated DNA damage via the detection of 8-hydroxydeoxyguanosine, References 352,354 (Chapter 5).
24. An MDA assay of lipid peroxidation, Reference 67 (Chapter 5).
25. Detection of diene conjugates and hydrocarbon gases in lipid peroxidation, Reference 67 (Chapter 5).

APPENDIX I (continued)

EXPERIMENTAL TECHNIQUES USED IN OXYGEN RADICAL STUDIES

26. Application of fluorescence microscopic technique for studying hemoglobin oxidation in erythrocytes by oxygen radicals, Reference 412 (Chapter 5).
27. Using fluorescence-activating sorting of antimyosin antibody-labeled cells for studying cellular damage by oxygen radicals, Reference 419 (Chapter 5).
28. Measurement of cell damage by oxygen radicals via the change in cellular membrane potential, Reference 421 (Chapter 5).
29. Using the repair enzyme activity for identification of oxygen species participating in DNA damage, Reference 469 (Chapter 5).

APPENDIX II

ABBREVIATIONS

agg IgG	Aggregated immunoglobulin G
ANS	1-Aninino-8-naphthalene sulfonate
AQS	9,10-Anthraquinone-2-sulfonate
AZQ	Diaziquone, 3,6-diaziridinyl-2,5-bis(carboethoxyamino)-1,4-benzoquinone
BCNU	*N,N*-bis(2-chloroethyl)-*N*-nitrosourea
BHA	2-Tert-butyl-4-methoxyphenol
BHT	2,6-Di-tert-butyl-4-methylphenol
BLM	Bleomycin
BSA	Bovine serum albumin
BZQ	2,5-Bis(2-hydroxyethylamino)-3,6-diaziridinyl-1,4-benzoquinone
CGD	Chronic granulomatous disease
CHO cells	Chinese hamster ovary cells
Con A	Concanavalin A
CQ	Chloroquine
DG	1,2-Diacylglycerol
DHAQ	Mitoxantrone
DHF	Dihydroxyfumarate
DMF	Dimethylformamide
DMPO	5,5-Dimethyl-1-pyrroline-*N*-oxide
DMSO	Dimethylsulfoxide
DTPA	Diethylenetriaminepentaacetic acid
EDRF	Endothelium-derived vascular relaxing factor
EDTA	Ethylenediaminetetraacetic acid
ESR	Electron spin resonance
EBV	Epstein-Barr virus
Fd	Ferredoxin
FMLP	*N*-Formyl-methionyl-leucyl-phenylalanine
GABA	Gamma-aminobutyric acid
GAPDH	Glyceraldehyde-3-phosphate dehydrogenase
GM-CSF	Granulocyte-macrophage colony-stimulating factor
H-7	1-(5-Isoquinolinesulfonyl)-2-methylpiperazine
HAQ	Ametantrone
Hb	Hemoglobin
HCQ	Hydroxychloroquine
5-HETE	5-Hydroxy-6,8,11,14-eicosatetraenonate
8-HOdG	8-Hydroxydeoxyguanosine
HPD	Photofrin II, hematoporphyrin derivative
HPLC	High-performance liquid chromatography
HRP	Horseradish peroxidase
ICRF-187	Bisdioxopiperazine
IFN	Interferon
IP_3	Inositol 1,4,5-triphosphate
KMBA	2-Ketothiomethylbutyric acid
LDH	Lactate dehydrogenase
LDL	Low-density lipoproteins
LPS	Lipopolysaccharides
LTB_4	Leukotriene B_4, 12(R)-dihydroxyeicosatetraenic acid
4-Mal-TEMPO	4-Maleimido-2,2,6,6-tetramethylpiperidinooxyl

APPENDIX II (continued)

ABBREVIATIONS

MCLA	2-Methyl-6-[*p*-methoxyphenyl]-3,7-dihydroimidazo-[1,2-α]-pyrazin-3-one
MDA	Malondialdehyde
4-MePyBN	4-(*N*-Methylpyridinium)-*tert*-butyl nitrone
metHb	Methemoglobin
MG	Methylguanidine
MHAQ	(1-Hydroxy-5,8-bis(2-[(2-hydroxyethyl)amino]-ethylamino)-9,10-anthracene-dione
MP	Meracrine
MPP^+	1-Methyl-4-phenylpyridinium ion
MPPNHOH	*N*-Hydroxyphentermine
$MPPNH_2$	2-Methyl-1-phenyl-2-aminopropane
$MPPNO_2$	2-Methyl-2-nitro-1-phenylpropane, phentermine
MPTP	3-Methyl-4-phenyl-1,2,3,6-tetrahydropyridine
NBT	Nitroblue tetrazolium
NDGA	Nordihydroguaretic acid
NK	Natural killers
NMNH	Reduced nicotinamide mononucleotide
NPGB	*p*-Nitrophenyl-*p*′-guanidinobenzoate
OAG	1-Oleoyl-2-acetylglycerol
OXANOH	2-Ethyl-1-hydroxy-2,5,5-trimethyl-3-oxazolidine
PAF	Platelet activating factor, 1-*O*-hexadecyl-2-acetyl-sn-glycero-3-phosphorylcholine
PARC	Poly-L-arginine
PBN	α-Phenyl-*tert*-butyl nitrone
PDD	4-β-Phorbol-12,13-didecanoate
PHA	Phytohemagglutin
PHSTD	Poly-L-histidine
PIP_2	Phosphatidylinositol 4,5-biphosphate
PMA	12-*O*-Myristate 13-acetate
PMNs	Polymorphonuclear leukocytes
PMS	Phenazine methosulfate
PUFA	Polyunsaturated fatty acids
ROS	Reactive oxygen species
SCE	Sister chromatid exchange
SDS	Sodium dodecyl sulfate
SIIS	4-Acetamido-4′-isothhiocyanostilbene-2,2-disulfonate
SOD	Superoxide dismutase
SR	Sarcoplasmic reticulum
TBA	Thiobarbituric acid
TNF	Tumor necrosis factor
TPA	12-*O*-Tetradecanoylphorbol-13-acetate
TTFA	Thenoyltrifluoroacetone
VP-16	Semisynthetic derivative of podophyllotoxin
XO	Xanthine oxidase

INDEX

D

E

F

G

H

I

N

T

U

V

W

XYZ